Deptford, Toronto and Kingston

Charles Aird. his Career.

born 12 December 1831. the eldest child of John & Agnes Aird. Christened at St Georges Church Southwark. My father & Mother then living in a small six roomed cottage, in a yard occupied by the old Phoenix Gas Compy for Stores & Stables in Wellington Street Blackfriars Road a very squalid part of London. with a knacker's yard adjoining for the preperation of Cat's meat from dead horses. At that time I remember there used to be seen going about in all Town's Catsmeat barrows with choice morsels on a little wooden stick for the domestic pussy, Such a thing I do not think is seen anywhere now. My father was then an Outdoor Foreman or Inspector of that Gas Company. I cannot remember how he became connected with the Compy but I have often heard him say, he was born in Scotland in Rossshire & when a lad about 16, he came up to London with his father & mother

A page from the diaries, actual size.

Deptford, Toronto and Kingston

The early years of Charles Aird, Victorian Engineer

Introduced by Peter Stirling-Aird

Edited by Laura Hone

The Grimsay Press
2005

The Grimsay Press
an imprint of
Zeticula
57 St Vincent Crescent
Glasgow
G3 8NQ

http://www.thegrimsaypress.co.uk
admin@thegrimsaypress.co.uk

First published in Great Britain in 2005

ISBN 1 84530 020 3 Paperback
ISBN 1 84530 021 1 Hardback

Peter D. M. Stirling-Aird (1915-2004)

In 1995, at the end of a dinner celebrating my father's eightieth birthday, he started to tell the fascinating story of his grandfather's early life and of the meteoric rise of the firm John Aird and Sons. My stepsister, Caroline Dawnay, had read some of my great-grandfather's diaries and at this dinner encouraged my father to get them published.

Over the next few years, I transcribed these diaries and became increasingly interested in all they contained. My father was a meticulous editor and proof-reader and mercifully lived long enough to correct all the copy, as only he could identify various archaic and family references. Sad to say he died in July, but before he died, he knew the diaries would be published and in the exact form he wanted. This fact gave him great pleasure and satisfaction.

I would therefore like to dedicate this book to my father's memory – to the memory of a gifted and extraordinary man who was rightly proud of his forebears - John and Charles Aird.

Laura Hone,
October, 2004

Charles Aird (1831-1910)

Contents

John Aird (1833-1911), younger brother of Charles Aird, and son of John (1801-1876), created Baronet 1901.

List of illustrations

Preface

My grandfather, Charles Aird (1831-1910) then living at The Glen, Eastbourne, with his wife and four of his unmarried daughters, started in 1908 to compile a diary of his active life. Also, as he remembered it, that of his father and grandfather. It is now copied, word for word, as he wrote it down in his old age. My daughter, Laura, has undertaken to transcribe the seven books of this diary. They give a fascinating background to life in the last two centuries and show what can be achieved by a Victorian family with tremendous drive and determination.

Edward Meldrum's book "From Loch Ness to The Aird" (printed by John Eccles, Inverness, in 1987) gives an interesting background to the Aird family, although no connection is proved with the current family whose history we know only from the early 18th century.

During the 13th century the new feudal families of Byssets, Fentouns and De L'Airds held the fertile lands of the valleys of the Nairn and the Ness. The land south of the Beauly Firth was, and is, known as The Aird and the family had taken this as their surname. In a similar manner, my first wife's family, the Stirlings (also Anglo-Norman followers, and in 1100 then called de Strivelin) could have derived their surname from Stirling Castle and can show their direct descent to the present day.

John Bysset, Lord of the Aird, founded Beauly Priory in 1230. Dourie Castle, 13th century stronghold of the Fentouns (subsequently Beaufort Castle) passed by marriage to the Frasers of Lovat in 1511. The chiefs of both the Frasers and Chisholms (originally incomers) gradually absorbed the lands of the lesser lairds. Alexander Chisholm, second son of Sir Robert, who was Constable of Urquhart Castle in 1359, married Margaret del Aird of the Aird, Lady of Erchless in 1368. To the east of Erchless Bridge on the A831 is Hall-hill and the earthwork at the top was most likely the earliest motehill castle of the del Aird or de L'Aird family first mentioned in 1296, possibly of Scoto-Norman origins like the Byssets and Fentouns, but adopting a local territorial surname.

To put Charles Aird's diaries into perspective and to identify the family members in his narrative, I have summarised the family tree and the parts played by some of its members. Basically, it covers the first forty years of his life. He retired from the family firm in 1876, having suffered a major stroke, but did not start his diary until a few years before his death. Unfortunately it stops soon after he became an active director of the family firm. Having previously obtained much experience in managing and expanding, amongst others, the Kingston Gas Works, he was in a unique position to control John Aird's ever expanding involvement within the gas industry.

In the 18th Century, the Airds lived in the Black Isle and went through a "period of depression" after the '15 uprising when father and eldest son were not able to see eye to eye. The property was divided and this son became a tenant farmer.

The known family background, according to the Ross & Cromarty Register of Births, recorded David Aird of Teanuald who was born around 1740 and who married Roberta Munro of Foulis. They had seven children, of whom Robert was the fourth. Robert, born in 1773 at Rosskeen near Alness, married Elspet McRitchie. She was the daughter of John McRitchie, sometime of Templand and who afterwards farmed on the Black Isle. He was born in 1709 and his wife Elspet Junor, born in 1749, lived to 98. The McRitchie tombstone, with family details, is at Avoch churchyard.

From early records in the Ross and Cromarty fiche, Robert and Elspet had five children before John. Janet and Jean were born at Avoch in 1791 and 1793 and Elspeth, George and Alexander were born at Rosemarkie in 1794, 1796 and 1799, but they all remained in the north and faded out of this history. [As an interesting footnote; many years later at John Aird's funeral in 1876 the completely unknown brother Alexander turned up out of the blue. The family fitted him out in full highland dress, gave him some money and sent him back up to obscurity in the Highlands. However, his brother had remembered him in his Will]. Robert and Elspet both must have been of sturdy and independent stock. They farmed at Insch near Avoch, but some seven years before Waterloo they went to London to find employment, partly walking and she with no language but the Gaelic. They arrived in 1808 bringing their youngest son with them, and Robert Aird obtained work with Hugh McIntosh the civil engineering contractor. He then became a ganger (foreman), but was

John Aird (1801-1876), founder of John Aird and Sons,
after his marriage in 1831.

killed in an earth fall leading his men to safety whilst engaged in excavating the Regent's Canal. A London solicitor, James Lemon, wrote in 1901 to say his father was present when Robert was killed. The ground at the side of the cutting had moved and he called in his men to lead them back, but stumbled and fell when moving backwards himself and was crushed to death. His 16 year old son, John, was thus left to provide for the family and initially achieved this by working as a "cat's meat man" i.e. taking a small barrow round selling horsemeat from a knackers' yard as petfood. John's mother Elspet comes into the family history from time to time.

John (1801-1876), whose eldest son was Charles, features largely in the diary. His brother Alexander married and had a daughter Ann, who in turn married into a family by name of Munro. This could have been the same Munro family who helped John when he obtained work in the Phoenix Gas Company of London.

In 1831 John married Agnes Bennett, daughter of Charles Bennett of Lambeth who died in 1869. Charles Aird's diary tells us a lot about his father John, who must have been an extraordinary man. As a totally uneducated, virtually illiterate boy, he taught himself to read and write in the evenings after work. He had to find employment to provide a home for his mother and worked initially as a stonemason in Bromley by Bow south of the River, before joining the Phoenix Gas Company. John and his wife then lived in a cottage inside the works next to a knackers' yard and after ten years he rose to become initially an inspector and finally the works' Superintendent. He then began to obtain outside work in his spare time and finding that there was a demand for laying water and gas pipes, he started to take contracts in London and elsewhere; for example Moor Park for Lord Ebury. So successful was this venture that he dropped his connection with the gas company and, at the age of 39, formed the firm of John Aird and Son, Constructional Engineers with a workforce of 100 men.

By 1850 his company and Mowlems had become the major contractors in the country. Two small portraits of John and his wife Agnes have been handed down in the family through four generations, otherwise there is no physical reminder of that early generation. Not only must he have been extremely able, with endless energy, but also farseeing

Agnes Bennett, daughter of Charles Bennett of Lambeth,
married John Aird in 1830. They had ten children.

and withal a kindly man. After training, he put his sons and sons-in-law into various businesses, all designed to service and assist the contracting enterprise. Even when they were children, his sons were sent out on a Friday night to different pubs (the gathering points) with the wages for each gang's weekly pay. The company was latterly split into two parts, John Aird & Co. based at Queen Anne's Gate, Westminster, responsible for constructional engineering and John Aird & Co. at Belvedere Road, Lambeth, with the responsibility for all the gas side.

In Scotland John Aird joined up with Sir Arthur Lucas to build the Ardrossan Dock and the West Highland Railway. Later on the Westminster Construction Company was formed to look after government contracts. The manager of this company, J.M. Summers and his brother O.C. Summers set up on their own after Airds finally broke up in the early 20th century. The following obituary from the ***Elgin Courant*** of April 14th 1876, gives a good summary of John Aird's career.

> THERE died in London, on Sunday last, a gentleman connected with the North of Scotland, whose life has been a sufficiently remarkable one to entitle him to more than a short newspaper notice - Mr John Aird, the head of the firm of Messrs John Aird & Sons, the great London contractors. He was a man who took rank with the late Mr Brassey, the great railway contractor, though the work of his firm was more in sewage, gas, water, harbour, and other undertakings of that nature, than in railways, although latterly he had also very considerable railway contracts. Deceased was a native of Ross-shire, but left Scotland a mere boy. When the late Mr William Munro, who built the fine house looking up the High Street of Elgin at the Little Cross, and spent the end of his days amongst us, some twenty years ago, was engineer of the Phoenix Gas Company in London, Mr Aird turned up a young lad among those employed in connection with the works. His superior knowledge, skill, and character, attracted notice, and he was, after a time, placed in charge of some departments of the work, and then entrusted with contracts on his own account. Ultimately he left the Gas Company, and, when contractors were asked to estimate for works that he saw his way to undertake, the young man from the far North competed for them, and, having his early friend, Mr Munro, to stand by him

Charles Aird (1831-1910) when an old man.

until he became better known, he got into business. At first his transactions were on a small scale, but, as it became known that his fixed principle was to execute whatever was entrusted to him in the most substantial and perfect manner, the sphere of his operations rapidly extended. From one point to another his business rose to be one of the largest in the kingdom, The London contractor, who had earned a reputation for failing in nothing he undertook, was encouraged to look at works of importance and difficulty in the provinces, on the Continent, and also in America. He took gas and water into Berlin, where he laid down some eighty miles of pipes, and so satisfactorily did he carry through these very important works, that the Government presented him with a magnificent testimonial and the freedom of the city, in acknowledgment of their satisfaction with his work. He took water also into Amsterdam, a work of very great difficulty, and had large contracts in Rotterdam, Schiedam, Hamburg, Archangel, Singapore, Copenhagen, and other places, At Copenhagen he carried a tunnel under an arm of the Baltic seventy feet below it, and lost a large sum of money by it, as it turned out to be of sulphurous stone, in which the men could work only an hour or two a day. But he never failed to finish what he undertook, whatever it cost, and thus he was often preferred to others when estimates came to be considered though he was not the lowest offerer. In Ottawa, and at other places in America, extensive contracts were entered into by Messrs. John Aird & Sons, for, when his large family grew up, he took two of his sons into partnership with him, and they have in recent years chiefly conducted the business. Friendships led them even to Scotland to undertake one or two small contracts. The Buckie Water Works were executed by them, and so satisfactorily was the work done that we believe we are right in saying that not a shilling has been required to be laid out on it since. To the village of Urquhart and to Innes House Messrs John Aird & Sons brought water from Lochnabo, and, though the works have been in operation for perhaps twenty years, no repairs have been required. The firm would in all probability have been the contractors for the great railway bridge across the Firth of Tay but for the condition insisted on by the promoters of it that, with the other work, there should also be joined certain responsibilities connected with iron. Gas and water works they erected without number. In

> London they had a large share of the works going on, such as the Thames Embankment, gas, water, and sewerage operations. It has been said there is hardly a street in London in which they had not laid down either gas or water pipes. At Christmas of 1874 a reckoning was made of the people in their employment in various parts of the world, and they were found to be considerably over 30,000. The strength of the firm of John Aird & Sons was brought into special notice by the money crisis of 1867. Sir Morton Peto, Bates, & Co., who had then perhaps the largest contracts of any firm in the world, succumbed under the monetary pressure of that memorable time, and Messrs. John Aird & Co., in the great majority of cases, took up their contracts and finished them, and made by them the fortunes that the original contractors lost. In the midst of all the responsibility of a world-wide business of such gigantic proportions as we have indicated, the gentleman who has now been gathered to his fathers at the ripe age of about threescore and ten was as serene, and calm, and happy as any man could be who had nothing to think of except how to enjoy himself. Most gentlemanly in every way, he had a nature literally overflowing with kindness and goodness of heart. And much of his warmest feelings were towards the North of Scotland, of which he was ever mindful, as was shown in the frequency and liberality of the contributions he made to the Elgin Museum through his friend, P. MacBey, Esq of Darliston. In the course of his very extensive works, whenever anything was found of special interest, it was duly forwarded to Elgin to enrich its already interesting museum. Those who have had opportunities of meeting the deceased on his visits to the North have the most kindly recollections of the old gentleman, and every person he knew in the North had in him a warmly-attached friend on going to London.

Charles Aird's diary mentions his other brothers from time to time; his father John quite a lot, and of course much of his own family life as well as the events in his own business career. He was first articled to a well known surveyor – Henry Austin (brother-in-law of Charles Dickens), but this was not a good start in life. His diary discloses that Austin gave him no technical training for three years and he had to occupy his time

Nellie Aird (1861-1947), eldest daughter of Charles

by making occasional drawings, or copying plans. Eventually he went to Canada to assist in surveying a route for the proposed Canadian Grand Trunk Railway. He then returned and had various jobs in the gas industry, eventually becoming manager of the Kingston Gas Works and buying Pope's villa in Twickenham in which to live (rather to the concern of his parents who thought he was over stretching his finances).

He wanted to manage more works, but with all his background his father felt it was time for him to join the family firm, using his training and experience to supervise the gas side. He was 35 years old when he became a partner. In Charles's time and subsequently, the firm built many gas plants – Oxford, Rugby, Eastbourne and Birmingham are examples in Britain – and the firm also built abroad; notably Palermo, Rome and Milan in Italy; Ottawa Parra in Brazil and Behia Blanca in Argentina. In England they built the Becton gas works, then the largest producing plant in the world.

John Aird's eldest son Charles (my grandfather) married Ellen Noakes (1843-1945) in 1860. We have miniatures of them both painted by their daughter Edith who had gone by herself to Paris to study under Millais – not quite the done thing for a girl to do in those days! Charles and Ellen went on to have 11 children, my father Kenneth (1871-1941) being the ninth. Only three of his daughters married: Margaret (1866-1938) to Sir Alfred Dent, Founder of the Chartered Company of North Borneo; Mabel (1869-1956) to the Rev. Westgarth Duval and Maye (1871-1950) to Ernest Jörgenson. These eleven children only produced five grandchildren between them. I well remember my grandfather Charles's wife Ellen: she once asked me, a little boy of six if I would like to see her brush her teeth; she then removed them all from her mouth whereupon I fled, terrified, from her bathroom. We used to stay with her every year at East Grinstead, surrounded by staff and her four remaining unmarried daughters. She lived to be 102, and became a bit vague at times. Once she had a long talk with my mother, who had married her son Kenneth some 30 years before. On parting my grandmother said you must come and meet my son one day, you will like him!

John's second son, also John (1833-1911), married Sarah Smith the daughter of Benjamin Smith of Lewisham, in 1855 and had nine

Ellen Louisa Noakes (1843-1945), wife of Charles Aird (1831-1910),
about the time of her marriage

children. In due course, when his father died he became the driving force in the family firm, concerning himself initially with the construction and engineering sides. He was very much a "front" man and his greatest achievement was in constructing the Aswan Dam, plus its two barrages down river, in Egypt. This was first opened in 1902, enlarged in 1912 and doubled again in 1931; for this work he was created a baronet. His eldest son John (1861-1934) married Ellen Renton and I remember them both before the War; he was a kind and gentle man. Cousin Nellie was quite firm and formidable, but she used to give children's Christmas parties at the Rubens Hotel in London to all the young family. My sister and I used to go - a highlight being the digging for presents in a huge wooden bran tub. His next son Malcolm (born 1872) married Nellie Dummett; and of his daughters, Sarah married Basil Ellis, whose family worked in the family firm. Jessie married G. Abernethy; Kate married Alfred Bean; Ada married Louis Miéville; Vida married Edward Anderson and Dorothy, Spyridion Mavrojani. Charles's diary mentions his brother John from time to time when growing up and afterwards in the firm.

Of the rest of the elder John's family, Alexander (born 1836) married Janet Keith and was with the firm for a short period; also living in Germany. Agnes (1837-1939) married Henry Gill, a director of the Berlin Water Company (built by Aird's) and they lived at Potsdam. Jessie (1842-1918) married Sydney Hargrove, a London solicitor and he, together with the following two generations, became the family lawyers for all the Aird descendants. In 1872 Joseph (1845-1906) was given the Wellington Tube works by his father in order to supply the huge quantities of pipes, both water and gas, needed for the Aird contracts. Although apparently of quiet disposition, he developed the works at Tipton near Birmingham into a concern with world-wide outlets, eventually also specialising in small bore tubing. My father had hoped to have been offered this firm when his uncle died, but instead was given 1000 shares, being one fifth of the capital. The latter increased to a value of several millions under the control of a high-powered family named Turner. In 1974 Wellington Tubes combined with Gibbons Dudley, a large manufacturer of furnace and associated bricks.

Of the remaining children, William (1840-1917) married Blanche Stock and emigrated to Adelaide; Henry (1849-1914) married Martha

Margaret Dent (1866-1938), fourth daughter of
Charles Aird (1831-1910) and friend.

Henry Aird (1849-1914), sixth son of John Aird (1801-1876), emigrated to Australia.

Rutherford and emigrated to Melbourne. Walter (born 1852) emigrated first to Australia, then to Oregon. He married Leah Bailey in 1876 and had two daughters, one called Violet, before emigrating to America in 1882. George, the youngest died when only a year old.

John Aird and Sons reached their peak during the second half of the nineteenth century and continued up until the Great War. The firm had become (after Thomas Brassey) the largest contractors in the country – if not the world – carrying a labour force of over 30,000 employees with additional help as necessary. At its height, the Aswan Dam project required an additional 20,000 local labourers. (It is interesting to record that a hundred years after the building of the Dam, the Old Cataract Hotel at Aswan issued a special commemorative menu, reproduced on page 398, for its centenary. This included dishes named after those responsible for the Dam's construction, together with other more recent notables. Amongst those mentioned were the Khedive, Lord Cromer, Earl Kitchener, Howard Carter, Agatha Christie and John Aird).

Apart from their gas interests, John Aird and Sons laid and built most of the pumping stations, mains and reservoirs supplying London with water - including Hampton and Staines. They laid the water and gas to the Great Exhibition of 1857 and again when it was rebuilt as the Crystal Palace. The firm constructed the water works of Southwark, Grand Junction and West Midlands. Abroad, the water works in Berlin, including the laying of 80 miles of piping, the water works in Rotterdam, Amsterdam, Copenhagen (which needed a tunnel 70 feet below the Baltic), Hamburg, Altona, Schiedam, Brunswick, Simla, Calcutta, Cagliari, Riga, Archangel and Moscow were all constructed by John Aird. They built the first ever outfall sewer in Copenhagen and, in conjunction with Brassey, they built the unified sewage system for London plus the Thames Embankment and part of the Metropolitan and District underground.

In the 1860s, as Lucas and Aird, the firm began to take railway contracts. These included the Thames Valley railway; work on the Midland, Metropolitan and Great Eastern Railways, including Liverpool Street station and an industrial railway network in Hull. In 1885 they built the West Highland Railway, which ran from Glasgow across Rannoch Moor, through Corrour and Roy Bridge to Fort William. To cross the

OLD CATARACT
Aswan

100th ANNIVERSARY
1899 OLD CATARACT ASWAN 1999
Hotel Sofitel

LES POISSONS - *FISH DISHES*

DUKE OF CONNAUGHT	Le damier de rouget et filet de sole au pistil de safran, aubergines et tomates aux herbes *Red mullet and fillet of sole with saffron, compote, of squash eggplants and tomato with herbs*	**62.00**
KHEDIVE ABBAS HELMI	Le filet de loup de mer et sa petite compote de tomates à l'anis, accompagné de légumes du jour *Sea bass fillet with tomato compote in anise with vegetable of the day*	**56.00**
HOWARD CARTER	Les pétales de saumon cuisinées à la sauce verte, semoule aux petits légumes *Salmon petal cooked in green sauce, semolina with vegetables*	**60.00**
LORD BENBROCK	Les gambas mijotées aux courgettes et sauce aux crustacés, mitonnée de petits légumes *Jambas with courgettes, crustacés sauce and simmered small vegetable*	**85.00**

VIANDES - *MEAT DISHES*

AGATHA CHRISTIE	Les pigeons farcis au blé des Nobles de l'Ancienne Egypte, petits légumes acidulés *Stuffed pigeons with nobles wheat in ancient Egyptian style with acid vegetable*	**61.00**
JOHN AIRD	La fricassée de blancs de volaille aux crevettes et banane gratinée, riz basmati et poêllée de légumes *Fricassee of chicken breast with shrimps, gratinated bananas, basmatic rice and fried vegetables*	**59.00**
MR. FAVARGER	Le carré d'agneau rôti aux herbes et petit jus de tapenade, pommes Darphin et tomate confite *Roasted rack of lamb with herbs, tapenade juice Darphin potato and Confit tomato*	**63.00**
LORD MOUNBATTEN	Le Rossini de filet de bœuf en papillote croustillante, fricassée de légumes, jus nature au cognac *Beef fillet Rossini in crusty light pasta, vegetables fricassee, natural juice with cognac*	**75.00**

Menu from the Old Cataract Hotel, Aswan

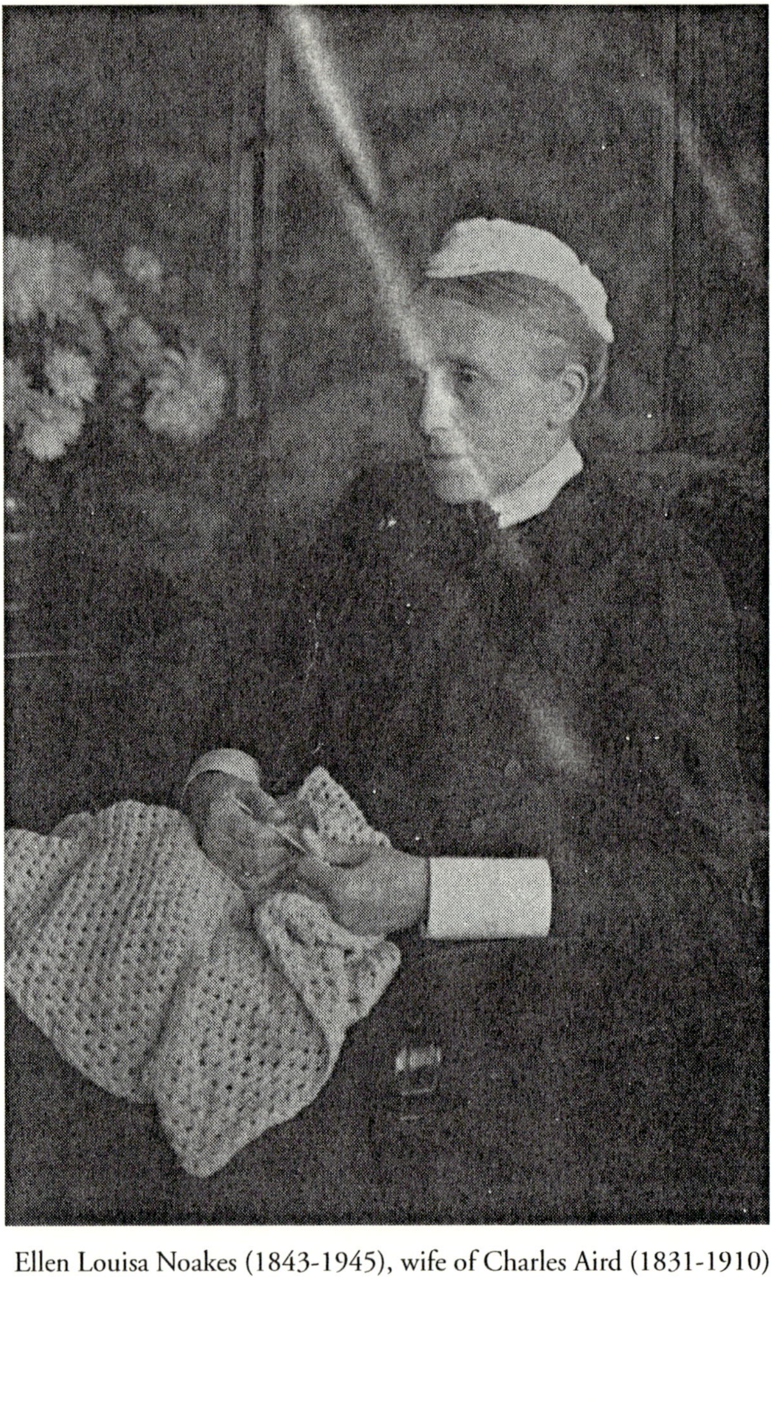

Ellen Louisa Noakes (1843-1945), wife of Charles Aird (1831-1910)

marshes, they used the old Roman method of great bundles of brushwood covered by many thousands of tons of earth and ashes.

Many docks were built. The Millwall Docks in 1863, the Royal Albert, the West India and Tilbury Docks in the Port of London; Portsmouth, Southampton, Newport and Newhaven dock schemes; the Ardrossan Dock in Scotland; the Royal Edward Dock, the Avonmouth and finally the Singapore Docks. This latter project was their last major contract and, due to inaccurate survey work, led to huge losses.

The firm worked with other contractors, particularly the Lucas Brothers, with whom they carried out railway and civil engineering contracts. It is interesting to note that some of the great contracting names of today – Wimpey, McAlpine and Mowlem – all started working with John Aird. Wimpey and McAlpine started as subcontractors and Mowlem repaved the streets after Aird had laid the gas or water mains.

In 1910 John Aird (the son) retired, having suffered a stroke. Charles Aird had already retired in 1876 and the next generation lacked the same interest, drive, single purpose and inspiration that brought the firm to such heights. Other firms were expanding and competing and there was a huge financial loss at Singapore, a million pounds being involved; so the family finally decided that the firm should be liquidated. It is interesting to note that subsequently, a number of Aird's senior managers with their own staffs, carried on as contractors in the same fields and each under their own name; some of them well known to this day. William Press, Avant, G Wilson in Holland, E. E. Jeavons, Pearce, R. Robinson, O.C. Summers and Whittaker Ellis are examples.

The two names that stand out above all others in this saga are John Aird the Elder (my great grandfather) and his younger son, also John. The former was an uneducated boy who, having to provide for his mother and with nothing but sheer determination, educated himself, found a job, started pipe laying in his spare time, founded the company and then developed this into a large international concern. One of his greatest assets which gave him the advantage over others was that he always finished his contracts and kept to his stated costs; in other words he kept his word.

The latter (the younger John Aird) also had great determination and single purpose. He continued the expansion of the firm from 1876

William Aird, (1840-1917), fourth son of John Aird (1801-1876), emigrated to Woodville, Southern Australia when his father died and his brother Charles had a stroke, until he himself had a stroke and retired in 1910.

"The Master Builders", a book by Keith Middlemas, published by Hutchinson in 1963, is a study of the four outstanding contractors of the nineteenth century and one of these was John Aird and Sons. The "West Highland Railway" by John Thomas, reprinted by Redwood Books, is the story of John Aird's construction of the line from Glasgow to Fort William and ultimately to Mallaig.

To complete our side of the family history; following on from Charles Aird, his son Kenneth decided that the new electrical industry was where his future lay. However, his father, with all his Aird background, put him into the gas industry and he was trained under Henry Hack of Birmingham, then the foremost gas engineer in the country. In 1887 his uncle John Aird had joined the Board of the old Gaslight and Coke Company of London, whose governor was Sir William Makins. The latter had been asked to advise about the future of a gas appliance firm whose partners could not see eye to eye with each other. This firm was called Glover and Main, and Sir William Makins arranged to install his son Paul and my father (John Aird's nephew) as directors of this concern which involved the manufacture of gas cookers, fires, meters and catering equipment. They had factories in Scotland, where my father first started as the Works Director, and London, plus seven branches throughout the country, Ireland and Holland; with another works at Croydon later on. Apart from making shells and army equipment in the two wars, including sub contract work on aircraft during the second, the domestic appliance side expanded so much that the output of cookers alone had reached five thousand a week by the mid thirties. Kenneth Aird spent his whole life with the firm, ending as Governing Managing Director and Chairman.

I myself, the fourth generation to be connected with the gas industry, joined Glover and Main as a management trainee in 1933 and, apart from six years war service, progressed to becoming Executive Chairman of the group. The previous Chairmen had been Sir Ernest Hiley and then Colonel William Glover. By the time the firm was taken over by Thorn Industries in the 60's (basically an electrical appliance group who wanted a major stake in the gas industry), we had expanded the two

Kenneth Aird (1871-1941), second son of Charles Aird (1831-1910), with his Argyll in 1914, at Maryfield, Bridge of Allan.

original companies into a group largely centred in the gas industry. This included George and William Glover, Braddocks, Sawyer and Purves, all meter companies. Main Morley, who manufactured washing machines, Main Vitreous Enamel, Diaphragm and general leather manufacturers, Main Gas Cookers, Fires and Refrigerators; and the two largest catering firms in the country – James Stott and Benhams – with the heating and ventilating company connected to the latter. The era of the specialised companies, some with an almost family interest in their workforce, was now giving way to the multinationals.

PDMSA,
July 2004

1
Childhood (1831 – 1841)

I was born on 12th December 1831, the eldest child of John and Agnes Aird. I was christened at St George's Church, Southwark, as my father and mother were then living in a small, six roomed cottage in a yard occupied by the old Phoenix Gas Company for Stores and Stables in Wellington Street, Blackfriars Road. This was a very squalid part of London, with a knackers' yard adjoining for the preparation of cat's meat from dead horses. At that time I remember these men used to be seen going about in all the Town's Catsmeat barrows with choice morsels on a little wooden stick for the domestic pussy. Such a thing I do not think is seen anywhere now.

My father was then an outdoor foreman or inspector of that gas company. I cannot remember how he became connected with the company, but I have often heard him say he was born in Scotland in Rosshire. When he was a lad of about 16 he came up to London with his father and mother to work on the Regent's Canal; Mackintosh, the contractor, having that work to construct.

My grandfather was one of his gangers over the navvies employed and my father drove one of the horse trolleys. Unhappily, during this work my grandfather was killed, with several other men, by a fall of earth in one of the large cuttings. This left my father the task of supporting his poor old mother who could hardly speak a word of English, nothing but the Gaelic. I remember the old lady well in my child days and many years afterwards, living in her cosy little room near Wellington Street and dying there at 84 years of age.

How my father managed to succeed in this task, I do not know, but he did. He had never the chance of any schooling, but taught himself to read and write and would sit up, night after night, reading any books on any subject he could get hold of. His old mother never tired of mentioning this, she was so proud of him and considering in those times there were so few books published, no board schools, free libraries or Education Acts, it is simply marvellous that he should have qualified himself in this respect, to take the position in the great and busy world that he did – and so successfully too. But then he was one of those few that Dr Smiles in his books 'Self Help', 'Character' and others, has so pleasantly and faithfully represented and was born also one of Nature's noblemen.

Returning now to my own childhood days, my mother was the daughter of Charles Bennett, a well known veterinary at Vauxhall; my parents for some reason preferring to name me after him and not after my father. (He went to Russia for some Russian nobleman and died there). At this time I cannot remember much of my father, being away early and late at his work. I can, however, very well remember that on the day Queen Victoria was crowned, when I was about five or six years old and my brother John two years younger, my father put some seats in one of the Gas Company's vans and drove my mother, myself and some few friends about the streets to see the Illuminations. Soon getting tired of it, I went to sleep very comfortably on the floor of the van. I think about this time my parents had some fear that I might not be reared, being so frail they feared I would fall to pieces. They often told me this years afterwards and that is probably why they took me to Margate and bribed me with macaroon biscuits, to be dipped in the sea.

Weston Green, Thames Ditton is one of those untidy irregular bits of common one meets with so often in Surrey, with its patches of gorse, its dirty pond and flocks of geese quacking about – but still pretty and picturesque with its surrounding little cottages. When I was about seven years old, we visited an old friend of my mother's, a Mrs Freakes, living in one of these cottages with her husband, a corn chandler and miller with no shop, but three little bins in the front parlour filled with various kinds of meal to sell. This kind old friend, seeing that my brother and myself were not doing much good running about this yard and the street, like little street arabs, suggested that we should be sent to a little village school she knew of at Esher, about a mile or so from Weston Green, and she would look after us. As we knew nothing whatever of schools or education, my father thought well of the idea and took us to Esher and placed us in the care of a Mr Pounds. There were only our two selves to board and live in the little house, the other few little children being from the village. The school room, as it was termed, was a small wooden structure like a barn in the little bit of garden at the back.

Esher - what a pretty, ideal village this was 70 years ago; no church then erected on the beautiful green, as level as a billiard table. On one side this was bounded by the Portsmouth Road, on the other by the Squire's wall of the manor house, with at least a dozen peacocks on the top of it, sunning themselves and spreading their beautiful tails and uttering their strange cry. The remaining side had cottages of various sizes, the little chandlers' shop with its one little window, ornamented with glass bottles containing several sorts of sweet stuffs, jumballs and biscuits called Parliament. Then the blacksmith's shop, the little village inn with its wooden benches outside for the villagers to drink their beer, smoke their pipes and watch the cricket match. Then comes the little village school. Such villages have been depicted so well in Bulwer-Lytton's books, 'My Novel', 'The Caxtons' and 'What Will He Do With Me?'

I learnt my ABC here and having, I believe, a lot of my father's character about me, it was not long before I could read any book – in fact devour them - whenever I had the chance. Literature was then very different to what it is now and books for boys very scarce.

At this time I little dreamt how closely in later years I should be associated with all this part of the country. My brother John and myself, I think, remained here about a year, with occasionally a visit to our kind old friend Mrs Freakes, remembering well the seed cake always ready for us, with some ripe gooseberries from the little back garden. I must mention also my father coming to see us and taking us to Hampton Court. I have never, never forgotten this, going over the Palace and Gardens, but my delight, being taken on the river for a row, was almost too much - the beautiful cool, clear, rippling water – so clear I could see the little fish playing about when the boat was stationary. During the whole of this life there has never been a greater lover of nature than this poor old man.

Soon after returning to London, we had a most unexpected treat, being taken to a theatre, near to the old Surrey. At that time I had very little idea what a theatre was like. It can therefore be very easily imagined what a novelty and astonishment it was to us. It was arranged that we should have tea first at the Leopard Coffee House for working men close by and we then thought it an additional treat to sit at the wooden benches and drink our tea in china pint mugs and eat two large slices of bread and butter an inch thick. I can even now fancy the strong smell of sawdust and oranges about the theatre and hear the click of the boxkeeper's key opening the doors to find room for us. Before going, my mother tried to persuade us to go to bed in the afternoon to sleep and be quite fresh for the evening, but that was a failure. I remember on this occasion we were accompanied by a very great friend of my father's – Mr Joseph Quick, of whom much more anon.

I can also remember being taken once to the celebrated Vauxhall Gardens shortly before they were closed and built over.

Some time after this event we were all greatly excited to learn that my father had been appointed Engineer and Manager to the Phoenix Gas Company works at Greenwich. To our little minds at that time it seemed to mean undreamt wealth and luxury. Although those dreams were never realized, it meant a great change and rise from our then surroundings, but even then I am sure we were happy. I must have been a strange child for I remember taking walks in the early morn in the New Cut and Lambeth Marsh and watching the many hucksters and costermongers with their barrows and stands on both sides of the road, and feeling so happy and joyous with the lovely scent from the wallflowers and daffodils they had for sale as well as the many little plants for gardens such as pansies, daisies and myosotis.

Before proceeding further, I should mention that the chief works and offices of this company were situated at Bankside by the River Thames. There was another carbonizing station by Vauxhall Bridge; the Gasholder Station being in Kennington Lane. The Chairman of the company was a Mr Shuter, then residing near Bexley in Kent. This gentleman must have been the grandfather of the present family of Shuters, the well known Cricketers, Johnny Shuter having been Captain of the Surrey county eleven for many years. On consideration, possibly he may have been their great-grandfather, but one or the other. The Chief Engineer was a Mr Munro, a little old man, a Scotchman, and the Secretary was a Mr Charles Pridden. All these gentlemen evidently had a very great regard and opinion of my father and shewed him great kindness, and when later on, a committee of the directors came to Greenwich to inspect the Works, they always had a kind word for us boys should they happen to come across us.

At last the time came for our removal to Greenwich. There was then no direct railway service, only as far as Deptford; the railway seemed to me to be worked with ropes and I could not make out where the horses were.

The gas works were situated in the very poorest part of the town, as most gas works are, on a tongue of land about eight or nine acres, facing the River Thames on one side, and surrounded at the back by the River Ravensbourne, the mouth of the river coming into the Thames here.

The only street to enter the gasworks from the town was Thames Street, running parallel with the river and on each side with poor little houses for work people, mostly from the gas works. At the dead end of Thames Street, there were large gates and a side gate with a bell. How well I remember the sound of that bell (the gates always being shut at 6 o'clock) longing for my father's return home when away.

Just within the gates stood the manager's house and office, alone, looking of course more important than the home we had left, but with very little more accommodation for a family. But it was a great change and it took some time to realise and settle down to it all.

Some distance from the manager's house, there were two little cottages very similar to those we had just left. One was occupied by a young man, just married, named Taylor, at that time the only clerk in the office, except the Collector, a Mr Rutherford. The other cottage was occupied by the Day Foreman, a Mr Swanston; the Night Foreman was a Mr Stevens, for in a gas works it is always both day and night work. Gas must be made. This man always had the keys of the gates during the night.

The other principal men I can remember being Still, the retort setter and bricklayer; Neal, the blacksmith; Crisp, the gas fitter; Winch, the foreman of the coke yard and Mitchell – jack of all trades and messenger. It was not very long before we boys were great friends with all these men. I should mention that in addition, there were probably 50 or 60 stokers and others.

I remember very well we had not been there very long before I had induced the blacksmith to make me an iron hoop and trundler, the

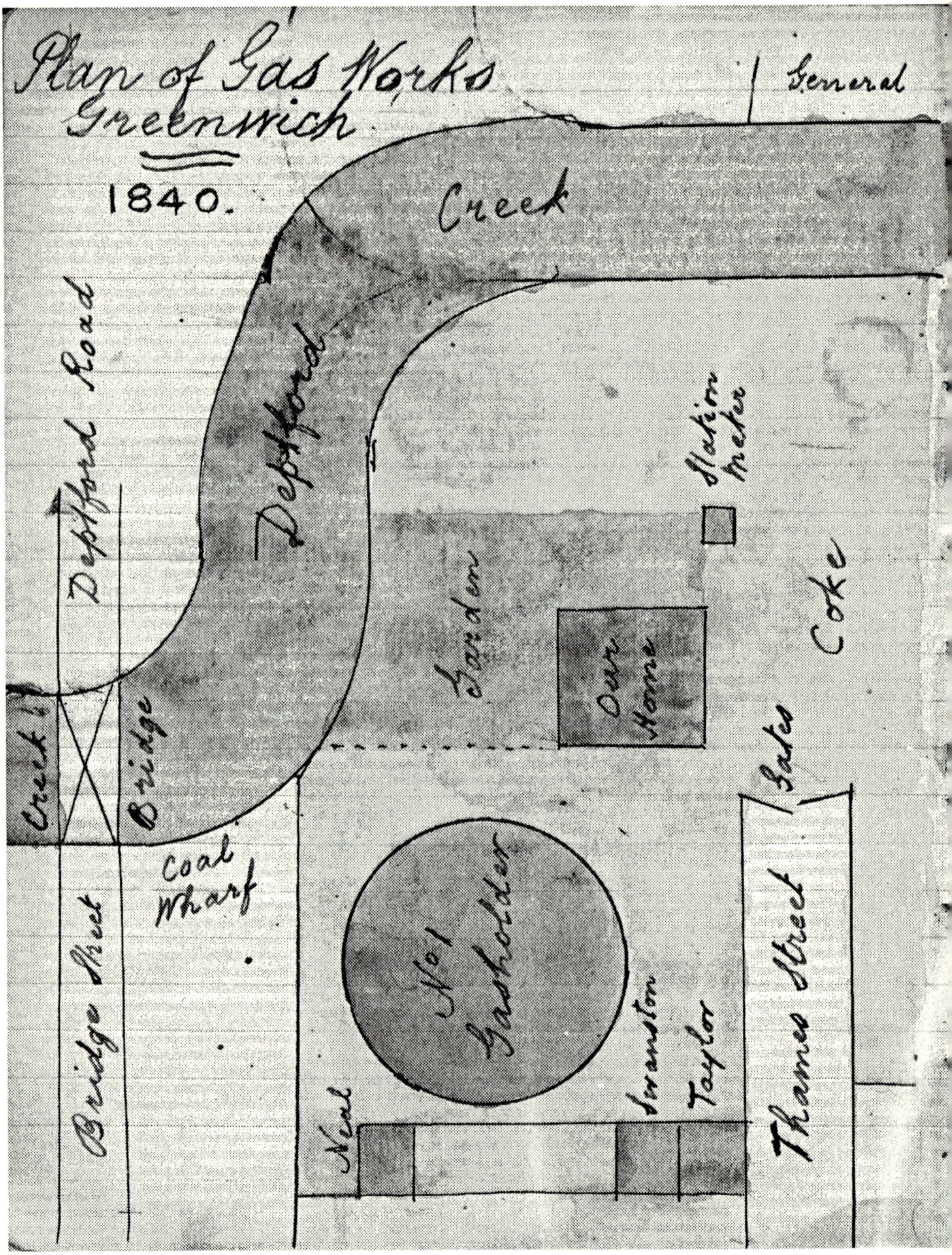
Plan of Gas Works
Greenwich
1840.
General
Creek
Deptford Road
Deptford
Station meter
Garden
Our Home
Coke
Creek
Bridge
Coal Wharf
Bridge Street
No 1 Gasholder
Gates
Thames Street
Neal
Swanston
Taylor

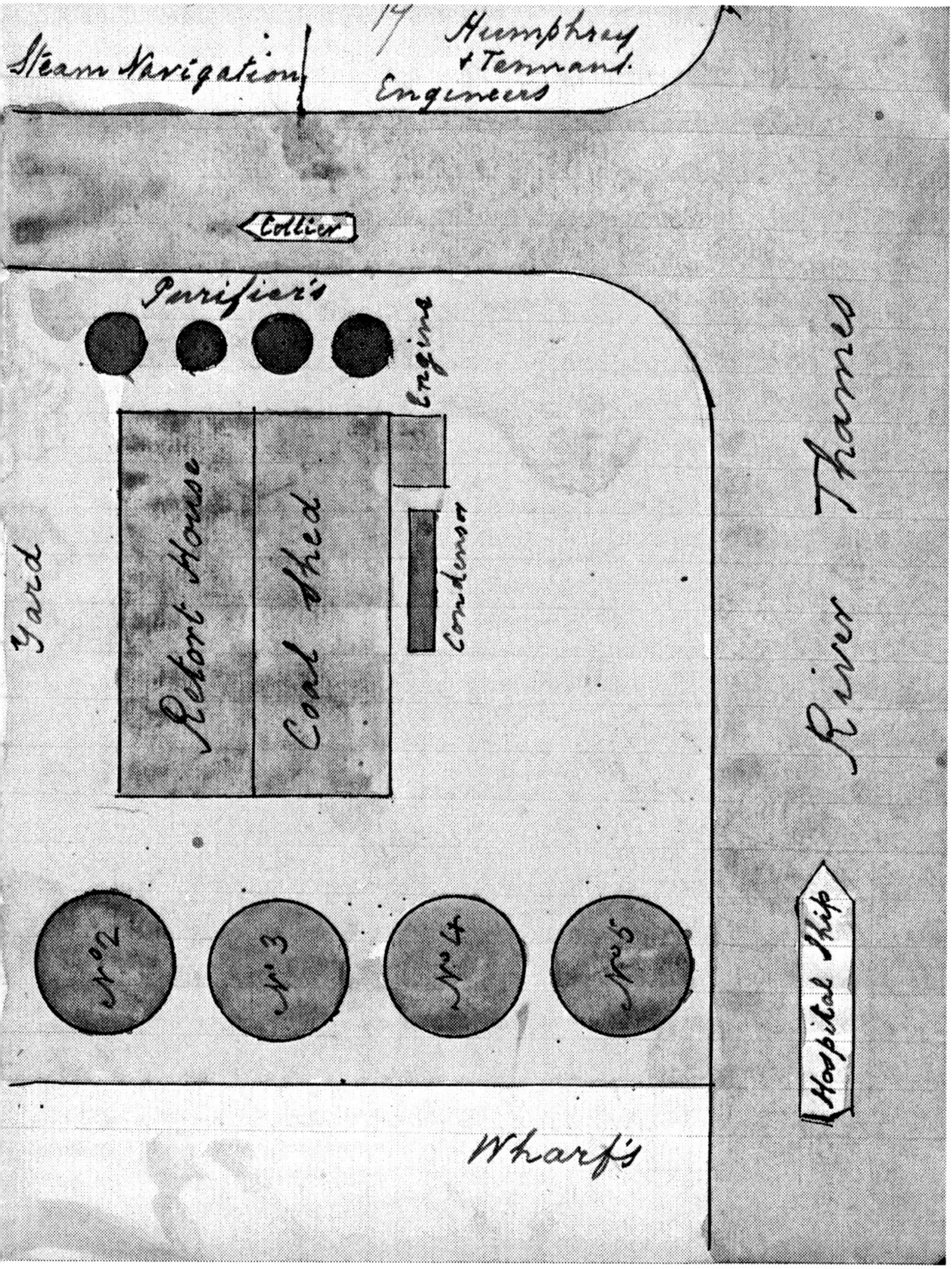

Plan of Gas Works, Greenwich, drawn by Charles Aird

greatest delight I had ever had.

Between the manager's house and the little cottages, there was a large gasholder, and from the side of this round the company's ground and boundary was a timber campshed or quay wall alongside as far as the mouth of the River Ravensbourne by the Thames. Barges could be moored at high tide and at one part near the mouth some strong piles were driven for the colliers, or Geordies as they used to be called. They were ships with two masts bringing 300 to 400 tons of gas coal to unload into the sheds. I remember it amused us very much to see them unloaded; four or five men jumping together from a raised platform on to the deck with a rope, bringing up a large basket or tub filled with coal.

From this point, facing the River Thames there was only a mud bank or beach at high water. Although the water was dirty, it was somewhat exhilarating to see the large vessels going to and fro from the sea to the Pool and to the East and West Indian Docks, that then existed. The old Dreadnought hospital ship was moored in the middle of the river, just opposite, and looking across to the other side was the vast tract of flat wasteland called the Isle of Dogs. How little any of the Aird family then thought that John Aird and his family would, years afterwards, be so closely connected with the construction of the two largest docks in the world on that ground – the Millwall and the Tilbury.

Gasworks as a rule are considered somewhat objectionable to neighbours, the smell in the manufacture being rather strong, but it is really not unhealthy and we did not mind it. Indeed, many children would be brought to the Works to be cured of whooping cough. But I must say it did try us very much at first to find that when the tide went out and there was low water, the shore and the mud in the River Ravensbourne (or Deptford Creek as it was then called), was covered with the most horrible objects – some indescribable: dead dogs, cats and cattle, any number. On two occasions I remember two dead bodies being picked up in the creek, but we seemed to get used to it.

We were very friendly with the captain of one of these colliers bringing coal – Captain Griggs (of course a Scotchman). His eldest son was his Mate and he had two boys at the large naval school opposite the Hospital. There was then a large model of a ship by the roadside for the boys to practice upon. We often went to see these boys, taking some few little things in the way of cakes and sweets from my mother. In return, they gave us some very pretty things worked with different coloured beads. I have often since wondered what has become of those boys. We were always running about these ships, as they came alongside to unload, and many a time have I enjoyed a cut of the salt junk and a hard ship biscuit – the sailors' food.

The wonder is we did not become sailors. I know both my parents at that time were very anxious about us, fearing we would get drowned. The river was all round with no railing or protection whatever.

A little way beyond the manager's house there was a circular building, containing the large station meter and the governor, which regulated the pressure in the district. On the right going on, there was a large space of ground for the storage of coke from the retort houses – in the summer, almost a mountain high; in the winter almost none for sale. On the right again from this space were two very large, long buildings, the retort house and coal shed. At the end of these buildings, near the collier ships were the purifiers, condensers, tar and ammoniacal liquor tanks. At the other end, facing the manager's house, were four smaller gasholder tanks and holders. I think this gets rid of all the Works (illustrated on pages 40 and 41) and now let me try and picture our new house or home, where we had to live the next seven years.

I am afraid all the above writing may seem to others very prosy and uninteresting, but 'thorough' is my nature and I cannot help it. Just inside the front door of the house was a hall, another door in front of this one, being a private door, leading in to the house. A door on the left opened in to the office – a small room, although it then looked to me large. It had a desk by the window that would accommodate four clerks, but in my time it never had more than three.

A door from this office led into a smaller room at the back for my father's use in which were the photometer and other gauges and tests for illuminating power. I used then to wonder to myself, when looking on, where he [my father] had picked up his chemistry, although of course at that time I did not know the meaning of the word; but I could not then understand it. Later on it was to me a marvel how he became so well acquainted with so many scientific subjects.

Another door from the office led into the house. Crossing the passage at the foot of the stairs, a door opened into a room called the Board Room which was used by the directors of the company when they came down to inspect the Works. At other times it was used by the family, but only on special occasions.

The stairs led to the first floor. The bedroom facing the yard over the office was my father's and mother's; another smaller room behind this was a spare room. Over the Board Room was a room dignified by the name of the Drawing Room, but considering that at that time we had no personal friends or acquaintances, it was of little or no use to us. As far as my memory goes, it was mostly used as a kind of hospital, when going through our childhood ailments, which were many, goodness knows. Also it was used for my mother to rest and pick up her strength again after her confinements. At that time the family consisted of Charles, John, Alexander, Agnes and William. Later on at intervals of about two years, there were Jessie, Joseph, Henry, Walter and George, who died in my arms when only about 12 months old.

There was another small bedroom at the end of this hospital or drawing room and two 'attic' bedrooms on the floor above in one of which my two brothers John, Alex and myself slept. There were two little cribs or bedsteads in the room with little planks round the four sides which dropped into sockets in the corner posts to prevent our tumbling out. Well, boys will be boys all the world over. Early in the morning we would begin and imagine we were rival ships, rig up sails with the sheets, pull out all the planks for masts and try to make a gangway from one bed to the other and then the combat would begin. The result was that all that the planks would fall flop on the naked floor, there being no carpet, and making the most awful noise with my father and mother sleeping underneath. There could be but one result to this and in a few minutes afterwards, my father made his appearance with a slipper we had tasted before, and gave each of us a good beating I fancy I can feel even now. Yet a few mornings afterwards, we would find on our pillow, when awake, a particular kind of bun we had a fondness for – such was our father. I am afraid that, when boys, we must have been a great trial to him.

On the ground floor at the back of the house and at the end of the passage there was the kitchen, really our living room, with an ordinary kitchen range, a common dealwood dresser with little shelves for dishes and plates and hooks for little jugs. Most curiously under the dresser there were two very handsome boxes which I firmly believe now were real Chippendale. They were about two foot high and one foot wide, the lid being lifted, there were little interstices for about a dozen knives and forks in each box. I need not mention that such things were quite out of place and character there; and how my father and mother possessed them and where they have gone, I do not know. I only know that they were used on some special occasion and it was my duty afterwards to carefully rub the knives over with mutton fat and put them away.

At the end of this room, there was a large window looking out on to Deptford Creek and also to Deptford Bridge. Along this window there was a large shelf which, when raised, formed a table for ironing and when not in use for that purpose, made a useful table for me to work at. In the further corner of the room was a large copper with a fireplace under, generally used for washing days, but at Xmas time it was used for boiling the Xmas puddings.

Under the stairs there was a cupboard or pantry, and a door opening into a small covered courtyard only a few feet long. On one side was a cupboard for fresh meat and on a ledge, a stone pitcher with a little cup for anyone thirsty. A door on the right led into a little scullery with sink and gas stove for cooking. At the back of this, a place for the man Mitchell, the man of all work, to clean knives and forks, boots and the like. At the end of this little courtyard, a door led into what was called the garden. A little path led in front to the campshed at the creek. Here there were some few willow trees with a number of what we called 'musk bugs' generally upon them (a large beetle smelling very strong of musk). Right and left of this little path there were some few cabbages planted and a few untidy earthed up celery beds and that was all. A more miserable piece of ground I have never seen. After a little while, I thought I would try and see what I could do to make a garden and finding that Winch the coke foreman was really a gardener, I got some few lessons from him and became quite interested in my work, sowing seeds and later on helping my mother in her housekeeping by supplying lettuce, radishes and little spring onions of which there was a large consumption in our family. The four seasons – spring, summer, autumn and winter - seemed then distinctly defined, not as now mixed up altogether, and I then always associated the 14th February (St Valentine's day) as the first day of spring and commenced my gardening operations.

I think I have now described the home where we lived the next seven years of our lives, and were seemingly quite contented. To many it might seem a very hard sort of life and no doubt it was, but we had never known any other and never dreamt then seeing any other. My dear mother worked away here as she did in London, keeping the house clean and in perfect order, doing all the cooking, and what a wonderful cook she was! She did all the lighter portion of the washing and ironing with no regular female servant, indeed no others to help her, but Mitchell, the man of all work, who, in addition to all boots and shoes, knives and windows; scrubbed all the floors (there were no carpets, except little pieces in the bedroom, the board room and the drawing room and a little piece up the stairs, and a piece of oilcloth along the passage). He also dealt with all messages and did all the family marketing.

Of course at times when my mother was unable to work, one of the men's wives would come in to help, but that was always a very rough time for us, and a young girl, a niece of the foreman Swanston, acted as nurse.

Every Saturday night Mitchell started out with a large basket and a list of things required. I was sometimes allowed to go with him with a penny to spend myself. This most frequently went in the direction of a certain little pie which I then thought delicious. But looking back seriously to those days, the marvel is that I lived so well. I remember well the first Saturday I went with Mitchell marketing. He went into a large shop to buy some cheese and butter for the next week, and, to my great astonishment the man behind the counter serving us was Mr Taylor, the clerk in my father's office. I learnt afterwards that, being friendly with the proprietor of the shop and Saturday night being such a busy night, he always volunteered to lend him a hand. That was John Taylor all his life.

I might mention that the five most important events in the year for my mother occurred in the summer and in the winter. At the time when all fruit was taken to the market to be sold and made into jam, she would get up at six o'clock in the morning and, with Mitchell and myself, go to the market (quite a large one, I can remember) and carefully select the fruit she required, which was sold in the large bushel baskets. This being sent home, it was then made into jam of various kinds and placed into little gallipots of different sizes. It was then my work to put a little round piece of paper on each pot, carefully covered with another piece tied very tightly with the name of the fruit written on it. This little job occupied several weeks and I need not say that the sickly smell about the house at that time was not particularly pleasant. My mother was always very proud of her jams as well she might be, and I know I always did my very best to make it a success in the covering up.

The other event was a great one for her. Weeks before Christmas days she would purchase and prepare all her Xmas dried fruits – raisins, dried figs, citron etc. Here again I was called in to help; stoning the raisins, chopping the meat and suet and other little matters for the Xmas puddings and mincemeat – all taking some time.

The day before Xmas day every one of the family had to stir up the pudding – what an exciting time that was and to celebrate the occasion I remember every one of us before going to bed had a wine glass of hot elderberry wine (made by herself) and a little piece of toast. But the real exciting time was the next morning. My mother would get up at 4 o'clock and only at my earnest wish wake me, for I was always to be the stoker of the little furnace under the large copper, in which the two puddings were to be boiled, and to keep a regular and steady heat. One pudding was for our family, the other for the men - with a glass of beer or anything else they preferred. It did not, however, need this little Xmas gift to gain their good wishes. One and all, their families had the greatest regard for my mother. She had then a wonderful old receipt for some ointment; a certain cure for rheumatics or any aches or pains and was always administering to their wants; successfully as I myself then had reason to know.

Their regard, I was going to say almost love, for my father was the same. How often I have wondered since, how different things are now between men and master.

I have thus tried to sketch my mother and her housekeeping. There was not then very much money to spend, but she worked hard, I know, to make it go round. She was always happy and contented and when her day's work was done, she tidied up, as she would say, and sit down quietly for the afternoon with her needlework and her tea. At that time there were no callers or society or any idea of such a thing, for really then we had no friends or acquaintances. Later on things in that respect improved a little. We seemed to live very simply and without any regular time for meals. On Sundays we did have our meals together, then a joint of meat; in the week, anything that turned up, but always wholesome and good. Wine for our own family was simply unknown, as a rule water. By and by when I left school and had to work, then about 13 years old, a glass of beer for supper with bread and cheese and lettuce. There were some few bottles of sherry kept for the special use of the Directors when they came on a visit of inspection.

I can very well remember when they came shortly after Xmas, a decanter of sherry being sent into the Board Room with a dish of hot and cold mince pies; how proud and pleased my mother was when they afterwards complimented and thanked her. Such mince pies as hers were not made every day, but the men always had a kind word to say to her and old Mr Munro, the chief engineer, would frequently stay on and remain to have a cup of tea with her. What a good kind man he was. Many years afterwards when earning my own living he came to see how I was getting on.

After all this writing, I am still not more than 10 years old and it was then soon after our arrival at Greenwich that I really began to know the worth of my dear father. When a boy gets to the age of 10, if he has anything in him he would be able to form some sort of opinion of things about him. My father at once set to work. To put things at the works into shape according to his own views and make many great alterations. In the meantime, until he could decide upon something better, to keep us out of mischief and danger of getting drowned he sent us to a little day school for boys and girls in Bridge Street, kept by an old woman. I am certain I never learnt anything there and was very much more at home teaching myself.

It was at about this time I think my father suggested to his Directors there should be another clerk in the office, to take special charge of cash, men's wages etc. This was assented to and a friend of Mr Taylor was engaged – a Mr Ingleton, poor dear old fellow. How different in every way to his friend Taylor. The one always on the move, always cheerful and smiling in fact almost like a Harlequin - when speaking to you he would keep moving about rubbing his hands like a shopwalker, when asking what is the next article; such was Taylor, a dear friend and help to the Aird family throughout his life. Ingleton in manner was the very opposite to Taylor. Short, sturdy with a very short neck. I never remember ever seeing him run or walk quickly or smile. In fact the most matter of fact individual that ever existed. His handwriting was most formal and clerk like; as for figures, he was simply wonderful. He would add up miles and miles of figures of wages lists £., s., d., with great rapidity and with certainly never a mistake. This was dear old Ingleton. When my father decided to resign his position as Manager and make contracting his future work, both these men volunteered to go away with him and serve him during his lifetime, which they did truly and faithfully in every sense of the word.

I have written of Mr Quick being a great friend of my father's, when living in London. He was now very frequently with us and it seemed to me they must be engaged together in some business transactions; I had understood in some way that my father had received permission from the Gas Company to take little contracts for pipelaying work outside his duties for the company.

Some many months after coming to Greenwich, my mother was in great distress one morning, for my father really seemed to be lost. She asked me whether I thought I was man enough to go to London to try and find him. She explained that he had arranged with Mr Quick to meet him on Sunday morning with a gang of men at Kew Bridge to cut out some large mains and make certain alterations; a Sunday being selected, as less pumping for the District would be required. Everything had been arranged with all necessary castings on the ground, and all would be finished on the Sunday evening, that he might be home that night. He never came. How I had longed for that gate bell to ring.

It was on the Tuesday or Wednesday morning that my mother asked me to go and although I was then only a little over 10 years, of course I said yes. I had never been alone in London before. However, off I started. I was told to go by the railway to London Bridge (there were very few other railways then and no telegraphs of any kind); to go over the bridge and ask for the Bank of England. When there, I would see several omnibuses about and find one with Hammersmith and Kew Bridge on it and get into it. Well, I remember getting safely to the Bank and seeing an omnibus with the right address going along. I ran after it shouting to the conductor to stop.

After a time, the man seeing I was in earnest took compassion on me and stopped the bus. Having told him where I wanted to go, he explained that he was on his way home, having just come from Kew Bridge, but that another one would be there presently. After a little while it came, I embarked and after a tedious journey which I though would never end, I got to the water works close by Kew Bridge with its tall chimney and standpipe. Entering the large gates, there close by, I saw my father and Mr Quick and a number of working men, all seemingly wet through and in a most dirty condition. Their astonishment and surprise seeing this little chap running up to them, so joyishly, cannot be expressed.

It seems the work which was to have been completed on the Sunday, had only just been done in consequence of some defect in one of the castings. However, all's well that ends well. With my father and Mr Quick we went over to a tavern at the foot of the Bridge, washed, had a good meal and then I took my prisoner straight home to Greenwich again and with great joy handed him home to my mother.

I should have mentioned here that Mr Quick was evidently a well-educated man, a few years younger than my father. He had served his time in the shops of the then great engineering firm of Maudsley, Field and Co. in the Westminster Road. He was a widower at this time, living with his sister Annie and his two children Joseph and Lizzie in a small house in Park Street, Bankside, Southwark near the Barclay and Perkins great brewery. In their future lifetime, no two families could have been more intimate. Mr Quick was then the Engineer of the old Southwark and Vauxhall Water Co. then a very small concern and in some similar capacity for the Grand Junction Water Co., Sir W. Clay being Chairman.

Some little time after this incident, a matter occurred to me, which might have been rather awkward. My mother had gone with the younger children to Margate for a change, leaving my brother John and myself at home with my father. I do not know how it happened, but some dispute occurred between us over some childish matter. One of us must have had a knife in his hand at the time, for in a minute the whole top of my left thumb was cut right off – not the end of the thumb, but all the flesh below the nail. There was of course a great outcry and the doctor was sent for. In the meantime I looked like bleeding to death. However, Dr Keeble arrived at last, put the piece of flesh carefully on again and sewed it up. But not liking the look of it, he had the arm strapped to the bed and Mitchell and another man took turns to keep it bathed during the night. I did not mind a bit about the pain for myself, but I was so sorry to see Father so worried about it and my brother so punished. However, after a time it healed up, but plainly left a scar.

I must now mention two events that occurred annually during our time at the gas works; one being the then notorious Greenwich Fair held during the first three days of Easter week. As children we knew nothing about such a matter, but when Easter came and we saw and heard the great preparations made for it, we could think of nothing else. Two or three weeks before Easter, there trooped into the town every caravan, every cart and booths of every kind; every gipsy and costermonger in the Kingdom coming with them.

The principal site of the fair was the whole length of Bridge Street and the streets running parallel with Thames Street, the street into the gas works, and within a stone's throw of our windows. At the back of our house we looked right onto Deptford Creek Bridge over the Ravensbourne. All one side of Bridge Street, the caravans and carts were pitched close up in front of the poor little houses, with tents and booths of every kind erected in front, and during the three days of the Fair, with a great display of toys and gingerbread in all of them.

On the other side of the street there was a large untidy piece of vacant land; I have no doubt left vacant for the purpose. This ground was allotted to the great dancing and drinking booths and always the then celebrated Richardsons acting show, menageries and boxing booths with most of the prize fighters in the kingdom and many others too numerous to mention. One of them, the Crown and Anchor booth, was immensely large. One half was for eating and drinking with wooden benches fixed in the ground; the other half had a wooden floor put down for dancing. There was a great big orchestra in the middle, mostly of trumpets and drums, evidently to make the more noise the better. Two great chandeliers and the same wooden benches for drink – it was a most evil place.

Bridge Street being thus filled, then all along London Street, Nelson Street and Park Street the entrance to Greenwich Park; wherever there was a scrap of ground it was taken advantage of and filled with a booth of some sort, or the game of three throws for a penny at Aunt Sally, coconuts or thimble rig. This extended right through the Park to the Gates on to Blackheath, passing the Observatory Hill where hundreds of foolish people tried to break their necks running down the hill. On Blackheath, the scene was extraordinary. Every knackered or broken-kneed horse or donkey in the kingdom seemed to be collected here and let out to the dirty mob to race – a most extraordinary scene.

When Easter Monday came, the town seemed to be inundated with the whole of the worst part of London and certainly the most evil half; wholly given up to scenes of disgraceful character. Of course we boys at the time knew nothing about that. My father and mother seemed to hate the place and would only let Mr Swanston, the day foreman, take us quietly round in the early morning, before the fun of the fair began (as they used to say). What we did manage to see, during the several years we were there, created great excitement with us, particularly seeing the actors, in their play-acting costumes, strutting about the outside stage of Richardson's Show, and the constant loud cry of "All in to begin". There was also all kinds of hideous painted pictures of the various monstrosities to be seen inside and also pictures of various wild beasts inside the menageries, but it all gave us plenty to think and talk about long after. Living so near with deafening din, noise and discord of music with every booth playing something different was most distracting and there was no getting away from it all.

Some few years afterwards when we had left the gas works I was induced by a young fellow, with whom I had become acquainted, (Robert Munro Christie, then a student at Putney Engineering College and a nephew of Mr Munro, my father's old friend and master) to visit this fair with him. I then fully realised what an evil, wicked, place it was and never went near it again.

The other annual event to mention was a different thing altogether - the Greenwich Waterman's Regatta, which we all looked forward to with pleasure. The watermen on all the different stations of the River Thames were a fine lot of men and the same name was always to be found there, family after family. I well remember the name of 'Doust' appearing every year and curiously, many years later, I came in contact with a Greenwich Doust elsewhere. The prize was always a new boat- a wherry – and the course was to row from the Steamboat Pier round the Dreadnaught hospital ship moored opposite the gas works. There were generally about a dozen competitors rowing in heats, each man wearing a jacket and cap of very bright colours to distinguish them. There was always great interest and excitement in the town, particularly on the riverside.

All those watermen along the river then must have had a much better time of it than their successors have now. There were, in 1841, very few railways about, and those worked mostly by endless ropes. There were no underground railways or even dreams of such locomotion then. It was not until 1847 that the great railway boom or madness began with railway schemes everywhere which were so amusingly depicted in Theodore Hook's book 'Gilbert Gurney'. In this book an old gentleman, a landowner, comes down in the morning to his breakfast. Looking out of the window to his astonishment he observes a lot of men at work on his lawn and park, levelling and surveying with chains and driving in pegs. On enquiring what it all meant, he was told it was for a new line of railway to be made and run through his lawn, and it would also be necessary to take off a corner of the house! But back to the watermen.

At this time there was a fine fleet of paddle wheel steamers (the screw propeller was not then known) running up and down the Thames, all manned by these men, the boats always full of passengers and running every half hour between London Bridge and Woolwich, stopping at Limehouse, Greenwich and Blackwall. How many, many times in later years did I travel on these boats; each boat being called Waterman No. 1, No. 2 and so on, there being 12 altogether I think. Every one of these river boats, both below London Bridge and above, were engineered by John

Penn and sons, then the great marine engineers, indeed almost the only firm, and they then did all government work. How different now.

The steamers running above London Bridge were smaller, called Citizen No. 1, 2 and so on, again about 12. They ran as far as Chelsea and intermediate piers and were always well patronised. In addition to these there were three larger steamers going as far as Herne Bay and Margate, The names of them were 'Red Rover', 'City of Canterbury' and the 'Prince of Wales'.

These steamers were very popular with the London Cockneys and known as the husbands' boats. Margate was then always the Cockneys' or Trippers' paradise. I know well what delight it was to me to go there and revel in the beautiful sea air, although my stay was only two or three days at a time, or perhaps a week when with my mother; but certainly not more than four visits the whole time of my childhood for my father could not afford the time. He only had the weekend (as we call it now), with my brother John and myself he would always stay at a little inn called the Fountain. In size, this little inn was really not more than a private house. It was at the end of a row of houses, fronting the inner harbour and on an open level space called the Steine (I think), with the only hotel then the White Horse or Hart in the middle and with a row of bright green little wooden huts on stilts for bathing. The Fountain was small indeed. Entering the room there was a little passage, the narrow stairs just in front and a bar room on the right. On the first floor above there was a sitting room, the only one in front, and a room at the back which we occupied. But the well-fried sole, chops and tart could not be beaten. We knew nothing then of luxury and were very happy and content with what we had. Afterwards we would have a drive in one of the outside cabs to the well-known St Peter's Tea Gardens, about a couple of miles inland where for a small charge one could indulge in music, dancing and tea. Oh what a tea! At my age it was never to be forgotten. To select the particular arbour first, order the tea, to see the waiter presently coming with a tray with a large, crusty new cottage loaf, several fresh pats of butter, a large plate of shrimps or prawns and watercress: what more could a little fellow like myself want?

Once or twice I came with my mother and my greatest delight

was to spend the whole day on Jarvis's Jetty, an old broken down wooden landing pier quite green with slimy moss, seaweed and barnacles. The stone pier was not then completed and at high water, the middle of the jetty having sunk down, people at the sea end of it could not get to land again without being taken off, until the tide turned. However, my great delight was fishing for crabs and other fish from this jetty with its beautiful blue/green sea and transparent water and seeing both large and small fish at a great depth.

My mother would be occupying some small, quiet, cheap lodgings for about a week. In the evening we went into one of the bazaars in the High Street. This was a shop of great length with toys of every description on each side at the end of a large room. There were three musicians on a raised platform and a large table in this room for raffling fancy articles with dice. There were workboxes, tea-caddies, trays, tea and coffee pots and no end of rubbishy things: I don't know how many things my mother won. There was a good deal of what one would call the mosaic element about these bazaars. Of course everybody in Margate would wear what used to be the Margate slippers – yellow shoes.

Such were the few incidents of my visits as a boy to Margate. They would have been few and not worth mentioning had my father not been feeling ill and knocked over by work and so tried to get a few days rest. On this particular visit it had been arranged that after our bit of dinner at the Fountain we were to drive to the St Peter's Tea Gardens, hear the music and have tea. But when the time came he said he felt too ill to go. At this I began to sulk and shew my disappointment very disagreeably and have no doubt my brother John helped me. I mention this incident to shew what selfish, wicked, unfeeling creatures children can be and although so many, many years ago, how many times have I thought of it and felt disgusted and ashamed of myself. However, instead of giving me a thorough good thrashing for my foolish cruelty, my father simply said he would try and manage it and he did. That was always my dear father; always too weak and kind at such times, but I am afraid selfishness with the young is part of human nature.

This visit was my last for many years, the next one being an errand of deep grief and sorrow. Some 30 or 40 years later I went once more for the last time to try and regain some strength after a short and desperate illness, but alas, the dear old Margate of old had gone. Although I was then able to take a suite of the best rooms in the Cliftonville Hotel, the old charm had quite disappeared and the place seemed filled with hospital patients. The little Fountain Inn was still there after all the many years, but the quaint little bathing huts had all gone and a massive stone sea wall had just been constructed round the inner harbour (now called The Parade) by John Aird. How curiously this world moves on sometimes. This reference to my very last visit to Margate bring to my mind a very pretty little essay Louis Stevenson wrote in his book '*Virginibus Puerisque*' entitled "Ordered to the South".

We had now been a little more than two years at the gas works and in that time my father had made great additions and improvements. I took the greatest interest in of these, for the work seemed to fascinate me and I think my father seeing that, did all he possibly could to explain and teach me.

One piece of work was to lay a large pipe across the muddy Deptford creek which I thought a wonderful job, little thinking that one day I should be superintending the laying of pipes almost large enough to walk through from one side of the River Thames to the other. There was also an iron outfall sewer, a mile and a half out into the sea.

I have already said what a clever mechanical draughtsman my father was, although without education, and I have every reason to be grateful for the many useful hints at the time. I have had several of his drawings in my possession for many years and I now think one of my sons has them. What with writing most of his letters for him for his signature, and helping a little with his drawings, I was kept pretty busy and venture to think that I was even then of some help to him. I could very well see that his pipe laying work outside, apart from the gas work, was increasing by leaps and bounds. At that time he was not only laying mains for his own company, the Phoenix, but also for the Kent Water Company of the same district, Mr Morris being the engineer and a very singular man, but also a very kind one. My father laid pipes for Mr Quick of course [the engineer of the old Southwark and Vauxhall Water Co.] and also for Mr Wicksteed, the engineer of the East London Water Company. It was always a mystery to me then how he managed it all so well, but it was simply a question of organisation and training men for that particular work.

At first blush, pipe laying would seem a very simple and easy outdoor work, but I do not know any work that requires more care and judgement, with gas particularly. Making the joints, either with lead or what are known as turned and bored, was then quite an art. My father's very best and most reliable men as jointers were nearly always Irishmen who would have died for him, and yet these men one and all were the greatest blackguards, drunkards and scoundrels that ever breathed. The frightful language they made use of was simply awful, yet one could always rely on their good work. Many would be away on the drink for a week, and having spent every farthing they had, would come back and be quite steady for a month. But my father seemed to know exactly how to deal with them. What with helping him with his letters and drawings, and my own little pursuits such as my garden, my white mice, my silkworms (for I then had a great many and any quantity of the pretty yellow silk) I was quite busy, and could not be happy when idle. I was always reading, even at every meal, and was still boy enough to enjoy a good trundle with my hoop or a game of marbles or peg top with my brothers.

I must have been a strange sort of mixture at the time, something of an old man and a child. My brother John then (c. 1843) aged about 9 years old was quite different. So far as I can remember he was a somewhat dull boy, very hasty tempered, had not the least idea of doing anything to educate himself and never cared for reading. One disagreeable peculiarity he had was to walk about in his sleep. Many evenings when we were all sitting, in what was called the Board Room, engaged in something or other, he would be laying fast asleep on a couch. Presently he would begin to talk or grumble to himself about something and get up, walk across the room and try and get out through the wall. Failing in this he would quietly go back to the couch again and continue his sleep. Yet with all this dullness and want of effort as a child, in days to come he left me far behind in the world's race. Not from reading or writing a good hand, or being clever with his fingers, but simply from extraordinary force of character and of mastery over all about him. Even men of great position in the busy world who had far more culture and education than he had, would be glad to associate with him and seek his opinion and advice. Even my dear father in those days to come, could never summon up courage to oppose his views.

The difference between the two of us was simply that he was always in the front of everything and meant to keep there, but I always preferred a back seat. I would work my life out as I did to do my duty and faithfully carry out any work in which I was engaged. I never wanted to be or cared to be a great or prominent man. All through my life somehow there was always a strange diffidence and want of confidence in myself – a very unfortunate and unhappy trait in any man's character.

2
School (1842)

The time had now come for serious change, my father having decided to let us have some real schooling for a year or two. There were at that time many first class schools in Blackheath, both for boys and girls, but the terms were too high for my father's means. He did, however, find a good boarding school in the town very near our home. In a narrow passage, only for pedestrians, leading from London Street into Greenwich Road, passing the parish church St Alphege called Straight Mouth, an open space suddenly appeared on the right hand side on which stood a large house of some age, with a large carriage drive in front. An annexe had been added on the left hand side of the house for a schoolroom with a passage from the front leading into it, distinct from the house. On opening the door, one stood in a large room, with a double desk in the middle; there was a door at the right end communicating with the house. Two desks stood on raised platforms for the masters and at the other end there was a lavatory. By the door of this there stood a tall wooden cage with a very large monkey called "Jacko" who was fortunately very good tempered. Close by there was another door leading into a large playground, then with open meadows all round, but also with building going on and the new railway being completed from Deptford into Greenwich.

This school was carried on by an ex-military officer, a Major Bevan, and his son George, who was about 30. They were perfect gentlemen and exceedingly kind. The housekeeping was done by the eldest daughter as the Major was a widower, and there were two other daughters plus a Mr Tomkins, a stockbroker, past middle age, who was a friend of Mr George's and idolised by the boys.

This sudden change from the wild kind of life I had been leading since babyhood to stern school rules, with schoolfellows of quite a different grade to myself, was rather a trial, but in a very few days it all seemed to come round quite naturally. This school was distinctly a boarding school and no day boarders were taken, but my father by some means induced the schoolmaster to make an exception in our case. The boys, about five and thirty in number I think, were much older than my brother John and myself and all of good families. Instead of shunning us as day boys, they really seemed to make pets of us. I can honestly say that the something less than two years I spent at that school was the happiest time of my life.

Curiously, about hundred yards further on in Straight Mouth there was another school, under a Mr Knightley, where my brother Alexander was wholly educated, the result being that in intellect and genius he had more than all the Aird family then put together. However, great intellect and genius is of no earthly use in this busy world without application and hard work, as was so sadly proved in his case later on.

When we had made a start and settled down, Mr George the son, together with many of the boys who had never been to a regular school before, were somewhat surprised to find that I was not quite such a fool as I looked. One of the first classes of the day was a dictation class, each boy having a large slate and Mr George dictating sufficient to fill both sides of it. They were then stacked up by his desk or rostrum and by the afternoon he had reviewed them and the boys were at liberty to take them back and learn their mistakes by a large figure on them. On this little chap's slate there was a round '0' only. I do not venture to say for one moment that this was always so; but sometimes with grammar I was not so very bad and I put that down to reading so many books. Not superficially, but noting and learning as I read. Not only light reading (novels and stories) but history and geography were quite as interesting to me. There can be no question that any boy really fond of reading can teach himself a lot.

I think I did fairly well with sums at this school and there was one work I revelled in and that was mapping. The room in the house next to the school room was a large room where the boys dined, and when at liberty the large tables were made use of to make large maps of the different countries. I took much interest in this and did my best to excel.

There was also another little branch of the education there that I took to easily. I could make and work out the latitude and longtitude of any place on the terrestrial globe. This was very interesting, but never of any use to me.

Latin and Greek were not thought of, but there was French if there was time and the opportunity to learn. I rather took a fancy to it and the little French master did his best to get me on; had I remained longer at the school, or had the opportunity when I left to follow it up, I should have spoken it very well. As it was I could read and pronounce it well and can hear now the little man shouting out "*parfaitement*" when reading to him. But to translate French into English required a lot of dictionary and the verbs I hated. However, I must have got on a little with it, for many, many years after I had in my possession a part written out for me by the French master from Molière's *'Le Bourgeois Gentilhomme'* – a female part – which I learnt. This was for some private theatricals being got up by the boys, but for some reason or other, it never came off. And now, there was even one more extra and that was, of all wonders, a dancing class. This child, after all his rough, strange life, was to learn dancing, and in the presence of ladies!

Fortunately, it was not until I had been some few months at the school that I made my debut, so I had by that time got a little courage, but never shall I forget that first lesson. The dancing room was the boys' dining room and the room where I learnt the mapping lessons. The tables had been cleared and some seats placed for the ladies plus any visitors, and the dancing master, a Mr Beale, first appeared. He was a tall thin man with a little fiddle (I think called a 'kit') having only one or two strings to it. I remember the poor man looked half starved.

Not more than a dozen of the boys attended the class I think, but when the three daughters of the house came in – Miss Sarah, the eldest, who seemed to be about 30, Eliza, and the youngest, Miss Sophia - being the first time that I had associated with ladies, I seemed to get in an awful state of confusion. But after a time it seemed to come all right and I took very great interest in it. One lesson, Miss Sophia was asked by Mr Beale to take me as a partner in a set and I felt myself lost for ever as I fell deeply in love for the first time. At that time it seemed to be the practice at dancing lessons to teach boys to dance the sailor's hornpipe. I took to this very kindly and became very proficient at it, but my father hearing of this used to make me dance it for him almost every evening.

Being only day boys, my brother and myself walked home to our dinner, but if sometimes it was inconvenient (washing day, house cleaning or something) we were allowed to have the meals sent to us at the school.

Therefore, at about 1 o'clock the man Mitchell, or one of the O'Brians (my father's favourite men) would appear at the school door and we received the little basket.

What a dear mother we had to work and slave for us all as she did then. What could be more dainty and appetising than the basket's contents. Sometimes a couple of delicious meat pies just made, or perhaps sausage rolls, a couple of bread rolls or clean slices of bread. Perhaps a lettuce and some cake all carefully wrapped in clean white little napkins. When the resident boys had gone in to the house for their own dinner, we thoroughly enjoyed ours.

I can well remember the names of many of the boys and, many years after, met several face to face; curiously not one of them recognised me. There were two brothers named Cutler; two named Haigh; one named Woodman and one, Dick Rosier, a great favourite.

At this time my mother and many others used to say and think I was too much of a fop to be of much use in the world. Probably they were right; I must admit that after my introduction to the three daughters I looked at things in a different light and tried to make myself as smart as possible.

At that time my brother John and myself (and also Alexander who then joined us at the school) wore a dark holland blouse and trousers, a leather belt and a large white linen collar, together with a Scotch glengarry cap with thistle and feather. My brother John's was pulled on anyhow, but my own was at a captivating angle which took me some time to perfect.

How boys can risk both health and lives by buying and eating indigestible fruit and other things! When the three of us were walking back to school in the afternoon, we would pass through a side street (Roan Street) where there was a poor little shop with a shutter outside on a few loose bricks. On the shutter for sale were some hard green apples in parcels of half a dozen. We never could pass this shutter without trying to ruin our health.

The boys at the parish schools in this street, when seeing us coming along would shout out after us "Here come Matthew, Mark, Luke and John", although there were only three of us.

One of the boys at our school was marvellously clever at making little models with cardboard, principally of stagecoaches, carriages, and dogcarts, but many others. They were simply exact with wheels, springs and shafts all painted and varnished, with the name of the coach "*Tally Ho*", "*Wonder*" or "*The Times*" neatly printed behind. This work simply fascinated me and I determined to try my luck in the evenings at home and during the vacation, nothing like the length of vacations now at schools. However, this modelling work, which required immense patience and time, after a little while became quite within my power, and I produced coaches and other things so strangely real, and so far beyond those of my friend and school fellow, that they got quite talked about. My dear father and mother were simply astonished; there was hardly a single friend or acquaintance of our family who did not possess one. I know it was terribly hard work and for several years after I had to yield to the coaxing of many a pretty face for a specimen.

By this time my mother and father were very much respected in the town, not only by the many shopkeepers but also by residents. After a few months at school quite a little friendship seemed to have sprung up between Miss Bevan, the eldest daughter, and my dear mother.

One thing was quite certain that neither in manner or appearance was there ever the slightest sign of vulgarity or ill breeding in my parents. And another thing, they always all their lives kept things perfectly straight as they went along and paid all books and bills weekly and I am proud to say that this boy has always religiously followed their example.

Greenwich at that time seemed to me always a bright and cheerful place. After a visit to the Griggs boys at the Naval College I would frequently cross over to the Hospital and walk along the terrace, watch the steamers on the river, have a look at the pensioners at their dinner and the Painted Hall with the Nelson relics and sometimes walk in Greenwich Park. But the Park then always seemed to me to be so very untidy and that all the debris and rubbish from the last fair had never been cleared away. The three celebrated taverns for fish dinners "Trafalgar", "The Ship" and the "Crown and Sceptre" close by always interested me as I had heard and read of the great dinners and parliamentary whitebait banquets given there. There was one other equally good tavern there, at Blackwall called "The Artichoke" and a smaller one at Gravesend of the same character called "The Falcon". When my brother John and I were a little older, my father, once or twice on a summer Sunday would take us there by one of the steamboats to have some whitebait. At that time also, Roserville Gardens and their great attractions there were in great vogue.

I need not say how much we enjoyed such a day's outing. Many, many years after this, it was my fate and duty to report myself every fortnight for about four years to the Royal Engineers' Commanding Officer at Gravesend, (the great and never to be forgotten General Gordon), the entrance to his quarters being within only a few feet of what remained of the poor old Falcon – a few pieces of timber. It gave me something to think about and look back to – those younger days.

I think then at that time, my two best friends among the tradespeople were old Mr and Mrs Beaver, who had a stationer's shop and a very old fashioned circulating library in Nelson Street. Nearly every Saturday evening I would have a long chat with the old couple about books and other things which seemed to please them very much. Their library was not very much of a one, consisting mostly of the old standard works in three volumes very much battered about by that time. I had read and knew almost by heart all Scott's work, Fennimore Cooper, Captain Marryatt, Levers and many others.

It was when chatting with my two old friends the Beavers that the idea of a circulating library for the men of the gasworks occurred to me.

My father then had no library, except some miscellaneous magazines on mechanics and other odds and ends of books which he kindly allowed me to make use of. John Taylor, the clerk, lent me a few and with a few of my own I made a start, although I remember my time was then pretty well filled up. However, in a little place at the back of our house, behind the scullery where the man Mitchell cleaned the boots and the knives, there was a circular window to admit light, looking into what was wrongly called the garden. Getting some black paint and a short ladder I painted in large letters the word "Library" round this little window. All the men when told what the object was took it very kindly, but I am afraid looked upon it as a joke, or a bit of fun of Master Charles's. I had also fixed a little bell outside for them to ring if they wanted to change a book, but custom very seldom came. It is also more then probable that not more than half the workmen could either read or write.

The funniest part of the whole matter was that the Directors of the company one day going round the works noticed the word "Library" painted on the wall. They asked the foreman, who was in attendance, what on earth it meant. I afterwards learnt that when explained they were very much amused.

The only newspaper I had the opportunity of reading then was the Sunday Times and in that paper every week Harrison Ainsworth's works were published such as 'The Tower of London', 'Jack Shepherd', 'The Miser's Daughter' and others.

Charles Dickens's works were then being published monthly, but this way of reading books was very irritating and one had always to refer back to the previous number to get into the drift of the story. However, being now at a school where I had to learn, I had not much time for pleasure reading.

My father, I think, must have made the fortune of the poulterer and butterman - Russell, whom the clerk John Taylor helped on Saturday nights, as it was the practice every Xmas to send to everyone with whom he was associated in business a turkey, goose or a brace of pheasants. At that time there were perhaps a dozen; in a couple of years more – hundreds. For many, many more years to come he had the order, always under the direction of Taylor, to send away properly addressed, until it got too much for Russell and he retired from business.

I will just mention one or two other of the principal tradesmen and return and finish my school days which were unhappily not to be for very long.

The watchmaker's shop was kept by two brothers, both very well known men named Bennett – John and William. Years later John was knighted (Sir John Bennett) and had the well-known shop in Cheapside with the large clock above. He was an Alderman in the City of London Council, but in consequence of his eccentricities and radical speeches, when his turn came to be Lord Mayor, the Council passed him by and he was never elected. William, the brother, was a well-known poet and published some very pretty verses.

There were of course many other good shops, but I will just

mention one. When an opportunity occurred, my mother would, when in the town to do a little shopping, take me with her and go into West the confectioner, to give an order or pay an account, and would have a friendly chat with the daughters. I have already spoken of the three Miss Bevans, but the beauty of these three Miss Wests (particularly the eldest one, Miss Annie) even to a child like myself not yet 12 was bewildering. I remember well her saying to my mother one day in reply to some little chaffing remark "Oh no, Mrs Aird, I am waiting for your son Charlie there to grow up and will marry him if he will have me", but I was too young then to reply to this properly.

Annie West was, I think, the most beautiful woman I had ever met in my life. She was very fair, rather full figured, with fine face and eyes. Her hair was curly, parted in the middle and bound up in a knot at the back. It was my good fortune many, many years later to have the opportunity of buying (I think at Christie's auction rooms) a very beautiful life sized marble bust by the American sculptor Power, of Hermione from Shakespeare's 'Winter's Tale'. It was almost a speaking likeness of Annie West, but unfortunately it is not now in my possession.

Judging by the number of young men of good position who seemed to haunt the shop and make themselves ill with bath buns and soda water or lemonade, these girls were much admired, but I have never once heard an evil word said against one of them, neither did I ever hear of their being married, but I am certain they were always respectable, good girls.

Speaking of those residing in the town whom we well knew, I would like to take this opportunity of referring to Dr Keeble and his dear wife, both then over middle age. The doctor was very prim and precise; tall, clean-shaven and with hair always cut very close. He was always taking the greatest interest in our family, particularly in myself. I think finding that I was a great reader of books as he was himself, he used often to ask me to let him know of any interesting book worth reading as he had not time to seek them himself. I remember recommending 'The Romance of War' by James Grant and how pleased he was with it.

I know well how much indebted we one and all were to him for bringing us safely through the many childish ailments we all had to fight. At that time when I was about 12 there were then four others – John, Alex,

Agnes and William. Afterwards there came Jessie, Joseph, Henry, Walter and George, the last dying from croup when only 10 months old suddenly in my arms.

Dr Keeble wished to be my younger brother Henry's Godfather and when he died, left him £500. They never had any children themselves.

Somehow or other I seemed to be almost always running in and out of their house. If I wanted a tooth pulled out, the doctor would merely go to a drawer, take out a pair of pincers and all was over in less than five minutes. The charge would only be one shilling, not 10/6 or even a guinea as it is now.

Now, what can I say of Mrs Keeble. As far as I was able to judge, one of the dearest women on earth and one of Almighty God's angels for all she did to help me in days to come of the greatest sorrow. There are no words to express my gratitude and feelings. In every sense of the word she was the most perfect lady it was ever my fortune to associate with. She was very tall, as tall as her husband, very thin and straight as a backboard, but what a loving face and manner and so exquisitely but so quietly dressed. In conversation she was always calm, quiet and precise like her husband. Whenever she heard that Master Charlie Aird was downstairs, she would come down, find me chatting with the doctor and would join in so pleasantly and as interested as we were.

How my father and mother did love and reverence her. Unhappily I am not able now to remember many little incidents that would shew what a beautiful mind and character that dear lady possessed, but when the time comes, I shall have to refer more fully to her great love and help.

I have often and often wondered why, throughout my life, all the medical men I have met have shewn such interest and regard for myself. It has always seemed so very curious and will be referred to again later on.

Our schooling was going on very quietly and pleasantly and so was time, for I seem to feel that I was getting away from childish amusements.

Mr Morris, the Engineer of the Kent Water Company for whom my father was then doing all his main laying, mentioned to him one day that if I cared for, or knew anything about fishing, there were some good fish in his reservoirs. This was great news for me, and so nearly every half holiday found me at the waterworks with the rod and tackle my father kindly gave me. The only fishing I had known was fishing for crabs off Margate jetty, but where there is a will, there is a way, and with Mr Morris's kind hints I soon made a start, and in after life I don't think there was ever a greater lover of the piscatorial art than myself – from a crab to a salmon.

About this time, Mr Wicksteed, the Engineer of the East London Water Company, for whom my father was also laying mains, wrote to him that Lord Ebury of Moor Park, Herts, wished to see him about an improved supply of water to the estate. My father very kindly offered to take my brother John and myself with him for a little change, as it could be managed in a day. A day never to be forgotten. Except to Margate or the Roserville Gardens at Gravesend we had never been away before.

On arriving at Moor Park and entering the great gates, we found his lordship and his steward walking in the park, the most beautiful and largest place one could dream of. Lord Ebury suggested that his steward should take charge of us and shew us round while he and my father went over the ground of the proposed new water supply. Some little time after middle day they returned to the mansion and we were then shewn into a room by ourselves and a most ample, and to our young minds a most delicious, repast was served.

It was many, many years before I saw his lordship again. Returning home, my father, who was ever anxious to explain such matters to us said it was a question whether to bring the new supply by gravitation, or pump the water. I do not remember which system was adopted, but the work was successfully carried out by him. On how many similar occasions was he called in to advise on schemes of far greater magnitude than this?

I have mentioned that Miss Sarah Bevan, the eldest daughter of our first school master very kindly came now and then to have tea with my mother. Sometimes old Mr Munro and his niece Jessie Christie (a real Scotch girl), would run down from London for the same purpose, making things then perhaps a little cheerful and bright for her. On one of these occasions Mr Quick happened to come. I do not know whether he had been making enquiries from my father or not, but curiously he was found to be present whenever Miss Bevan came. Afterwards and later on I found out that he had proposed to her but was not accepted, the lady being married some little time after to a Mr Butler, an auctioneer and surveyor in the Old Kent Road. However, Mr Quick did not seem at all broken hearted. He very soon after this married the eldest daughter of a Mr Glover, who had a large tailoring business at the corner of St Thomas's Street in the Borough.

Before coming to the time when all my happy schooling was to end, I will just mention one or two little incidents that occurred in the meantime. I have already mentioned that in the house where we lived there was very little furniture, and that of the very cheapest. There was not a single picture or print or ornament of any kind and yet the house was always the picture of comfort and cleanliness and things seemed presently to alter and improve on this.

There was a tailor in the town in a small way, who had won a prize in the Art Union of London, a well-known and appreciated institution at that time. Any subscribers paying a guinea had a very first class engraving and the chance of drawing a prize at the annual general meeting of the Society, ranging from £150 to £10. I think there were about fifty prizes of various amounts. The prize was not in cash, but the winner could select any picture of that value from the Royal Academy or any other exhibition of pictures.

This tailor had drawn a £150 prize and for a tailor had selected a most beautiful picture, an Italian scene near Naples by a well-known artist named Herring. The picture was large and so beautifully framed – a square, but with the picture in a circle. As he had no use for such a beautiful picture, the tailor decided to raffle it at a guinea a chance and after a time got the requisite number of subscribers. I think the outcome was decided with dice, but I am not quite sure. My father, then knowing very little about pictures took one share to oblige the little tailor, and lost. Someone near him offered his chance for two guineas; my father accepted the offer, won and took the picture home. This was his first introduction to anything in the way of the fine arts. There will be much more to say about this picture later on if I am spared.

About this time there was a man named Orchard keeping an eating house at the corner of Straightsmouth by the church and he had an invalid brother struggling to be an artist. My father, hearing of this, saw the poor fellow and gave him a commission to paint a picture from Sir Walter

Scott's book '*Old Mortality*' – the murder of Archbishop Sharpe. This brother was more than delighted, set to work and produced in due time a picture that everyone was more than pleased with; my father greatly so. I do not think it was exhibited in London, but it was worthy of being so. The model who sat for the painter was a man named Murphy, one of my father's gangers – a great big Irishman with a cast in his eye. Poor Orchard, after this painted, in crayon, half life size portraits of my brother John and myself separately. My brother has his now in his possession, but mine has long since disappeared – I know not where.

About this time a little incident occurred in our home that was thought wonderful. My sisters Agnes and Jessie, being about nine and six years of age, attended a little day school in Park Street kept by two maiden ladies, the Miss Todds, who had a brother then engaged at the Observatory. (Curiously, many years after, when my younger brother William went to Adelaide in South Australia, he found this same individual there as Astronomer Royal and they became great friends.) He is spoken of in the book ' The Voyage of the *Baccante* with the two young Princes'. [Charles Heavitree Todd was the first man to link Australia with telegraph poles and wires from north to south. He gave his name to the Todd River; Alice Springs was named after his wife, Alice.]

My dear sister Agnes then shewed great aptitude for music and my father, hearing of this, gave us all a very pleasant surprise. I have already remarked that there was not much in the way of furniture then at our home, the two pictures looking rather out of place, but one day we were all invited to go into the drawing room (or hospital as it was sometimes called) and found a piano which was a present from my father. I can see him now standing in the room looking so proud and happy. We all thought it a grand piece of furniture then, but looking back now it was the very funniest little thing imaginable. Where he picked it up goodness knows. It was a little square mahogany box about five feet by two feet by one foot deep standing on four little thin legs. My father presented this piano to my sister with great pride and evident pleasure and asked her to play something. The little woman at once climbed the stool and played some air without any music before her in a way that certainly astonished me.

As the years went on there were not many ladies who could excel my dear sister Agnes on the piano. When the day came for her marriage with Mr Gill, the English engineer and director of the Water Company in Berlin, my father was then able to give her the most handsome grand piano from Erards that could be purchased. When the time came for my many business visits to Berlin, how proud I then felt to sit and listen at their little parties and note how the many German guests appreciated her beautiful playing on it. How I am digressing and galloping on into years to come! Before going back into childhood days again, I must say a few more words while they are on my mind. How devoted and proud my father was all through his life to my two sisters Agnes and Jessie and how devoted they were to him in return. This was at a time of trouble and sorrow that would have crushed many other men.

During the months of my schooling I did not see so much of my two friends, the clerks Taylor and Ingleton, but only now and then during the short holidays. On an off day we managed to get together again and see something. Ingleton was always so quiet he never seemed to care for a holiday but Taylor was very different. Always on the lookout for one, he seemed to have many friends in Deptford and when our family left the gas works he took a house in New Cross Road and lived and died there, or in that same district. Many and many a time did he take me in to Deptford dockyard, which at that time was fully occupied with government work – principally making biscuits for the Royal Navy. I remember very well how interesting it was to watch the different processes. Now, I think the dockyard is only used as a gigantic abattoir for the slaughter of foreign cattle.

One day poor John Taylor got into disgrace with my father and mother. There was to be a great event at Chatham: a review, sham fight and mining operations, and he wanted very much to see this. He asked my father for a holiday, explaining that there was an excursion steamer leaving Greenwich pier for Chatham very early in the morning and returning that same evening. His wish was granted; but when he asked that Master Charles should go with him, the parents distinctly objected. They gave way eventually, however, on the distinct understanding that we returned the same evening. We embarked with a number of other passengers and arrived in good time, but I soon found that the affair did not interest me. It was really hard work rushing about from one point to another to get a good view and prevent ourselves being blown up in a mine, to say nothing of the awful noise from the guns; this gave me a headache and deafness for a fortnight afterwards. As it happened, worse luck, there was very little food or refreshment to be got. I know I was very glad to get on the steamer again and look forward to the jolly supper I meant to have when I got home. But the old adage still exists *"L'homme propose mais Dieu dispose"*. We had hardly been underway for half an hour when a thick, dense fog came and after a little while the steamer was aground and no one seemed to know where. It certainly was a bad look out for my proposed supper, and unhappily with the large number of passengers on board, there were no provisions of any kind for them. At all events, I painfully remember none came in my way. Poor John Taylor, with all this was like a lunatic, remembering the position he was in with my father, but the poor fellow could not of course help the misfortune. I shall never forget the night I went through in the steamer's cabin with nothing to eat or drink and the awful stench of so many people in it. When the next day came round, however, things looked a little better and when the fog lifted a bit, a tug or tugs towed us off, but it was in the afternoon before we found ourselves at home again. I need not attempt to relate how painful it was to all, or what my dear father and mother must have suffered during our absence. After a little time the soreness passed away, but there were no more Reviews for this child.

I remember that my father at about this time took it into his head on one, or maybe two occasions to engage a 'one horse shay' (as they used to call them then). This had a very civil and respectable cab driver he knew something of and, together with his sons, drove all the way from Greenwich to 'Appy 'Ampton races – the Cockney's hunting ground. That was indeed a happy day, my mother having carefully and generously packed a hamper for us all. These races at that time were thought a good deal of – almost as much as the Derby, and our delight to see the pleasure boats again on the clear rippling river was very great.

We did not take any interest in the races, but the novelty of the scene altogether gave us great pleasure. We thoroughly enjoyed our picnic luncheon and, leaving early, enjoyed our long drive home.

Some months after this my mother, seeing how much we had appreciated this little event, thought she would like to take us all on another little trip in the same direction, so arranged with her old friend Mrs Freakes to meet us at Hampton Court Palace and spend the day.

I do not quite remember how we got there, but I think it was by omnibus or coach as I don't think a railway went there direct at that time. However, with our lunch nicely packed in a little basket, we managed to meet for a picnic in Bushey Park afterwards.

Having walked about the beautiful gardens, seen the great vine and tried to get through the maze, we started to walk through the rooms and see the pictures. As we were so many, to keep us together, my mother unfortunately made us take hands. I took my sister Agnes. Going from one room to the next, she suddenly turned round to speak, not looking where she was going, and struck her face against one of the sharp heavy hasps of the door, cutting her face badly. With the attendant's kind help with water and bathing it, we bound her face up, and, finding a doctor's surgery near, had the cut carefully strapped. Afterwards we had our little picnic in the Park, but the day was unhappily spoilt. However, we got home again and the cut soon healed up.

Looking back, I find I have omitted to mention that my mother had been suffering for some time past with very acute pains in the head. The only remedy when an attack came on was to apply leeches behind the ears, and this was always my duty somehow as no one else would undertake it. The poor dear sufferer would never let anyone else do it for her, but I never shirked it, although a most unpleasant and painful task, as can be imagined.

Some time after this, someone of the family (I rather think it must have been my mother) was not in good health, and Dr Keeble told my father that it was very desirable that all the family should get away from the gas works for a month or so for a change. My father therefore took a little furnished cottage kind of bungalow, with a garden full of fruit, in Blackheath Park. It can very well be understood how delighted we all were at this. There was one curious thing, when we all arrived at our new quarters, the lady who received us, the owner I suppose, to my great astonishment, was in bare arms and bare neck and shoulders. I remember thinking it very horrid.

Soon after we had settled down, we made the acquaintance of the then great surgeon at St Thomas's Hospital, Dr John South, who was living in Blackheath Park, near this house which my father had taken for six weeks, I think. He was a great friend of our own doctor, Dr Keeble, and hence the introduction. He was a singular man, dressed something like a Quaker, but always full of humour, and became, together with his wife and daughter, a very pretty and most ladylike girl about my own age, our dearest and best friends for many years afterwards. Some years later the daughter married one of the officials at the hospital – I think the Secretary.

At the same time, Mr Quick followed suit and took a furnished house for a few weeks on Blackheath, bringing his new wife, his sister and his two children, plus a couple of his wife's younger sisters, making altogether a large party.

I need not say we made the most of this great change, although for so short a time in our lives, and we lived out of doors; playing cricket on the heath being our great pleasure. Unfortunately, one day we persuaded father to join us; he went in to bat and was struck on his leg with the ball, causing a sore place that inconvenienced him many weeks before it would heal up.

During the time we were staying there, there was a great event to take place – a fancy fête on a very large scale. This was at Charlton Park quite near the beautiful place of Sir Thomas Maryon Wilson, with tournaments, concerts and many other attractions. Of course our clerk John Taylor was there, very busy seemingly; he never missed anything of the kind if he could help it. I found out that he was one of the singing choir of the then great teacher of singing Mr John Hullah, who had been engaged for the occasion to sing some glees and madrigals. Let me say here, that in music there is nothing more beautiful of that sort, than Sir Henry Bishop's compositions.

It being a most beautiful summer's day, my father and Mr Quick made up a party and took tickets as it was for some great charity. All the great people in London seemed to be there, I had never seen such a crowd of grandly dressed people before in my life. I remember very well, the heat being so great, the excitement and crush around the refreshment tents, the frantic struggles of the gentlemen of our little party to get some ices for the ladies, which they did succeed in getting, but without any spoons. Consequently, they had to collect all the half crowns and florins they had in their pockets for the purpose.

Pleasure and happiness cannot last for ever and we all returned to our home at the gas works and settled down again. When at Blackheath and walking about there, I could not help noticing what a sedate, old fashioned conservative set of people the residents looked. As if they had been born, lived and died in the same quaint old houses for centuries, mostly wealthy City merchants and professional men.

My father's solicitor (Mr Herbert Sturmy) lived in Aberdeen Terrace, his London office by St Saviour's Church in the Borough. I became well acquainted with Mrs Sturmy later on. If ever there was *une grande dame*, that lady was one.

I have mentioned the word conservative. This reminds me of the curious silent way in which my brother and I became politicians. Whenever my father and Mr Quick met in the evening, after whatever business or work they had been engaged upon, they would have a game of cribbage or chess and discuss politics. We used to sit and listen to their arguments which at times were sometimes very violent – Mr Quick being Conservative or Tory and my father, a Whig or Liberal. From that time to this day, some 65 years, my brother John and myself have never been other than staunch Conservatives.

At about this time, their game of cribbage soon developed into a four handed game – my brother and myself joining. Mr Quick was always my partner and John was my father's – always. The intensely exciting evenings over these games of cribbage for many years are almost indescribable - Mr Quick thinking nothing of missing his last train to finish an exciting game.

It now wanted but two or three months to my birthday when I should be 13. If all was well we would spend our usual quiet little Xmas, our only guests being old Mrs Freakes and my old Scotch grandmother, the dear old lady never failing to bring with her a paper bag of bath buns.

Between this and then, some one of the family ventured to suggest a little evening party and dance on Twelfth Night (January 6th).

At that time that particular night was always celebrated in the following way. There was a large cake frosted with white sugar and several figured ornaments fixed on the top with a fairy in the middle. There were also always what were called 'characters' – sheets of coloured paper figures with a king and queen cut singly folded up, put in a vase or tumbler, handed round to be drawn and pinned on to the dress. Also what used to be called 'kisses' which were little sweets of barley sugar wrapped up with a little motto in two lines.

I remember my father and mother looked very serious at this suggestion (a dance), but after a time asked us to make out a list of friends we wished to invite. The list we made certainly did not run into numbers and permission was given.

Being our very first attempt at anything of the kind, it was rather serious. My father, when he once agreed to anything, always did his best for its success. He had the boards of the drawing room, the 'hospital', planed for dancing and the floor chalked, having unfortunately been told it was the proper thing to do. He had a large gas illumination, a crown or Prince of Wales's feathers (I forget which) fixed outside the window. For an orchestra we engaged my dancing master Mr Beale and a friend.

This was our first venture into, shall we say, fashionable life. That, however, is a misnomer for it certainly was never our intention. It was nothing more or less than a little offering to a few of those who had been kind to us and I am perfectly certain the little invitations were received in the same kind spirit. *"Con amore"*, the two words I have ever loved best since I was born.

Of course, Taylor and Singleton when they heard of it worked like niggers to help in the preparations, making long lengths of festoons of coloured tissue paper and rosettes to hang round the ceilings. Both were invited to the event, Taylor of course being the MC.

The little affair came and passed by in spite of two or three small *contretemps* that occurred, such as the chalked floors spoiling the ladies' dresses; the illumination outside the window being too fierce and blinding and having to be turned off; and our orchestra (Mr Beale and friend) being found half way in the evening in a state of intoxication from a rather too profuse use of the cold gin and water they had desired to be placed by their side. However, they were got rid of and a lady kindly volunteered to play the little piano with Spanish and country dances.

From the beginning to the end it was a bit of fun and a romp right through. The nice supper, of everything that was good that my dear mother had so carefully prepared, would of itself have made it a success. My young schoolmaster, Mr George Bevan, who came with his two younger sisters, had drawn the king character and certainly he kept everyone going and cheerful and although Annie West was not the queen, I could see he looked after her well. I fear most of my time was devoted to his youngest sister Sophia. It was a strange little party and a strange little medley of guests.

Looking back to what I have already written, I am sorry to find that during our schooling at Mr Bevan's I have made no reference to church on Sunday. When we first attended the school, Mr Bevan most kindly suggested that we should accompany the school each Sunday to the Parish Church of St Alphage, which we afterwards did regularly, but it was required, before we returned home to our dinner, that we should return to the school with the other boys and repeat the Collect of that day in French.

Some little time after I left the school, my father took a pew in the gallery of St Mary's church and I could but observe, in the pew opposite to our own in the gallery, a family of young ladies who were all very pretty and interesting. For some time I wondered who they were, little guessing then what was to follow.

Among the many friends my father had made since coming to Greenwich, were two gentlemen (brothers) George and John Smith who were architects and surveyors living at Croom's Hill. My father and Mr Quick had arranged together to invest their spare money in building houses, or house property as it is more properly termed. They had engaged these gentlemen to prepare plans for a terrace of 12 houses to be called Morden Terrace, half for the one and half for the other, at rents of £65 per annum. Rents then were much cheaper than they are now, labour and materials being so much cheaper. I remember a man in Greenwich named Hart had the contract to build them in the Lewisham Road, not far from Mr John Penn's Engineering Works. There was a good bit more house building about this district by my father and Mr Quick as time went on. When the work was finished, John Taylor collected all the rents and had entire charge of it all, until my father's death.

I may as well now refer to a building or Institution in the town that assisted me to gain instruction and knowledge for many years to come. This was the Literary Institute, a thoroughly well managed affair. It was a large and important-looking building, or seemed so to me. The great hall, for concerts, meetings or lectures was in the form of an amphitheatre, the platform being on the ground level, with doors on each side, and the audience seats rising in tiers in the half circle until near the ceiling, where there was a balcony on which to stand. Facing you as you entered the main doorway of the building was a large stone staircase leading to the upper seats of this hall, also to a reading room and several class rooms. On the ground floor was my delight, a splendid, well-managed library and class rooms for almost everything – French, German, dancing and a first rate debating and discussion society. I was always much interested in this, but never had the courage to speak, and I remember very well a young lad about my own age speaking called Henry James. He afterwards became a great lawyer, counsel and judge and intimate friend of King Edward and is now Lord James – such is fate. It is curious that the very last time he appeared as Counsel in a Law case was one brought by Mr Quick against a Mr Henry Whiting, one of his Directors, for defamation of character and libel. Sir Henry James, as he was then, appeared for Mr Quick and won the case.

My father had been made a life member, but being too busy to make much use of it, gave his ivory ticket to me and how many, many instructive and pleasant evenings did I spend there. The lectures were always high class and the concerts also. How well I remember a grand concert being given by the then Opera Company in London, with Mario, Grisi, Lablache, Tamburini and many other well known singers. How delighted my father and mother were, and myself. Never shall I forget when the Ethiopian Serenaders made their first appearance with their beautiful simple melodies, comic also. They were never tired of; years afterwards there came another troupe, also Americans, called the Christie Minstrels who planted themselves in the St James's Hall, Piccadilly and seemed to live and die there.

3
First job at 13 years old

Having now entered in my fourteenth year, my father told me that I had learnt as much as I could learn at school and could be of much use to him on the work going on outside. He therefore took me away from Mr Bevan's school, where I had been so happy, and put me as the timekeeper on a pipe-laying gang of men, laying a '12' main from Shoreditch Church to Hackney. The change was very sudden and strange to me, for I knew nothing of London life then. However, I was always ready to accept what was put before me and saying "Goodbye" to my kind teachers and friends, never met or heard from them again, which was curious.

For the next two or three years I would leave home by the very first train in the morning, sometimes taking with me some bread or meat and sometimes buying a roll and a sausage at a shop for my dinner. I would return home in the evening, report to my father and then, after a big tea or supper, write out letters for his signature or do some little plan for him. Occasionally I went to the Literary Institute in the evening.

My new work or occupation interested me and I can honestly say, although so young, I did my very best to look after my father's interests. I found out after a little while that there were many matters that very much required it.

I well remember first coming to the conclusion that I must keep friends with the ganger in charge and not to let him think I was there to interfere with him or suspect him. All the same, I carried out my father's wishes precisely and had the joyful satisfaction of finding, after some little time, that two items (the blacksmith's and plumber's bills) were being considerably reduced.

These two accounts in pipe-laying every week form a very serious item and arise in this way. Because of the work, if the road is very hard, the picks used by the men to excavate the trench get blunted. If the men are inclined to be lazy, they go to the store, knock out the shaft or handle and put on a new pick long before it is necessary. The blacksmith in the neighbourhood who has the order to repair and re-steel the picks, comes round and collects as many as he can find.

The same trouble happened with the plumber. If the new main was being laid near the houses, there was almost the certainty of the lead pipe supplying each house with water being pricked by the men in the trench. This flooded the work with water and necessitated no end of plumbers' work and lengths of new lead pipe. All this trouble and expense was greatly modified by simply pointing out where the service pipe was likely to be found.

Another matter of expense was the carting away of the surplus earth after the main was laid, and seeing that only the proper number of carts were charged for. But really it would take pages and pages to describe the many subjects requiring attention on such apparently simple work as main laying and no doubt very uninteresting to all others except for myself.

During this pipelaying work, when excavating the trench, the men would sometimes come across very old waterpipes of wood. Simply the trunks of trees hollowed out, but with the ends carefully planed to form a joint, they were very curious and very old.

I have often wondered how many thousands of pounds have been paid by the firm of John Aird for broken glass windows from chips and stones, the shop keepers not having been cautioned to put up their shutters for a little while. How many thousands of pounds for damage to horses, donkeys, carriages and drunken old women, tumbling into the trench entirely from their own fault, but putting the matter in the hands of some low, pettyfogging lawyer who soon works up a case. By sad experience, we soon found that it was no use resisting the case in court, as the jury invariably gave their verdict in favour of the poor old woman.

Another trouble and almost constant expense that used to irritate me were the large sums that had to be paid for boys' eyes. On all pipelaying works, pipes have to be cut and this is done with a diamond point chisel. Street boys, although cautioned and driven away would collect round the men doing this and presently would get a chip of iron in the eye, frequently destroying their sight.

This is where John Taylor always came in so useful. If, after any such accident (if it could in any way be termed such), he would get hold of the old woman, boy or others making claim for damages before the evil-minded lawyer came in and would generally be successful in settling the case fairly reasonably. But nearly every claim put forward has been dead robbery.

Every Saturday, having prepared the wage sheets, my father would give me the money required and when the day's work was done, the ganger and myself would go to the nearest public house and pay the men (numbering thirty, forty and often more), together with the various weekly bills. This was then the worst part of my work, for there were frequent disputes. Because I was so young and some of the men were great scoundrels, it was a little dangerous. However, soon after this an Act of Parliament was passed, making it illegal for wages to be paid in such places, and very properly, for many of the fellows would spend half their money on drink before going home.

There were at this time two or three other timekeepers in charge of other gangs as well as myself, but until the time came for my father to resign his position with the old gas company, his contracting work was in a somewhat primitive state. But I still jogged along fairly well with it.

During the several years that I went up to London in the morning by the train, I became much interested in the old Greenwich Railway and nearly all their men. Except for a short line belonging to the Brighton Company running as far as Croydon, there was no other line to London Bridge. It was then worked by the South Eastern Railway Company, as it is now (1908).

It always seemed to me a silly little line, only taking a quarter of an hour to do the whole journey and stopping at Deptford and Spa Road stations. The locomotives were little, old fashioned tank engines, with no tender, and I knew each one of them well, together with its number. Consequently, when attached to the train, as I always travelled second class and frequently in a half compartment, or coupé, when I had the chance I would chat with the engine driver during the journey from the window, and really imagine myself on the engine. My great excitement some morning was to find one of the large main line engines, with a tender, put on temporarily. But what I really wanted to say is this: how I watched the career and rise of some of those men on the line.

There was a young lad named Abbott, always running about the

Greenwich station and who was a kind of under porter. Also two guards named Richardson and Grinstead, guards of this quarter-of-an-hour line (really hardly worth building – and on an expensive brick viaduct all the way). For the fun of the thing, when I had the chance and the time, I walked the whole distance through New Cross, Old and New Kent Road and coming out by St George's Church in the Borough, the whole journey taking me about one hour and a quarter. A day seldom passed without my having a chat with each one of these men. No doubt by attention to their duties they were made guards on the main line, where I often afterwards used to travel with them; then they were made Station Masters at intermediate stations and finally the top of the tree – years after being made Chief Superintendants of the London Termini - Charing Cross, Cannon Street and London Bridge. This interested me, as in after years I very well remember reading in the papers that each one of these three men had on several future occasions been presented with valuable gifts by certain Royal sovereigns who had travelled on their lines.

As I got older I was very fond of long walks and exercise and with my two brothers I walked right round the Isle of Wight with nothing by way of luggage beyond a knapsack on my back. Afterwards, when I was older I travelled in a similar way several times in north Wales and afterwards through Switzerland.

One day I learned that my father was going to some large dinner at the Artichoke Tavern at Blackwall – one of the fashionable fish dinner places – and John Taylor suggested, being summer weather, that he should hire a waterman and his boat to come and fetch him. The idea was accepted and so arranged. It did not take me long to persuade John Taylor that it would be a great advantage to them if they took me with them. At first he was doubtful, remembering the trouble he had with my father after the sham fight expedition; however, it all turned out well. We got to the Artichoke Tavern all right, and as we were rather too soon, adjourned to the tap room for a biscuit and cheese and a glass of beer until my father came. The boat was ready and he jumped in, evidently much astonished to see me there, but thank goodness not angry. No doubt he understood it all and fortunately it was a fine evening. It is always a heavy pull from Blackwall to Greenwich, particularly if the weather is bad and many great steamers are passing, but it ended all right in our case. I little thought then, that a time would come when I myself would be an invited guest on my own account at this very same tavern, with the choicest of wines and luxurious food, but which I can honestly say I never cared for.

I would now take this opportunity of saying that from the day I could first remember my dear father, until his death, I never saw or heard him given to drink. He was not a teetotaller, but a most abstemious man. In early life he could not afford wines, but in later years he had some most choice wines; but for friends, never for himself. How well I remember dining alone with him at his home, only some weeks before his last illness and telling his butler to get up a bottle of some particular port wine. It was brought, and a superb wine it was, but we neither of us took more than one glass, I never having cared much for wine myself. Sometimes he would have a glass of whisky toddy before going to bed, but never more than one glass.

I also never saw him with a pipe in his mouth. It was the funniest thing in the world to see him and his friend Mr Quick, after some dinner party, try and smoke a cigar as the other guests were doing and enjoying. Having accepted a cigar, there came an anxious sort of look in their faces, a little uncertain about which end to light; and when lit, never letting go of the cigar. They would give some few puffs, blow out the smoke and finding there was no spittoon handy, give the whole business up as a bad job. I never knew them get beyond this on any occasion.

Now there comes a time when my father takes a very momentous step in regard to his future career and decides to resign his position with the old Phoenix Gas Company, feeling that he cannot now do justice to the company's work with the outside contracting work increasing so much. At the same time he really did not need to seek fresh work as he was applied to by so many other companies to carry out their pipelaying work for them. The name of John Aird was even then becoming well known in the engineering world.

Having sketched out and dictated to me a very nice and proper letter which I wrote out for him to sign, the letter of my father's resignation was sent and accepted by the directors in the kindest words possible. What a change to face now! How anxious we younger people all were. However, although it took some time to make all our arrangements, my father had been making his plans for our future home. He had decided to rent a house in Sandfield Terrace in Lewisham Road, some short distance beyond the Morden Terrace that I have previously referred to, but on the opposite side of the road. This particular house was nearer to Lewisham than to Greenwich and when the time came for our removal, I had a somewhat longer walk every morning to the Greenwich railway station.

I would like to describe the geography of his neighbourhood a little before going on. Sandfield Terrace number, I think, some half a dozen houses, and returning towards Greenwich on that same side of the way there were several other houses before coming to Morden Lane, exactly opposite to the first house in Morden Terrace.

The occupants of this Morden Lane, almost a private road leading on to Blackheath, were as follows. The first on the right in the corner was the family of Mr Benjamin Smith, a wharfinger in a large way of business at Deptford. The next house up the lane was occupied by Mr William Noakes, a hop factor in London, and the third and last house by a Mr John Ellis who was a solicitor in London. Coming down the lane again on the other side towards Lewisham Road there was Mr Trail, the magistrate's house and Mr Hartree, the principal partner in Mr John Penn's engineering works. Several years after this we had become very friendly and intimate with these and other neighbours.

Soon after taking the house in Sandfield Terrace (a temporary home for the time being), my father and Mr Quick commenced building the six houses immediately opposite – Dartmouth Villas – with the intention of both families living there when completed. They were to be larger houses with a third floor and much greater accommodation and the rent only £75 a year. What a difference compared to rents now!!

My brothers John and Alex were still left at school, John at Bevan's and afterwards at a good boarding school at Southgate, Alex I think wholly at Mr Knightley's. Alex seemed to be the most intellectual of any one of us and shewed great aptitude for drawing and painting. Had it been followed up, many thought he would have been a great artist, but my father, although at that time greatly pleased at him shewing this gift, seemed to have other views as to his future – as it happened, very unfortunately.

Among the clever and pretty paintings he did was a miniature, a most perfect copy of the beautiful Italian picture my father won at the raffle. It was exquisite and was given to our dear friend Dr Keeble.

Of course it took all the family some time to make all the arrangements as to moving and to settling down again in another new home. Everything seemed so strange and different and, for the first time, indulging with servants' help in the household. Poor old Mitchell and other such old friends were left behind at the Gas Works. It seemed quite to upset my poor dear mother.

Some few months before leaving the gas works I had prepared a plan for my father of a new setting of retorts. Most strangely, this very same plan is before me now at this moment, signed and dated December 1846, although I am now an old man in my 77th year. It is most curious how this one plan, out of the many, should have been preserved during my many wanderings and changes. I have also some few old books still that I made use of for the Workmen's Library. Looking at these now takes one's mind back a long way.

Having now arranged for a new home for us all, my father set to work to find a place of business in London for his work and finally fixed upon a little place in Emerson Street, Southwark, which was a little street leading from Bankside to Summer Street.

About the middle of Summer Street, opposite the grammar school, there were the new offices of the Southwark and Vauxhall Water Company. The offices were on the ground floor and seemed very spacious to me. Our old friend Mr Quick and his family were then living in the rooms above. How many times during my life have I been in those offices and rooms above, more particularly up that mysterious little staircase in the Engineer's department leading to Mr Quick's private room and drawing office and where, years later on, would be found young Joseph, Mr Quick's eldest son. The room below this one was generally occupied by the foreman of the turncocks – Restler – and his men; and also a very mysterious being who acted for Mr Quick as his confidential agent, in the same way as John Taylor did for father.

I am afraid I must plead guilty about that time of having a little chat now and then whenever I had the chance, with Mrs Quick's sister, Louisa

Glover, and with Mrs Quick's little two year old daughter, then a most precocious child known as Polly.

I think I have before mentioned that the chairman of the company was Sir W. Clay, Bart., MP for the Tower Hamlets and a very great man in the City. The Secretary, Mr Roseter, was also a great man in society and had strong friendship for my father. When he resigned the secretaryship some two or three years later, Mr Roseter, much to my father's surprise and delight, presented him with a large and beautiful watercolour drawing of Edinburgh by William Callow, which is now in my brother John's possession. But everyone loved and respected my father.

Somewhere about this time, my father, wishing to give Mr Quick a little surprise and pleasure, got Mrs Quick and her daughter to sit for their portrait to one of the leading artists. It was done and was a great success to everyone.

The little house in Emerson Street was a little place of about six rooms and reminded me somewhat of our first home again. There were three small rooms on the ground floor for the offices and three above for the caretaker, with a gateway at the side leading into a little yard large enough for a horse and cart. There were also a couple of sheds for stores such as lead, yarn and tools. "John Aird, Contractor" was painted over the gateway.

Of course it took us all some time to get settled down and reconciled to this great and serious change. The two clerks, Taylor and Ingleton, had both expressed a strong wish to follow my father and remain in his service, but he could not, of course, assent to this until the directors of the Gas Company had been communicated with. It was, however, so arranged and these two true and worthy men left the gas works and became my father's assistants, together with myself, in these offices and thus John Aird the elder, at the age of 39, laid the foundation stone of the great contracting business which has followed.

At this particular time my father's work was mostly pipe-laying for both gas and water. His name becoming widely known, the business soon greatly increased and drifted into other branches of work such as reservoirs and filter beds for water and gas holder tanks and retort houses for gas. Yet, with this good start, it is quite certain my father never, nor anyone about him, ever dreamt or realised what rash contracts, both at home and abroad, he was to see carried out before he died.

Some little time before we left the gas works at Greenwich I remember he took me with him across the Channel, his first visit and certainly mine, and a very rough crossing it was. I think it was to arrange for some work in Holland. I can well remember arriving at a little inn at Dover facing the harbour. There was then no Lord Warden Hotel, pier or breakwater and only very small steamers to cross in. I recollect my father looking very anxious and worried, leaving me alone in the coffee room while he went outside to ascertain if a boat would leave. It was blowing a regular gale then and pitch dark. In a short time he rushed back in great haste to myself and our bags, for I can even now recollect that our luggage was two large carpet bags. What a voyage we had! On the other side we had to land in boats. Poor dear father, what a bad sailor he always was, and in spite of the many times he was obliged to cross on business, he never got better.

Well, everything seemed settled down now. I continued my work of timekeeping and trying as well as I could to get many of the younger lads trained to be jointers and gangers and even advance further, if so disposed. It was evident, even then, there would be a great demand for such men. I remember well what great interest I took in this and how, many years later, many of these same young fellows remembered me. It is curious and pleasant to remember how many of all grades devoted all their lives to the firm. For instance, my father was standing by a trench one day when a young fellow came up and asked him if there was any chance of a job. My father looked at him and told the ganger to give him a pick and shovel to start him to work. That same young lad became a jointer, ganger and one of the firm's chief responsible agents in many parts of the world – Holland, Berlin, Copenhagen, Bahia and Para in the Brazils, Altona plus others I cannot now remember, besides great works in our own country. This man was James Eagle and the last I heard of him, not a great while ago, was that having completed, I think for the third time, another great contract for the Amsterdam Water Company, the Burgomaster presented him with a magnificent scarf pin.

Resuming again our start at the new office, Taylor and Ingleton, the clerks, had evidently settled down, quite content and reconciled to their new quarters, although so quiet. It was not very long before my father found out how useful Taylor could be in seeing that his wishes and instructions were properly carried out, more particularly on any work abroad. It was in that respect that he was mostly occupied for years, more than ordinary clerk's work.

The gas works being now of the past, I had a little more leisure in the evening after returning to my new home. I would read what books I could meet with, or walk down to the Literary Institute at Greenwich to some lecture or class in which I was interested. I have mentioned that John Taylor was very fond of dancing and at this time, although he had left the gas company's service, he still attended the dancing class there. Now and then I accompanied him, so in days to come I got very fond of dancing myself.

Some few years before this, like everyone else, I had been reading with the greatest delight and enjoyment every work, as it was published, of that great writer Charles Dickens. They were then published in what were called 'monthly serials' which was a most irritating system, as you had to wait a month for the next number, and then were obliged to read the previous one again to keep the thread of the story.

What an extraordinary mind and brain this man must have had. He enters into an agreement with certain publishers to write a certain book or story for a certain price in 12 monthly parts making, when completed, three volumes (as was usual then). He sits down to write and concocts the first number having no idea what he is going to write about in the next number. Further than this, he has been positively writing another novel of a different character at the same time and sometimes a short story or play as well. To write and concoct a plot from hand to mouth in this way is simply marvellous, as must be apparent to anyone who has read and so thoroughly enjoyed all his books, as I have done over and over again.

4
Articled to a Civil Engineer at 16

I am now 16 years old and up to this time my career perhaps has been somewhat uneventful; but one evening my father said to me that it had been suggested to him that I ought to have a better chance to be articled to some civil engineer. At first this sounded something grand and I was to be advanced to a different sphere of life, but I told my father that I was quite willing to go into anything he thought most suited for me. All the same, I was very much surprised and wondered why such a suggestion was made. It could only have been made by one person – Mr Quick. What could have been his reason? I don't think my father would himself have made it. However, after a short time I was told that the matter had been completed by the solicitors and that I was bound for five years to a Mr Henry Austin, a civil engineer with an office in Walbrook in the City. Mr Quick had some acquaintance with his gentleman, but my father did not know him from Adam. The premium paid was, I think, £400. A cruel thing for, as it turned out, it was money hard earned, absolutely thrown away and the best years of my life wasted. Why and how has ever been a mystery.

While this strange idea was being arranged I remember it being mentioned that the gentleman to whom I was to be articled was a brother-in-law of the then world-renowned author Charles Dickens, whose writings at that time I was simply devouring, and of course adding so much at the moment to my idea of the future. Mr Austin had been an old friend of Charles Dickens's from boyhood and had married his second sister Letitia.

The day however came when, with much sorrow, I said good bye to my two old friends Taylor and Ingleton, having to go through so much and so many years before we were united again.

I was taken to Walbrook by my father to be introduced to the gentleman who was to make a great man of my unfortunate self. I think it was only natural that I should have expected to find myself in an office full of work, excitement and business. Alas, what an awful disappointment!

Mr Austin had no work in his office whatever. Before referring any further to this very painful matter, I will try and give a sketch of the place or prison in which I was bound hand and foot to spend the best years of my young life. The office was situated in Walbrook on the right hand side about half way down between the Mansion House and Cannon Street – No. 10 I think. Walbrook is a very narrow street allowing only one vehicle at a time to pass, and never any sun. The front aspect of this building was an ordinary street front door and on the left of it was a very dirty little shop window said to be a stationers, with a few bundles of quill pens, some few pieces of sealing wax and few bottles of ink – nothing more. The window was always so filthily dirty you never could see anything. Never during the years I was imprisoned in this hole did I ever see or hear any soul going in to buy anything. Henry Austin's front door was always left wide open during business house from 9 to 5. Arriving in the morning and stepping inside there was a little square space of only about five or six feet; then a door at the foot of a very narrow stair with a little brass plate with 'Henry Austin, Civil Engineer and Architect' engraved on it and a bell alongside. To the left was a door into the shop. The upper panels being of glass, one could see at any time a gaunt old man sitting behind a counter waiting for a customer who never came. I was told afterwards that this person was the owner of the premises and thus Mr Austin's landlord. He lived alone in the upper regions and I do believe was the model of the old man who died from spontaneous combustion in 'Bleak House'. Much later on I frequently heard it stated also by those who knew that Mr Austin was Charles Dickens's copy for 'Pecksniff', and Mr Fraser for 'Tom Pinch' in 'Martin Chuzzlewit'. I so often wonder after, what rent was paid for the two rooms on the next floor used as offices.

Having rung the bell, some latch was pulled opening the door. I remember quite well my father and I groping our way up this dirty, dark stair and reaching the landing on the first floor, which was certainly nor more than ten feet square and almost pitch dark. There were three doors, one on the right, one in front and one on the left with the little stairs going on upwards. Waiting here a minute or two, not knowing which door to attack, the door on the left marked Private and facing the street, opened and a little old man asked us to walk in. This individual I afterwards learnt was Mr Austin's father and only clerk.

We entered the room which was very small, very dark and very dirty with no furniture other than three or four cane chairs, a little old desk in the middle of the room on a worn, ragged strip of carpet and a drawing board between the two windows. Here we found my new master Mr Austin. I had never seen him before and my father had only met him once at his solicitors' office to sign the deed and pay the money for my indentures. He was a tall, gentlemanly-looking man; very quiet and indolent in his manner. After a few minutes conversation he took us to the room on the right hand side of the landing called the Drawing Office, a room a trifle larger than his own with no furniture whatsoever except a long drawing table facing a window the whole width of the room. This window looked out on a courtyard at the back belonging to a wholesale fancy stationers and consequently had a better light.

There were two young men here at work on some drawing, one a Mr Alexander Fraser who, with his wife and family, became years after one of the most intimate friends of the Aird family. He was then quite a young fellow, only some five or six years older than myself. The other was a Mr Milner, the son I remember, of the Actuary of the Imperial Insurance Company and I supposed an apprentice like myself. The door opposite on the landing I afterwards found out was only a WC and a lavatory, but without any window or light. In the street again, there was a gateway next to the old stationer's shop, leading into the courtyard onto which our drawing office looked, and belonging to and forming an entrance to the business premises of Messrs Rowe, wholesale stationers. As the part of it facing the courtyard was a glass window also, one could see quite plainly the work people, men and girls busy making envelopes, fancy stationery boxes and many other things.

Having thus seen and heard all there was to hear, we parted from Mr Henry Austin with the understanding that I should commence work the following day. Although neither my father nor I had any idea of grandeur or knowledge of City offices (certainly I had none) we did expect something a little more cheerful than these. Although we were much surprised and disappointed, we thought it better to keep our opinion to ourselves, for the present at all events, and get some lunch. This we did by going to a place at the corner of Abchurch Lane and Cannon Street, known by the name of Chard's, which was a public house, but a side door led up to a large room on the first floor where there was an eating house. I can remember now, although this was more than 60 years ago, that our lunch was ham and peas and a glass of beer. How different such places are now to what they were then, all good and cheap and any number of them in the City. I believe that before I left the City I had tried every one of them.

They one and all had some interesting feature about them. For instance, take 'Izants', in Buckelsbury, just behind the Poultry where there were any number of these eating houses as they were called then, and not restaurants. This one I found to be exceptionally good and it is worth describing. A large handsome swing door from the street led at once into a long room, each side being partitioned off into boxes for four customers, with an open passage the whole length of the room leading to a slightly raised platform where the joints etc. were placed as they came up from the kitchen to be operated on by the carver. Having taken my seat, the head waiter , Robert - although very busy always seemed to me to have an extra eye somewhere about him to note new arrivals, for the swing doors were constantly going between 1 and 2 o'clock - would come at once to take the order from myself and several others. In a few minutes he would return with a pile of plates with flat tin covers over them – some eight or ten, as many as he could carry and give to each customer whatever he had ordered, never making any mistake. His memory was simply marvellous. But the great feature of the place to me was to see the proprietor, Mr Izant, and his son on duty, both looking like dukes, and certainly the very handsomest and most gentlemanly men ever seen and both dressed in the height of fashion. They could be seen every day driving up to the door soon after 12.00 in a splendid carriage and pair with a footman all the way from Clapham Common, where I afterwards heard they had a beautiful house. On their arrival they each put on a perfectly clean, bright, brown holland light blouse and strolled up and down the room collecting the shilling and the penny from their numerous customers. At about 4 o'clock the carriage would again come to fetch them. It all seemed to me very strange.

My father, sometime later, took me to lunch with him on separate occasions to two very celebrated places of this kind in Finch Lane leading out of Cornhill, called 'Joes' and 'Neds'. You descended a couple of steps from the street into a kind of cellar, a long bare room without the slightest attempt at decoration and with the usual partitions and rough tables for four persons. None could give more than a quarter of an hour for their meal. On the right hand side of the door there was a kind of counter upon which were placed some large dishes with any number of chops and steaks cut both thick and thin, or large and small. You selected any one of these that pleased you and saw it conveyed to a large grill at the end of the room, then in less than five minutes this was brought to you red-hot on a pewter plate with a splendid large kidney potato. After this you would say you had never eaten such meat in your life. It was curious to note that the frequenters of these two eating houses were mostly the wealthy men in the City – bankers, merchants and not the clerks, although the charge was the same to both.

My father was always very generous and kind (too much so I sometimes used to think) and at that time allowed me a few shillings a week for ways and means. Being always of a saving disposition, I did not wish to spend more of this than I could help and as time went on, and being left alone in this wretched office, I would frequently run out and buy a kidney and a roll from some shop nearby, stick a surveying pin in the kidney and toast it on the fire – when there was one. By the bye, there was also the celebrated Pimms, the oyster shop in the Poultry close by, where one could eat as many as one liked at sixpence a dozen, instead of 2/6 as they are now, with the danger of getting typhoid fever.

I must now put aside this perhaps uninteresting talk, and come back to the very serious position in which I found myself placed and endeavour to find out the object. If it really meant my advancement in life, why did Mr Quick (if it really was him) who was my father's greatest friend and companion throughout life - indeed I may say of every member of the Aird family; why did he not suggest my being articled to the same firm he was himself articled to – Messrs Maudsley, Field and Co? By working in the shops, he made himself a thorough, practical engineer. Putting all this aside, if for some reason this was impractical, why was this office selected? As time went on and by putting this and that together, I was obliged to come to the conclusion that my father had been made use of in some way. With all his great gifts, he had one great weakness which was to accept as gospel truth whatever was stated and he was too kind hearted to doubt anyone's word, particularly that of an old friend. However, the mischief was done and could not be undone.

The following day, at Walbrook shortly after 9.00, no one had arrived, the door being opened by some aged female from the regions above. However, Mr Fraser appeared soon after, then Milner and Mr Austin's father and clerk. During all this time I was sitting in the drawing office patiently awaiting instructions and chatting with the two others. At about 12 o'clock Mr Austin appeared, coming into the room hurriedly, evidently much disturbed and worried in his own mind as to what on earth he could give me to do. After fumbling about in some drawer he brought out a little plan of something or other, dead and gone, I felt sure, long ago and told me he would be glad if I would make two tracings of it as quick as possible. He then disappeared for the day. Before going, he mentioned that Mr Fraser was at work on a survey of an estate at Highgate, between the Seven Sisters' Road at Holloway and Highgate Archway (then all open fields and country), and he would be glad if I would accompany him whenever he went on the ground and do what I could to help him. This little work seemed to have dropped from the clouds quite providentially and it did not take me long to work out that Fraser was really the draughtsman in Mr Quick's office at the water works. He was allowed by arrangement to go to Mr Austin's office and help him in the matter, as he really had no draughtsman in his own office, so there did seem to be wheels within wheels after all. Yet this particular time, the year 1847, was the great railway boom when civil engineers were surveying and planning railways all over the country, and a leading man in Parliament, a great promoter called King Hudson, carried everything before him. But I can honestly say that during my time in Walbrook no railway work of any kind whatever entered Mr Austin's office.

So far as I could gather from time to time from Mr Fraser, who was the most simple-minded creature on earth, without one atom of chicanery and deceit about him, matters at this stage were thus. Mr Austin and Fraser were neighbours and friends then living somewhere in Lewisham or

Blackheath, but I never knew where or cared to know.

For some time before this there had been considerable agitation throughout the land as to the sanitary conditions of the country, there being great sickness and epidemics in all parts and much outcry, in the press and in pamphlets, against the water companies and drainage authorities. This agitation seems to have brought Mr Quick and Mr Austin together and in some way being kindred spirits, they took up this matter. Mr Austin backed up the agitation by writing, calling meetings and thus bringing his name before the public, being at the time very friendly with a very great man and authority on such subjects (Mr Edwin Chadwick). Mr Quick so worked the good cause that Sir W. Clay, his Chairman, and the chairmen of two of the other London water companies – the Grand Junction and the West Middlesex – all resolved to have plans prepared for the three companies together to build new works higher up the River Thames at Hampton, beyond Hampton Court. The plan was to pump the water from there and it was also arranged that Mr Quick should get out the plans for this work at his office, thus Mr Fraser was introduced into Mr Quick's office and became his assistant for the rest of his life.

The arrangement made between Mr Quick and Mr Austin at this time as to the survey to be made of Wainwright's Estate, which was to be completed in a couple of months, was that Fraser should divide his time equally. He should do three days in each office, or at all events as often as he could until the survey and plan of the estate was finished.

Putting all this together and thinking it over, it is my belief that in their interviews Mr Austin had mentioned that things with him and his office were very bad, but if he could only find a pupil and a premium it would keep him going. I am not of course in a position to assert this as fact, but can only surmise such from the matter before me. It is curious that many years after Charles Dickens's death in 1870 two volumes of his letters were published which were far more interesting than his biography by his greatest friend, Foster. They were, of course, the great man's own words. The very first letter in these volumes is one to Mr Henry Austin and in another written during his visit to America and from the Niagara Falls (some four or five years before my going to Walbrook) the following is an extract from that letter:

"I am sorry to hear that business is so flat, but the proverb says it never rains but it pours, and it may be remarked with equal truth on the other side that it never don't rain but it holds up very much indeed. You will be busy again long before I come home I have no doubt. Give my love to Letitia and to mother and always believe me my dear Henry.

Affectionately yours. C.D."

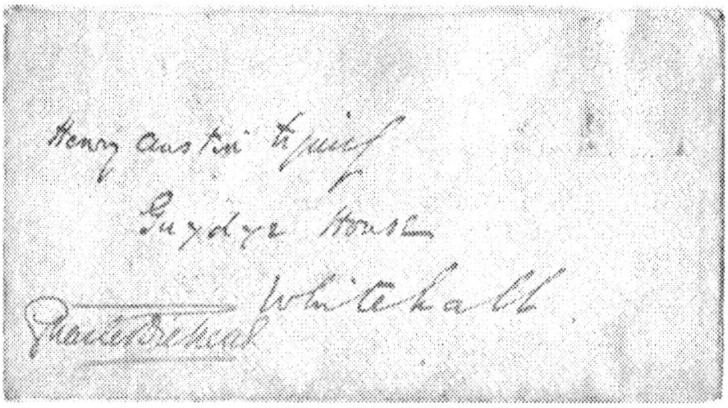

At this time Mr and Mrs Austin were evidently in charge of Dickens's children and also his mother. In another of the letters following shortly after his return home he writes:

"I return you the Niagara letter with many thanks for the loan of it. Pray tell Mr Chadwick that I am greatly obliged to him for his remembrance of me and I heartily concur with him in the great importance and interest of the subject".

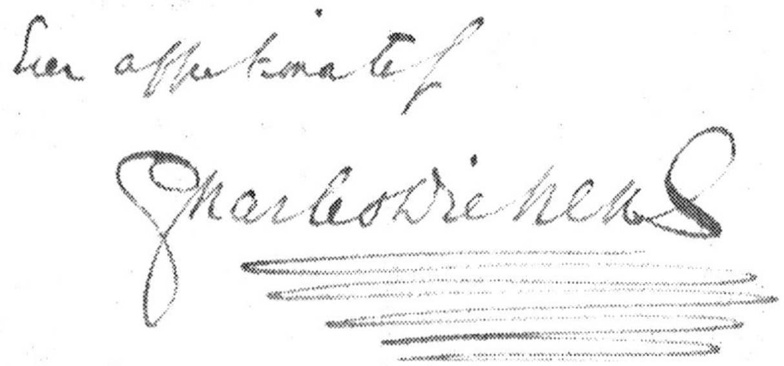

The "subject" in question no doubt being that great movement and pressure at this time going on for improvement in all sanitary and poor law measures. Later on Charles Dickens became, in addition to all his wondrous labours in the literary world, an active member in the movement, which ultimately tended so greatly to the advancement of his brother-in-law Mr Henry Austin. Poor man, what an extraordinary mind and brain. Always working at high pressure, always poor, never in really good health and yet almost supporting all the members of the two families – his own, the Dickens's and his wife's (the Hogarths) and his father at this time being in the Fleet Prison for debt – a ne'er-do-well – and several of his brothers much the same.

Well, coming back again to my daily life at the office, whenever Fraser went to Holloway to continue his survey there, the estate being all the land on the right hand side of the main road between the Seven Sisters' Road (then being formed) and Highgate Archway. It was then all country, meadows, fields and farmhouses. The work was, of course, mostly chain and rod work, but dear old Fraser, whenever the opportunity occurred would give me a wrinkle with the level and theodolite instruments, seeing I was anxious to learn what I could. I don't think I have ever been a coward, but there was one little part of this work I was not quite comfortable with at first. This was having to go several fields ahead with a red flag on a pole and fix it in the ground for a line. On such occasions it was generally a question of who was to win? A savage, bad-tempered bull, a savage farmhouse dog or myself? But as nothing serious did happen, I soon got over that little difficulty. How little I knew then of my future work at surveying and how this little bit of work would help me.

After a day or two's work in this part of the country, Mr Fraser would come to our office and plot his work on the plan or map he was preparing of Wainwright's Estate, a little job sent to Mr Austin by a friendly solicitor. During this part of this work, I used to put aside my absurd and childish tracing and watch and learn from his great ability. I found I could even help him a little, remembering how fond I used to be of mapping work when at Mr Bevan's school. When Mr Fraser had plotted the fields and meadow of his last outdoors work, it was necessary he should take out the acreage of each field, large or small, and before he finished his work and left Mr Austin's office, I found and he found that however irregular a piece of land might be, I could complete the acreage of it as correctly as himself.

I had always been very fond of figures and feeling then that I would much like to have some greater knowledge of square roots, equations and indeed mathematics generally, I asked my father one evening when home if he would let me go to Mr Knightley's school (where my brother Alexander was). This would be in the evenings for an hour or two to see what I could pick up in this respect. He at once cheerfully assented and for some little time I tried to do my best. I never saw anyone on those occasions but the old father (he was awfully clever). I remember sometimes by accident his son turned up, in appearance an ideal schoolmaster, and I have often thought since then that Charles Dickens, or his artist who made the illustrations for Dombey and Son, when he afterwards left for a large school at Brighton, made a model of him for Dr Blimber; it being an exact portrait. But I never remember seeing any of the ladies of the Knightley family.

Soon after these evening lessons came to an end and a little later my brother Alex's schooling there finished also. I heard later they had given up the school and taken a larger one at Brighton which did not prosper. Although my connection with this family was so slight, from that time to this, something about 60 years, there have been frequent appeals for assistance. That is the worst feature in charity. Only once begin and there is a certain unfortunate class who will stick to you like a leech. Although it seems incredible that such a thing should occur at such a moment and after so many, many years, while writing these lines, a servant comes into the room with a letter by post. Opening it, I find a letter from a Miss Knightley, again from Brighton, begging for assistance for herself and a sister.

During this survey work at Holloway Mr Fraser and myself journeyed to the ground by one of the well known blue 'Favourite' omnibuses and the old bus driver and myself became great friends. The fare was threepence each way. I paid for my own and Fraser for his, and during a short interval for lunch, I suppose a similar arrangement existed, for I remember well we separated and I adjourned to the nearest public house for a couple of Abernethy biscuits, a large slice of public house cheese and a glass of beer in the taproom. Mr Fraser may have had his lunch in his pocket. After several weeks of this I asked Mr Fraser what I should do about my expenses. I remember well how he laughed at this but suggested that I should speak to Mr Austin. That to me was an awful ordeal, for I dreaded more than anything coming in contact with him. It was certainly not from dislike, but I think a nervous feeling of fear and being so suddenly removed from one class of people and society to another so totally different. Curiously we never seemed to get much nearer to one another during the whole five years of my apprenticeship. No other member of my family except my father ever met him, neither did I ever know exactly where he lived and never met his wife during the whole of the time of my indentures. I understood he never had any family.

Very often Charles Dickens himself used to rush upstairs to see Mr Austin, seldom or ever finding him. Charles Dickens's brothers also used to call frequently, generally with the same result. Fraser, before he left had given me some little idea of these gentlemen. Poor Charles Dickens, what a time of life and worry he must have gone through with one or the other.

At times some lady or ladies would call and possibly one may have been his wife Letitia – Dickens's second sister, but I never had the opportunity of being introduced. There was only one exception – one of the ladies who called then most frequently was a Mrs Burnett, the eldest sister Fanny, who would sometimes stay and chat with me a little while and one day brought her son Willie. She called him then a clerk in the London and Westminster Bank, a very pleasant young fellow, a little older than myself, and we became rather friendly. The poor lady, I was much grieved to hear, died some little time afterwards.

Coming back again to the curious sense of fear which always seemed to haunt me whenever I had occasion to approach Mr Austin and which I could never entirely shake off, it was due in some measure to being now under authority and not a free agent as I had hitherto been. I can honestly say this, however, that during the whole of the five years I was under him, he never once had occasion and never did speak one angry word to me. He invariably addressed me as Charles, and yet I could never summon up courage to look upon him but as a superior person beyond me. He was always very quiet and placid in manner, one would almost say indolent. However, following Mr Fraser's advice, I prepared a little account and gave it to Mr Austin in his room, having in fear and trembling told him what it was. He remarked, although the amount was only a few shillings, "Thank you, leave it with me, it shall be attended to by and by", but I never heard any more about it. Poor man, there is no doubt at that time it must have been a desperate struggle to get along.

About this time, his great friend Mr Edwin Chadwick, with whom he was working up the great sanitary agitation by letters to the press and public meetings, was made a CB.

Mr Fraser finished the survey and, to my great sorrow, left, returning to Mr Quick. There was then in the office only old Mr Austin, the clerk, and the young fellow Milner who was very irregular in his attendance and seemed to be left to himself to do what he liked. The poor fellow always seemed to me to be ruining himself by dissipation and some short time after he disappeared altogether.

For many, many months after this, my dreary life in this office day after day was terrible. Mr Austin seldom turned up, his father came for a short time in the morning, but there was no post, no correspondence; simply stagnation. How well I remember those awful afternoons, nothing whatever to do but to take charge of the office and the bell which seldom or ever rang. I wandered about between the two rooms and stared out of the windows until 5 o'clock when I gladly left for home by the train to Greenwich with the certain prospect of the morrow being the same. I remember sometimes at the railway station at London Bridge, waiting for the train to start, seeing Mr Austin coming up the platform in his leisurely way also on his own way home. I would immediately take a frantic plunge into the compartment I always sat in, next to the engine, where I could chat with the engine driver and get out of Mr Austin's sight. On arrival at Greenwich I was half way home before he had got out of the train.

When we first came to Sandfield Terrace, the land all around was quite country, but soon after there came an extraordinary change, railways being made all over the Kingdom. One was a new line for the South Eastern Railway which had worked the old Greenwich line. They were widening the old line as far as Spa Road station and then branching off to the right and going through New Cross, Lewisham, Blackheath Park and then on to Woolwich. This line was called the North Kent Railway and was being pushed on rapidly. It crossed the Lewisham Road by a bridge overhead and the station was only a few hundred yards from our home.

It was expected that this railway would be open for traffic about the time that the Dartmouth Villa houses of my father and Mr Quick would be completed.

It was curious what a number of new friends we seemed to have now, both among our neighbours and in London also. Everyone seemed to like to know our family. My father's dear old friend and once master, Mr Munro, often came with his niece, Jessie Christie, to stay and my father and mother in return would stay with them in London at Kennington Lane. About this time London society had gone mad over a new singer at the opera, Jenny Lind, the Swedish Nightingale as she was called. My father took a box for a night as a little pleasure for Mr Munro, and I remember being one of the party. Unfortunately, it was an entirely new opera that night, which proved a dead failure and was never produced again.

Shortly after this, Mr Munro resigned his position with the old Phoenix Gas Company and was succeeded by a gentleman, a Mr Innes who also became a great friend, particularly to myself many years after. The directors of the old company allowed Mr Munro to remain in the house in Kennington Lane, and Innes and his family (wife and niece, a Miss Mynott) occupied the rooms above the offices at Bankside.

Speaking of Kennington Lane reminds me of the great agitation and alarm there was at this time over every part of the kingdom and fear of a revolution. Many, many thousands of evil-minded men had joined together for that purpose headed by a man named Cochrane and calling themselves 'Chartists'. Notice had been given by these men to hold a meeting on a certain date on Kennington Common. The then government and authorities were evidently alarmed and empowered the old Duke of Wellington, who was then Commander in Chief, to take whatever steps he thought necessary to crush the matter. Certainly the old gentleman did not mince matters, for he had almost every soldier in the kingdom quietly brought into London and hidden in suitable places, every man with a large quantity of ball cartridge, and also cannon. All the public buildings were fortified with sandbags and barricades with loopholes cut in the walls. I well remember walking about, seeing all these astonishing preparations with the greatest awe, for London was then nothing more or less than a fortified city in war. In addition to all these precautions, several thousand gentlemen were sworn in as Special Constables and the police (a recent institution and an idea of Sir Robert Peel's with their top hats and swallow tail coats) went from house to house and gave notice that all well-minded people should remain at home for their safety.

The eventful day came, my father and mother insisting that I should on no account leave the house and we all waited; but what a fiasco!!! The next day the news came that thousands and thousands of these foolish men assembled at Kennington with the wicked leaders Cochrane and Fergus O'Connor and before a word or a speech was said, a superintendent of police quietly rode up on horseback and in a friendly sort of way said "This meeting cannot and must not be held, and if you are wise you will leave at once. Disperse". They took his advice and did disperse, without a drop of blood shed. So ended the great Chartist meeting which was intended to follow the example of the great French revolution going on in Paris – it fizzled out like a squib. Had these foolish men refused the advice given, they would have been shot down like so many sheep, for the old Duke meant business.

A little beyond our home, then nearer to Lewisham, was a road leading straight up to Blackheath called Dartmouth Row – I think – but everything was Dartmouth, all the land round about belonging to the Legge family. On the other side of the Lewisham Road, there was a road called Loampit Hill, going straight on to New Cross, through a lot of brick fields and then on to London via the Kent Road. Near our home then there was a pretty little inn called the Roebuck. At the end of its garden the Ravensbourne (then a pretty stream) passed along to the old dirty Deptford Creek we had known so well. In the garden of the inn the stream had been considerably enlarged, making quite a nice little pool for bathing. Often and often during the summer months I would get up early, take a towel and enjoy a plunge before breakfast. My brother Joseph, who was then a very little chap, always liked to go with me, but never seemed to have courage enough to bathe. One morning, however, he said he would like to try, so I told him to strip and tumble in and that I would look after him. He took off his clothing but stood shivering on the bank instead of making a plunge. I jumped on the shore, got hold of him and threw him in the stream, but as it turned out it was a wicked thing to do, for he had a desperate illness afterwards from the shock, poor fellow. I well remember how anxious Dr Keeble and my father were about him and how my own grief and sorrow at the time, coupled with the miserable state of things at Walbrook, was very seriously affecting my own health. I seemed to be drifting into a state of melancholia, never speaking to anyone and shutting myself up alone in some room. My father and mother I know were very anxious and I think my father must have seen or written to Mr Austin about me as shortly after, three weeks leave of absence was given me.

My dear father, knowing how fond I was of wandering about the country alone, generously gave me the ways and means to spend a week in North Wales, walking through it with a knapsack only. Never shall I forget that little trip and others later on, with my father, and my brothers John and Alex revelling in the beautiful scenery. The ascent of Snowdon to see the sunrise and the charming and picturesque little inns at Beddgelert, Llanberis, Capel Curig and others, with the fresh caught trout for breakfast, from the little stream running through the garden.

Later, he also took me with him to some of his trips on the Continent. I think Rotterdam. We went, I remember, in one of the steamers running from London Bridge. I will mention a curious little incident that occurred during one trip. On my father's previous visit, the great man in authority on the work there had been speaking a great deal about Poultry. My father surmised therefore that he was a fancier and connoisseur of these birds and as at that time those enormous and ugly fowls, the "Cochin China" were the fashion, he had a couple of the finest prize birds bought. He gave a lot of money for them and took them with him in an immense cage as a present. But oh, the trouble and anxiety of keeping these two great ugly brutes alive was indeed no joke. But it was done and they were duly presented and accepted with thanks; but now the denouement! My father and myself were invited to dinner the following day and to our horror and disappointment, the two giant birds were placed on the table – cooked!

Having returned home from my pleasant little trip, there was just one day left before going back to Walbrook and misery. My father suggested that we should have a good walk together from London to Southgate where my brother John was then at school – The Eagle House

Academy – to see how he was getting on. I shall never forget that walk. It was in about the middle of April; a lovely spring day with all the gardens of the houses we passed most fragrant from the lilacs, laburnums and may, all in full bloom. What a dear lover of nature I was even then.

After this I made my reappearance at Walbrook and found matters just as I had left them with Mr Austin's old father the only one in evidence. I wondered who had been attending to the doorbell during my absence.

The next time Mr Austin came, I felt obliged to ask him for something to do. The poor man seemed quite troubled; what to suggest? At last going to some drawer he took out a large book containing lithograph plans of machinery. It was evidently an instruction book published in France and he said it would be useful to me to make a copy of them. I was glad to find afterwards that they interested me. Some of these drawings are even now in existence; also some of my father's work and my own when at the Greenwich Gas Works.

So the days went on. Now and then some friendly solicitor would send round a parchment deed, to have a little coloured drawing made on the first page illustrating the little piece of property referred to, either house or ground. I recollect Mr Austin asking me if I thought I could manage it, as it would be serious if the deed should be injured or spoilt. I could only reply that I would do my best and as there did not seem much difficulty about it to me, I suppose it was all right, for many other deeds of similar kind followed. On one occasion, to carry out this particular order, I remember it was necessary that this particular piece of property should be measured and planned first. I remember well it being some two or three shops in Old Street Road.

Mr Austin again asked me if I could manage it. It was done and as I heard nothing more, of course I assumed it had been done all right. I often wondered after, what he had charged for these little efforts of mine.

The same dreadful stagnation of any work in the office was still going on, being left all alone in the afternoon until 5 o'clock. Oh, what a dreadful, miserable time that was. How sick and ill and wretched I then used to feel, but to my great delight and surprise one morning a little change for the better came in this way.

What at home, I had heard Mr Quick one day telling my father of a scheme that was being talked of in the engineering world of carrying parcels underground in large iron tubes by what was called "Pneumatic Power", thus relieving the great traffic in the streets. The author of this idea was an intimate friend of Mr Austin – Thomas Webster Rammell, a civil engineer. This gentleman had, no doubt, seen Mr Quick at Mr Austin's suggestion, with the view to getting his permission to construct a model of the idea on the wharf of the Water Company at Battersea, because there would be a suitable length of ground to try the experiment there. Mr Quick having consented, Mr Rammell made use of Mr Austin's room to complete his plans and arrangements and would now and then ask me to help him a little, which I was always willing and glad to do, as he was always very kind and nice.

This went on for several weeks. The large iron castings were made and put together at Battersea, with an engine to pump, or rather exhaust, the air to work the model.

The day came when cards of invitation were sent out to all the great engineers, science experts and others interested to attend at the Water Works to see the idea in operation.

As my father had put the model together for the inventors, I also managed to be present at the trial. It certainly was very funny (to me) to see Mr Rammell in his immaculate costume and silk hat take a seat in the little trolley or car, supposed to be full of parcels or letters and running along inside the tube. The door at that end was shut, a signal given and before you could blow your nose, he was coming out at the far end of the tube.

Although many hundreds of the most clever and scientific men attended the trial and experiments with the tube, for some reason or another it did not seem a success. However, Thomas Webster Rammell was one of those men who put the motto '*Nil Desperandum*' before them and many years after this, a company was formed and the tube was laid under great difficulties by John Aird from the General Post Office in St Martins le Grand to Euston Square station. I will refer again to this again, when that time comes, if I am spared. [The Post Office tube line was finally closed in 2003].

In spite of the weary, sad life here in these offices time seemed to be moving on fast; quite a year-and-a-half having passed since I came to this place. But what great changes had occurred at home. The new North Kent Railway finished and opened and the six new houses finished (Dartmouth Villas); being occupied thus: No. 1, Mr Aird; No. 2, Mr Beater (silk merchant in the City); No. 3, Commander Popplewell RN; No. 4, Mr and Mrs Collasso (Dutch); No. 5, Mr Wilson (cooper and shipowner, Bermondsey) and No. 6, Mr Quick and family. For many years after these six families were almost like one, so friendly and happy as we were as neighbours. Among the many new friends we made was Mr Samuel Smiles, then Secretary to the South Eastern Railway Company, their offices being at London Bridge. In conversation he mentioned one day that he always walked across the road and got his lunch in the Coffee Room of the London Bridge Hotel. As often as I could, I tried to do likewise and sit at the table with him, for he was then a most pleasant and amusing man. Mrs Smiles was also very nice and very clever and I have no doubt helped him very much in writing his very popular books – 'Self Help', 'Character', 'Thrift' and several others. One of his daughters married a son of Mr Hartree, Mr Penn's partner. Mr Penn was living at a very pretty place called The Priory between the Roebuck Inn and old Lewisham Church. Mrs Penn was a strikingly handsome lady and always kind and nice. Mr Penn was rather an odd looking little man, but considered very clever. There were many young people and the sons, when they became young men, were great and well-known cricketers, playing in the Kent Eleven.

The Penn family were always very kind to us and would frequently ask some of us to their dances and garden parties. My two sisters would go and were always great favourites, but I never could be persuaded to do so – I never seemed to care for the society of others. Mrs Penn, however, seemed always pleased to come to any of our little dances.

When settled down at Dartmouth Villas, my father changed our pew from St Mary's at Greenwich to Trinity Church at Blackheath Hill, where there was then a very eloquent clergyman, a Mr Soames. I shall never forget a sermon he preached one Sunday morning. At that particular time, there was a great and serious agitation throughout the Kingdom and the cry of "No Popery" as the Pope was then parcelling out great districts and nominating Roman Catholic bishops over us all. This sermon, I remember, seemed to rouse every one of the congregation, almost to madness, but after much excitement, things calmed down as they always do.

We had in the meantime become acquainted with the family of young ladies who occupied the opposite pew at St Mary's and who, I am sorry to say, frequently took away my attention from proper subjects. They were the daughters of Mr Benjamin Smith, who lived in the house at the corner of Morden Lane, not many yards from our own home.

There were, I think, six daughters in this family, the two eldest married. The next two, Mrs Alfred Selby and Mrs Hoad, were married later. The two youngest – Amelia and Sarah were then too young, but always bright and merry girls and great favourites. Amelia was a particular favourite of my mother, who was always chaffing and teasing me about the poor girl. She was considered very clever on the piano. These two youngest ones were always together and full of fun and attended most of the evening lectures and entertainments at the Greenwich Institute.

Sarah, I remember, used to sing very prettily, one of her songs being "My mother bids me bind my hair". The youngest of the family, Sarah requires a volume to herself to describe her happy and bright life. With much sorrow, I must write that it was fated that poor dear Amelia should die of consumption some year or two later on; the two only sons Henry and Tom died quite young men.

When at St Mary's Church, I also remember well the family of the Gilberts who sat in the pew behind. At that time they were some four or five fine young men, and afterwards one of them became the great artist Sir John Gilbert. When these gentlemen became old and infirm, they behaved most kindly to Miss Chase, a lady residing with the family of Mr William Noakes [the future father-in-law of Charles Aird].

I would like to take this opportunity of mentioning the names of some few other neighbours and friends with whom we had now become associated. There were also some neighbours with whom we never did in any way become friendly, not from dislike, but simply because we did not seem to have time. We knew their names and that was all. There was a family in No. 1 Morden Terrace, (the first of my father's houses and opposite to old Mr Benjamin Smith's dwelling in Lewisham Road) occupied by a Mr Harris, a teacher of music as the brass plate on the gate denoted. He held some office as teacher at the Naval Hospital School at Greenwich and was rather a rollicking old dog. When I first became acquainted with him, he was a widower, but sometime later on he married his housemaid.

Someone had once remarked that I had something of a voice and I took it into my head to go to this old gentleman for a few lessons on music and the piano in the evening when at home, but it was time and money quite thrown away. Although passionately fond of music, I never seemed able to grasp the alphabet of it and if I could not play well, I did not care to play at all.

Mr Harris had two daughters and a son – rejoicing in the names of Marcia, Portia and Guido. They were very much more worth knowing than their old father. The boy Guido was mostly at sea, as he was a sailor. Marcia, the eldest, was always at home, while Portia was a good deal away in Paris and eventually married a Frenchman. I became well acquainted with them both, more so perhaps with Marcia who was, I think, two or three years older than myself. She was a good looking girl, not unlike in appearance to my old favourite, Annie West, with the same fine-formed head and pale, wavy hair; yet in manner more like a man than a woman with a deep base voice.

Marcia and I became friends and comrades at once and during the many years of our friendship, I was always Charles. I remember well at that time wondering how very forward this was, but soon our lady friends, both young and married, got hold of my Christian name.

Marcia Harris was a clever musician and having perhaps noticed that I was fond of singing, did what she could to help me. There was then a very pretty duet published called "What are the wild waves saying?" referring to little Paul in the book *Dombey and Son.* After practising this together some little time, she taking the second part and I the tenor, we rather pleased some few who heard us at our pleasant evening gatherings. Indeed, it was the custom then that whenever there was a little evening dance, there should always be a few songs introduced. I know I was never let off. I never minded singing alone to my father and mother when they asked me, but at the little gatherings, I disliked it very much and sometimes broke down from nervousness.

My repertoire was not a very large one; only half a dozen or so songs such as "Come into the garden Maud", "Molly Bawn" "The bloom is on the rye", "Then you will remember me", "Look always on the sunny side", "The minstrel boy" and some one or two others. Even my dear father and Mr Quick were not excused. My father with not a note of music in him could somehow make melody come from some of Burns's pretty little Scotch ditties; also Mr Quick with his one song, "The Englishman" with its boisterous chorus.

Marcia was always a dear friend and favourite to everyone, particularly to my mother and more particularly to Mr Innes's niece, Miss Mynott. They became inseparable friends for many years and Marcia, having arrived at past middle age, married an elderly gentleman in good circumstances.

There was also a family named Boustred. He was the magistrate's clerk and Emily Boustred was another friend and favourite during one's lifetime.

Now to the most favourite of all, our next door neighbours the Beaters. A nicer, more sociable and pleasanter family it would be quite impossible to conceive. The daughter, Miss Emma, was a perfect lady, very quiet, for whom I myself had the most sincere regard and admiration. The son Orlando was also very companionable and played remarkably well on the organ. Some friends from Bow in London called Horne frequently stayed with them. They were also very nice and we soon had a similar regard and esteem for them. The visitors I remember were Miss Sarah Horne and Mr Christopher, the brother. My father, who seldom made any remarks or reference to young lady friends, once called my attention to Miss Horne as being one of the nicest and most ladylike girls he had met. To which remark I very cordially agreed.

I remember driving all the way from Dartmouth Villas to Bow (an awful long drive) to an evening party at their home with my two sisters and a young lady who was then staying with us. She was a Miss Cochrane, daughter of Mr Alex Cochrane, one of the great ironmasters in the north, with whom my father was doing much business at that time. The party was well worth the long drive as it was a most enjoyable one and our hosts most kind. I remember my sisters scolding me afterwards for dancing so much with Miss Cochrane, but as this young lady had been so kind during our stay with us as to give me many lessons in chess when I came home of an evening, I thought she was entitled to some little attention from me.

In course of time, Miss Emma Beater and Mr Christopher Horne became engaged and married. I remember calling upon them afterwards in the Bow Road, where they were then living, to pay my respects. Curiously, from that day until a day fifty years afterwards, I never met or heard of them again, until my doctor, attending me, spoke of them as patients. It was altogether a most astounding circumstance.

I will mention now one other family near to us, with whom we had a long friendship – a Mr John Paterson living on Belmont Hill, Lee Road, not more than half a mile from Dartmouth Villas. This gentleman was a City man with a very friendly wife and family, always fond of those evening meetings and gatherings that we ourselves and immediate neighbours so much delighted in. Among the many we met at their home, were three ladies in some way related to Mrs Paterson – Miss Goring, Miss Betsy Goring and Miss Morton. The first two of these ladies were life long friends, in later years more so with my brother John's family than perhaps my own, but then these ladies seemed to make quite a kind of pet of myself and I was always 'Charles' until they died. Miss Goring married a Mr Darling, poor lady, a very unfortunate marriage and the old story – she had money and he had none. Betsy Goring never married. She had a little cast or twist in the eyes, but no disfigurement, and was a lasting friend with everyone. But her singing was never to be forgotten as she had a most lovely voice and was always willing and ready to oblige. Her voice remained beautiful even when old, and until she died I can hear her singing that pretty song "The Brook" - "Men may come and men may go, but I go on for ever – for ever".

At about this time I became very friendly with three young men living near to us, sons of Mr Mumford the miller. The eldest, Sam Mumford, went on a pedestrian tour to Ireland with me. I recollect he must have therefore been a favourite. As a rule, I disliked travelling with anyone but my father. The other two brothers were Charles and Edward. They were also great friends of Mr Benjamin Smith's sons, Henry and Tom.

Looking back to that time, when I was about 18 years old, I have often thought what a silly fool I must have been to be so particular, and waste so much time over dress. Although my dear father allowed me to get what clothes I required from his own tradespeople, I was never extravagant, and I cannot remember that he ever once had occasion to complain or reproach me. However, I can remember, before going home on a Saturday night, going to my father's hatters - Greacher – near St George's Church in the Borough and sitting in the shop for more than an hour while the poor fellow in charge (Gill) was patiently trying to iron the brim of the hat flat enough to please me. I would then carefully carry it home with me and call at the hairdressers on my way to get my hair curled.

Dress or costume was then very different to what it is now. Swallow tail dress coats were then very seldom worn at evening parties by people of our class until several years later. But frock, morning coats reaching down to my ankles, a bright coloured velvet plush waistcoat in the day time, or a white watered silk at a party with a long gold chain spread about it and a silk handkerchief tied round the neck with a large bow were some of the clothes I wore.

Trousers were cut very tight to the leg with straps and wellington boots while shirt fronts were very much embroidered. It was also the fashion for ladies to wear a crinoline and a very outrageous fashion it was.

Can anything more ugly or unsightly be imagined than the clothes I wore? Yet I remember young Joe Quick, Mr Quick's eldest son, thinking me an awful swell. He was four or five years younger than myself and was always asking me to shew him how I tied my bow. He thought this so 'debonair' as he termed it, but thank goodness I am glad to say that I got out of these foolish ways as I got older.

There was also a fashion then all over the kingdom of sending valentines on 14th February – St Valentines Day. Some weeks before that day every fancy stationers shop and toy shop seemed cleared out and stocked with nothing but valentines of every kind from a penny to five guineas – vulgar and beautiful. Some of the West End shops had a grand show with some of them being hand painted on white satin. The shops were simply crowded with people trying to make a choice. I remember very well being one of the foolish. But what a waste of money, for every care was taken to prevent the recipient from knowing who had sent it.

During all this time, I was still journeying day after day to Walbrook and journeying back again, with the prospect of soon becoming an idiot. However, one day to my surprise Mr Austin came into the drawing office with a young-looking gentleman and remarked that he was to make use of the room as Mr Rammell had done. This man desired to prepare plans and maps for a scheme he wished to put before the Geographical Society for an overland route to India, instead of voyaging all round by the Cape. I need not say what joy it was to me again to have the chance of doing something that might be useful.

This gentleman was Lt. Waghorn of the Royal Navy, and it was not very long before he accepted Mr Austin's suggestion that perhaps I could help him. This kept me fully employed. His handwriting was not very clear or distinct, so I offered to make him a fair copy of the paper he had prepared on the subject he wished to bring forward. He seemed very pleased with this and with helping him with the plans and maps, I was quite in my glory.

All this work occupied many weeks and I need not say I was glad to hear afterwards that the paper, when read, was a success and was received very kindly. However, it was a long, long time before it was a *fait accompli*, but as we all know it has since been accomplished. I very much fear the poor fellow was never sufficiently remunerated or rewarded for being the pioneer of such a great enterprise and for being of such vast benefit to the trade of this country. But he will never be forgotten as the pioneer of this project and this insignificant individual, Charles Aird, will always remember with some pride that he himself had a little finger in that same pie, although a small one. It was not until some years later that any decided progress was made with the scheme, but I think I remember Lt. Waghorn having established an agency business in Cornhill under the name of Wheatley and Co. for the dispatch of goods to India via the overland route. The poor fellow, however, died quite a young man and only still as Lt. Waghorn, but a very fine statue of him was afterwards erected, I think at Bombay. All I can say of him during our short acquaintance was that he was a thorough gentleman and a thorough good fellow. Several years later I remember reading in some book the following: "England was never made by her politicians, but by her adventurers".

5
Joining the new Board of Health

Some two or three weeks after Lt. Waghorn had left our room in Walbrook, and I had returned to my life of stagnation, on leaving home in the morning as usual for the office, I purchased a newspaper to make things a little livelier. To my great amazement, I read the following news, copied from the London Gazette of the previous day:

"That the Government (Lord John Russell) having regard to the great sickness and epidemics that had so long been prevailing throughout the country, had decided to create a new government office to be called 'The Board of Health' and the offices to be at Gwyder House [later to be the Welsh Office]. The President of the Board to be the Earl of Carlisle, the Vice President, the Earl of Shaftesbury", then Edwin Chadwick Esq. CB, Dr. Southwood Smith and two others whose names I cannot remember (one a great chemist). But the name mentioned that interested me most was that of Henry Austin Esq. who was to be the Chief Officer and Engineer at £2,000 a year. Fancy, after all these years of poverty and struggling to get along to arrive at this. The news overwhelmed me altogether. There was also the additional information that there were to be ten civil engineers to be selected and appointed as government inspectors. The Board was to have full powers to go thoroughly into the question of proper drainage and water supply, particularly as to London.

How I got through that day I do not know. Mr Austin never turned up and his old father only for a short time in the morning. He did not refer to the subject with me. Curiously as fate had it, this was to be my last meeting with old Mr Austin. How I did long for 5 o'clock to come, to get home and to hear what my father had to say about it all. Well, when he returned from his work I found that he was very anxious as to what was to become of me in the matter. However, in the evening Mr Quick came round from No. 6, presumably to have our usual four-handed game of cribbage (for my brother John had now left his school at Southgate and was helping my father with his work). He very soon, however, came to the same subject we had on our own minds and the conclusion was come to that it would be best to wait to hear what Mr Austin had to suggest under the circumstances.

The following morning I left home by the early train to get to the office in Walbrook by 9 o'clock as usual. My brother John was with me on the platform waiting for the train to go to his work (my father and Mr Quick usually came by the next train). The train came up and my brother, seeing Mr C. Robinson, the Secretary of the S & V Water Company in the last compartment, joined him, and I went into a carriage forward (always second class then). This railway, when opened, had carriages very much larger than other companies' and carried 12 passengers in each compartment instead of six and was therefore double the size. The second class compartments of course, were not padded in any way.

Now the finale. The train, when about to enter the station at London Bridge, by some neglect or mismanagement, instead of going in alongside the empty platform waiting for it, ran right into another train coming out. The result was a great smash. My brother John and Mr Robinson being in the last compartment were fortunately only shaken, but knowing I was in a forward carriage, hurried up and found me lying in a mass of splinters and in appearance more dead than alive. I believe I was taken outside the station to Rouse the chemist's shop, across the road and just by the steps leading down into the Borough market. (I wonder how many times in my life I have gone up and down those dangerous, slippery stone steps).

After examination, however, finding no bones broken, but much cut and bruised with certainly concussion, more or less, the advice was to get me home as soon as possible and to bed, which was done. I have, of course, no recollection myself of this unfortunate time, but repeat what I was afterwards told. I do not remember any death having occurred from the accident, but many were very seriously injured. One, an old gentleman whom I can remember after all these years (60) was sitting on my left. From the position I was in when picked up, I must have been thrown right across the compartment and my head banged on the other side. However, on getting home I was at once put to bed and there I remained for quite a month, kept very quiet and not allowed to see anyone. I was under the kind care and nursing of my dear old friend, Dr. Keeble. When I was eventually allowed to get up, I remember being a very poor creature to look at; it had pulled me down dreadfully.

My first thought when strong enough was to learn how matters now stood as regards Mr Austin and myself. My father then told me that, having acquainted him with the serious accident that had befallen me, he had written a very kind reply expressing his sorrow, but trusting all would soon come right again. He had also said that should my father under the circumstances of his (Mr Austin's) new appointment desire that my Indentures be cancelled, he would be willing. However, at the same time he would be glad to have me with him for the rest of the time (3 years) and as no other suggestion was afterwards made, it was so arranged. A further communication afterwards came from Mr Austin saying that as he was now responsible for the proper formation and equipment of this large office, even having to select and appoint the ten government inspectors (what a piece of patronage) he would not be ready to receive me for some few weeks, but would let me know. Under these circumstances, my father suggested that I should take a little holiday abroad somewhere and I accepted his very kind offer.

Before going, however, Mr and Mrs Quick were giving one of their dances. They were always having London friends staying with them and frequently one or two of Mrs Quick's younger sisters with whom I was always glad to meet. What bright and happy meetings those were. At this particular dance I remember it was assumed that I was not then strong enough to go to it, but when the evening came I begged so hard that I might go round, if only for one hour, that I was allowed to do so. Never shall I forget that evening. It really looked as if some of them could have eaten me when I entered the room; I had no idea that girls had such strong feelings, may I venture to say of affection. At all events, it was a very happy and pleasant evening for me, chatting with my old favourites one on each side and more wonderful to relate, persuading one to take me just one turn round the room in a waltz and no ill came of it all.

A few days after this I started on my holiday alone, having taken a passage in a well-known steamboat at that time, the 'Baron Orsy'. This ran from London Bridge to Antwerp – a day and a night. In those times, no one could put a foot anywhere on the Continent without a passport and one in every way properly complete. Knowing this, and from the trouble my poor father had with passports when I had accompanied him across the Channel, I was very careful. But in spite of all my care, to my astonishment I was positively arrested because of some silly informality in the passport.

I was not allowed to leave the hotel until the British Consul had been communicated with, I do not know how many times, a great fuss made and fees to pay all round. At last having done Antwerp, a most interesting and quaint old city, I went on staying a day or two at Ghent, Bruges and other places including Brussels, then finally Paris. My first visit; eventually I knew Paris as well as I knew London. But on this first visit, how enchanting it all seemed to me.

I put up at Meurice's Hotel in the Rue de Rivoli. There was then no Grand Hotel, or Continental or Hotel du Louvre, but it was very comfortable, barring the dreadful smell in all the WCs. On the ground floor there were a number of quaint, square little windows with white lace curtains facing the Tuilleries Gardens opposite. These windows were in the Salon or Table d'hôte room.

How well I remember my first breakfast in that room. Seated by one of the little windows (the window was open, but the white net curtains were drawn to keep out the observation of the passers by outside) on a lovely summer's morning with a delicious sweet air coming in, the world seemed full of happiness. Squads of the French soldiers in their shabby blanket-looking blue coats constantly marched past as if in haste, and always drumming – drum, drum, drum; the city seemed alive with soldiers.

It was then in the early days of the Empire and every day while there, I would see the Emperor coming out of the Tuilleries Gardens for a drive. How bright and beautiful Paris looked then on my first visit and how much I enjoyed it. During the many times after that I came, it was always Paris - but never quite the same to me - and how little the French people dreamt then of the fearful bloodshed and humiliation that was to fall upon them later.

Returning home again, much stronger and better in health, I found a communication from Mr Austin asking me to attend in a week's time

Water colour - View of a Dutch town painted at the period - painted for his father by Charles Aird (1831-1910) when 18

Water colour - View of a Dutch town - painted for his father by Charles Aird (1831-1910) when 18

at a certain address in Whitehall as the arrangements in Gwyder House were not yet quite completed. During this remaining time I took it into my head to try water colour drawing for amusement and think if I had persevered I might have done fairly well. The very first attempt was a little drawing in sepia of a church, it was not so very bad. This is still in existence in an old scrap book of mine with a lot of autographs, several of Charles Dickens (see page 98) and one of Queen Victoria (below)

I also painted two other pictures in colours, reproduced in black-and-white on the previous two pages, small but larger than the one in sepia. How well I remember my anxiety about these as I intended them for a little present to my dear father. One was a water mill and surroundings, the other the River Scheldt near Antwerp with some boats. I remember well how my brother Joseph used to sit by me and watch as the little drawings progressed and often, many, many years after, speak of them. Poor, dear old Joe, he always seemed to take an interest in my movements. The drawings were finished and I framed them myself, then presented them to my father one Sunday evening. I have never forgotten the pleasure in his dear face, for he knew nothing of attempting such a thing. Those two little drawings never left him and were by him when he died. Afterwards these were sold at Christies (of course by mistake) having been mixed up with all his pictures when they were sold. [They, and the autographs, were salvaged and are still in the possession of the family].

In Scotland Yard, where the chief police offices are, at the far end there used to be then a quaint old house looking down on to a coal wharf and barges on the river. There was no Thames Embankment then, nor such a thing thought of, and when the tide was out, these barges were left dry on the dirty mud shore. All along the riverside to Hungerford

Bridge were similar wharves, but mostly for forage. This particular house belonged to the Crown and was evidently made use of for reserve offices in case of pressure of work, or during change of departments. Mr Austin had therefore no doubt made use of these rooms during his removal from Walbrook.

What a bit of luck it was for him that my unfortunate railway accident occurred at the particular time it did, keeping me out of his way. When meeting him as arranged, he explained that even now things were not quite ready at Gwyder House, but in the meantime I was to make use of one of the rooms and do some few odds and ends of copying and other little jobs.

At times I could not help feeling very miserable and unhappy as to my future. When articled to Mr Austin two years previously I was to have learnt a profession by which I could get my living. Up to the present, I had learnt nothing except the little I had taught myself. The past two years had been absolutely wasted. However, as no other suggestion had been made, I had to drift on and take my chance.

Having occupied these rooms for several weeks (they were less than five minutes from Gwyder House and exactly opposite the Horse Guards; you merely walked past the United Service Museum and Sir Robert Peel's residence where he died some little time after, having been thrown from his horse), Mr Austin told me that the following morning I should go to Gwyder House.

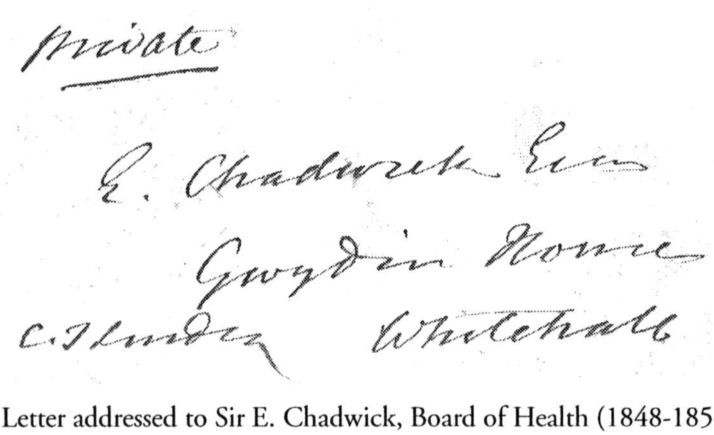
Private

E. Chadwick Esq

Gwydir House

C. J. London Whitehall

Letter addressed to Sir E. Chadwick, Board of Health (1848-1854) from the Bishop of London

I need not say with what interest and curiosity I had been waiting to see what the government office was like. The following morning, therefore, at 10 o'clock I nervously ascended the few steps and asked one of the many messengers that seemed to be about for Mr Austin's room. He had not arrived so I was therefore asked to take a seat in the waiting room. Some little time after I was taken to him. In the meantime I had been looking about me and taking it all in.

What a change from the two dirty, mean little rooms of Walbrook to this palatial sort of building, with its great staircase and its large, lofty rooms. It seemed so difficult for this young lad to realise that such a change should occur to himself. Mr Austin's room was upstairs on the first floor, adjoining the General Board or President's room. It seemed to me an immense room, handsomely furnished with official furniture – that is with writing tables, chairs, bureaux and thickly carpeted, with everything in the way of stationery. By the side of his table was a smaller one for myself, that I might assist him or prepare a copy or sketch of anything. Thinking it over afterwards, I really believe it was an act of kindness on his part to introduce me to all the others and shew that I was under his wing. Although without any official appointment then, I could not but notice that I was met and treated by all others in the different rooms as one of themselves.

What struck me greatly was the quiet and silence of the large building. The stairs and all the passage floors were thickly matted and all the rooms carpeted. There were never any loud voices or shouting. Every one seemed to pass from room to room in a sedate and quiet way.

Entering the front doorway, the waiting room was on the left, then two rooms for each of the two Secretaries – Tom Taylor and Alexander Bain. The first named was the well-known play writer and barrister; the second was the equally well-known writer and follower of Combe in physiology. Curiously I had been reading one of his books just before this and met him with much interest.

For some reason he did not remain for very long and was succeeded by Mr Macaulay, a brother of the great Macaulay.

There was a wide, long passage from the front entrance to the back of the building, where there was a large hall for the head messenger and his subordinates (six). I never could tell what these men did, except keep up the fires in the winter and distribute a fresh supply of stationery on all the desks. Judging by the immense stock of stationery Lang, the head messenger, always seemed to have stored in bureaux or cupboards in his room, what an enormous amount of money this must have meant for all the government offices.

The largest room on the ground floor was allotted to the clerks (four). This room was also carpeted and fitted up with large, massive bureaux with a desk or writing table in each corner of the room plus a large round table in the middle on which some of the morning newspapers were placed. The attendance book for clerks to sign (myself included) at half past ten was also in here. I remember that Lang used to come in and draw a line under the names every morning. This room seemed to be used as a kind of general room for the inspectors, although they had a room for their own use, but they were mostly in this room.

The Chief Clerk (Mr Walbridge) had a room to himself, the prettiest room in the building with a very beautiful sculptured marble mantlepiece. This gentleman and I soon became great friends. He was quite a gentleman and a connoisseur and collector. One day he asked me if I could make a little sketch of the marble mantlepiece for him. Well, I made a little try. I have now in my possession several autographs and old engravings that he gave me from time to time. Indeed so far as that goes, I am proud to say I was good friends with every soul in the building.

It is curious, but for the life of me I cannot now remember the names of the four clerks in the very large room, although I saw them and associated with them more often than any others. They also were gentlemen and quite middle aged. I remember I was much interested to see them arrive each morning, sign their names in the book and so quietly and sedately put aside their hats and overcoats, then always a friendly "good morning" to myself, although I was not one of them and only 19.

Referring again to Mr Walbridge. He was the most extraordinary snuff-taker I ever met. He seemed constantly to be forming a little pyramid of snuff on the back of his hand and then stuffing it up his nose. To take snuff in those times were quite the fashion; both my father and Mr Quick were guilty of the practice, but not to such an extent.

Having now tried to give an idea of the office in which I now found myself, I will try and give some idea of the *modus operandi* of the work being done, for it was very soon in full swing. As I have already stated, there was a Board of Commissioners. Mr Austin was Chief Officer with, I think, ten engineers as government inspectors under him – all well known men. Their names, so far as recollection goes were: Robert Rawlinson (afterwards Sir Robert); James Clarke; William Ranger; Henry Lee; Thomas W. Rammell and Alfred Dickens. The other names I cannot now remember. I was very glad to meet Mr Rammell again; Alfred Dickens was a brother of Charles Dickens.

When it was made known that the Board of Health had commenced work, several towns at once petitioned that an inspector should be sent to inspect, if necessary take evidence and report.

I remember that Croydon was one of the very first towns reported on. It was then Mr Austin's duty to select and direct one of the inspectors to make such a visit, and then so on with other towns until all were busy. Preston was another of the very first. The inspector, having made his examination and taken evidence if required, returned to Whitehall to write his report. When written (it was, of course, in the rough) it was sent downstairs to the basement where there was accommodation for some few copyists, or under clerks, who seemed to get their living by copying these reports in a good, plain hand – writing at so much a folio. The reports then went through Mr Walbridge's hands and then Mr Austin's. Now, this is where I seemed to come in.

The inspectors as a rule desired to illustrate their reports by little sketches or plans, depicting any serious evil of water supply or drainage they had come across. They nearly all liked to have a little reduced map of the town introduced in their reports when printed and this work I found devolved on myself. Mr Austin had referred to me to carry out their views and wishes which I did my best to do. When done, and to the satisfaction of the inspectors, they were sent to the lithographers to copy and include in the report when printed.

At the same time, Mr Austin would frequently come from the Boardroom and give me an order to prepare a sketch or plan of some subject in discussion, so there was plenty to do.

The reports when finished and printed with a paper cover only, were forwarded at once to the several towns with a request from the Board that the local authorities should at once take steps to remedy the very serious evils there pointed out. In a very short time these reports were needed for reference, and several of them covering different towns would be bound up together in boards, making a volume as large as a Temple Bar magazine. At one time I must have had at least fifty of them in my own possession.

Before going to another subject of the Board's work, I will just mention that in the middle of each day a waiter from the canteen at the Treasury office went round through each room for orders, but I do not think he got many from our department. Sometimes, maybe a chop or sandwich with a glass of sherry. It did not seem to be the rule then to go outside for luncheon and I know I was much too nervous and diffident to do such a thing myself. I therefore brought from home every day in my pocket a sandwich or a few biscuits to nibble with half a tumbler of water when the opportunity occurred. I rather think most of the officials did as I did at that time. There was always a large goblet of fresh water in every room with tumblers on a tray.

Fancy such a bit of refinement at Walbrook where an old, very dirty lead cistern dated 1670 was fastened to the wall by the large table in the drawing office. This had a tap in it and a dirty mug on it to enable you to get a drink or prepare colours.

The Board, having now with other matters arranged a good start with the Health of Towns enquiry, seemed to be anxious to commence with the investigation of the water supply in London. The outcry of the public at this time was getting very loud and startling. A Dr. Hassell of Richmond had written and published a pamphlet on the subject illustrated with a drop of Thames water magnified by a microscope. Certainly if we were drinking such awful things as shewn swimming about, it would alarm any old lady.

At that period there were very few places in London where one could go for a little amusement and instruction. I can only at the moment remember the Coliseum, the Surrey Zoological Gardens in the Walworth Road and, of all places in the world, that dead and alive place the Polytechnic in Regent Street. This was a building with a few little insignificant working models and a large tank of water just deep enough to allow a diving bell to be covered. Here a diver used to perform, but why I introduce this place now was that in a small room arranged as a kind of lecture room or theatre a lecture was given each day with these horrible drops of water magnified on a screen of enormous size.

By this time Mr Quick and my old friend Mr Fraser had completed the drawings and estimates for the new works of the three companies to be removed higher up the river to Hampton. An Act of Parliament had been obtained and a contract had been let to John Aird for the whole work consisting of three engine houses, reservoirs, filter beds and great mains, I think 30 inches in diameter. This work represented a large sum of money and was, I think, the first large contract of the kind my father had undertaken in England.

The contact for the pumping engines had been given to Harvey of the Hayle Foundry in Cornwall, Mr Quick having always had a preference for Cornish engineers. The only rival makers at that time were Messrs W. Simpson and Co., Grosvenor Works, Pimlico, London where the Victoria Station now stands. This firm had a monopoly for all government work similar to that which John Penn and Co. had for marine engines.

The railway company's arrival necessitated Messrs W. Simpson's removal and a very large new works were erected in Grosvenor Road near Chelsea Bridge. Although between fifty and sixty years ago, this firm is still in existence and prospering, whereas all the others such as John Penn and Sons, Maudsley Field and Co., Humphrey Tenerent and Co. have all passed away. It is curious that in after life this same Simpson family should be the dearest and kindest friends I ever had.

Referring again to this present water enquiry, there was no doubt that Mr Quick had the idea then that the government might be induced to purchase all the companies, place them under a trust or board, under a chief engineer and that that engineer should be Mr Quick. He was the first to give evidence and several other engineering experts followed. Mr James Simpson, the engineer of the Chelsea and Lambeth Water Companies and many others beside London experts came from the large towns in the north (Liverpool, Manchester and Glasgow). But there was an engineer then in London who seemed to be coming forward in leaps and bounds as an authority on such subjects – this gentleman was Sir Thomas Hawkesley. Mr Quick seemed to have a great dislike for him. As all the London companies, in consequence of the great outcry, were taking steps to move on to a better source for pumping their water, the Board at Gwyder House found the wind rather taken out of their sails when starting. But they persisted in the enquiry and eventually made a very useful report.

I remember very well my father at this time making known a very simple and useful idea which brought in for him a good bit of work along the River Thames, the well-known brewers Barclay and Perkins at Bankside being one of the first to give him an order. The idea was to sink a large iron casting or tube fitted with shelves or trays and filled with clean, bright shingle and sand, forming a good filtering medium. An iron pipe was then laid from the casting to the wharf and premises, by which quite bright and clear water could then be pumped.

In the early part of these memoirs, I have mentioned that my brother John and myself, when boys, often heard the rather warm discussions on politics between Mr Quick and my father. Mr Quick was a Conservative and my father a Liberal. We both became staunch Conservatives and have never altered our views all these many years since. Now being so near to the House of Commons and it being so easy for me to get an order of admittance to the Strangers' Gallery, I frequently availed myself of the opportunity whenever there was a good debate on.

Things seemed to be going on quietly at home. The younger members of the family of course, were growing older. My brother Joseph had just been sent to a boarding school at Eastbourne (in the old town) called The Gables. Afterwards he spent a few years at King's College School, and then to Lucas Brothers (builders) workshops at Lowestoft. At this time there was an institution called Putney College for training young men to be civil engineers. Several of these engineers in after life became great friends of myself and also of our family generally. I can remember the names of several – R. M. Christie, J.W. Gray, H. P. Stephenson, Alfred Williams, J. Vaudray and Henry Gill.

At about this time my father had another contract on the Continent to carry out, a new water works for Copenhagen. The engineer was Mr James Simpson of 29 Great George Street, brother of Mr William Simpson of the Grosvenor Engineering Works and with perhaps the largest practice in London of that class of work. This contract, and with so much other work coming in, all without competition necessitated John Aird's removal from the small premises in Emerson Street, Southwark to much larger ones in Belvedere Road, Lambeth between Waterloo and Hungerford Bridges.

The house stood by itself and really looked almost as large as Gwyder House. There were rooms sufficient for my father and brother John and at least 20 clerks of different denominations

How well I remember dear old Ingleton's pride at the advancement to such luxurious quarters and how many, many years he lived to enjoy it. Outside was a very large yard with outside offices for John Taylor and the Storekeeper. There were stables for many horses and workshops for carpenters above, and storerooms with a large wharf by the side of the river to unload barges. My father never required to move again. What an active strong man he was then; how well and easily he seemed to manage and arrange all this different work. Nothing seemed difficult to him and his name was so well known, there was hardly a single gas or water company in the country for whom he was not carrying out work.

My brother John was becoming a great help to him, I understood. He had tumbled into the pipelaying work so readily, that he had then almost the entire charge of it.

I had now been some three or four months at Gwyder House, my work being as I have described, the only alteration being that I was shifted from Mr Austin's room to a room to myself on the floor above. Mr Austin had been inconvenienced by the inspectors coming so frequently into the room to give me directions, so one morning he came into my room in his cold-blooded, silent manner, as if I had committed some awful crime. On this particular occasion, however, I had reason to open my eyes for he handed me a cheque from the Treasury, being payment of three guineas weekly to Charles Aird, draughtsman. Without any feeling he said a few words that he hoped it might lead to better things and also that it was the rule that all names on the staff were entitled to three or four weeks holiday, subject to the convenience of others. It is curious that during the whole of the five years I was under his charge his manner never changed.

I very well remember I could not say a word in reply, but he must have seen from my eyes how much I felt. When alone, I tried to realise it all and to remember that only nine years ago I was the little chap running wild at the gas works at Greenwich without any real education – it seemed hardly possible. However, returning home in the evening I passed the news on to my father and mother, and need not say how pleased they were and to know that in future I could earn my own living – and did.

It should be remembered that everything in 1849 was much cheaper then than now in 1908 when I am writing these lines. Rent, food, clothing and not so much temptation to spend money, plus income tax at 3d in the pound, not 6d as now in time of peace . Therefore from that day I started saving and started a little money box.

The Chief Clerk, Mr Walbridge, seemed to have the arrangement of the holidays of the several under-officials, myself included, so I asked my father when he was likely to go to Copenhagen again. I hoped that I might get my holiday at the same time to go with him. He was very pleased with the idea and so it was carried out.

I often used to wonder how my father used to get along, travelling alone on the Continent as he had to do, together with the passport nuisance and the different coinage, but he managed it somehow or other.

Having arranged a time for a second visit to the work at Copenhagen, I also got permission for my first little holiday to accompany him. It was a long journey in those times ; very fatiguing and disagreeable, particularly the very nasty crossing in a wretched small steamer between Kiel and Cönsor in the Baltic. As my father had some pipelaying work then going on at Altona, we journeyed first to Hamburg, Altona being on the other side of the river. I was greatly struck with Hamburg; such a fine important, wealthy-looking city. I was also much impressed with my first acquaintance with the Germans. The hotel we stopped at (Stroits) was the most comfortable I had ever been inside. I shall never forget the large tankards of Burish beer.

We left Hamburg by rail for Kiel and *en route* made friends with a German, evidently a man of some position who spoke English fairly well. Conversing, he suddenly pulled out a large cigar case and offered it to my father and myself. I declined, but my father foolishly accepted one, which he could no more have smoked than fly to Heaven. The old gentleman very politely struck a match and handed it to him, giving my father no escape. Therefore with a struggle, and a look of agony on his face, he made a start. To add to his misery, our new friend remarked what a very disagreeable practice it was for some smokers to expectorate. My father would go through any amount of misery and discomfort rather than appear rude or unkind, but it was no use. Had there been a spittoon about, he might have struggled on a little longer, but he had to give it up and apologise, remarking that the cigar was of delicious flavour, but rather larger and stronger than he was accustomed to. The old gentleman very kindly accepted his view of the difficulty and by the little twinkle I saw in his eyes, no doubt was amused with the pater's explanation. On arrival at Kiel we parted and commenced a most unpleasant long journey to Copenhagen. We embarked on the most wretched, miserable little steamer I ever voyaged in. The weather unfortunately being very rough and bad, the boat danced about like a cork on the top of the short, choppy seas. As I have said before, poor Father was always a bad sailor under any circumstances, but got along fairly well if comfortably placed and quietly left alone with his eyes shut. He therefore decided to go below at once and settle down.

I preferred to remain on deck, although very wet; but having borrowed an oilskin from one of the sailors, it was not so bad. After a time, however, I began to wonder how Father was getting on and thought it would be better that I should go down below and see. Worse luck I did. It would be simply impossible to describe the awful atmosphere below. There was a wretched small hole of a cabin crowded with foreigners and nearly all sea sick. When a foreigner is sea sick, it is a caution, as I had reason to know some two or three years later on. The place was so dark and the atmosphere so thick, it was at first impossible to find my father.

However, I persevered a second time and found him at the very far end of the cabin, lying full length on a hard horse hair couch, about the very worse place he could have pitched upon. But probably it was Hobson's choice with him. However, now comes the misery to both of us. Having at last found him I asked him quietly how he was getting on and whether I could do anything for him. But I had hardly got the words out, when I gave a great heave and was as sick as a dog. The first and only time I was ever sea sick. Coming from the fresh air into such an awful atmosphere, and tumbling about trying to find where my father was, would have knocked over the very strongest man. Unfortunately, the accident to myself quite upset my father and also made him very sick and very, very angry as he had been feeling fairly comfortable, until I had gone down and upset everything. Never try and do any good-natured action say I.

On landing at Cönsor we still had another long railway journey before reaching Copenhagen. The country was flat and uninteresting, but the journey came to an end at last.

There seemed to be only one really good hotel in the City (not to be compared to the Stroits at Hamburg), but we were both glad to get there anyway. The hotel was situated in the middle of one side of a large square or market place; with the quaint, very, very old long, low palace directly opposite. There were no modern buildings of any kind about and no striking public building. The people were all very quietly and plainly dressed with no Parisian fashions – nothing but home made tailoring. Also, there were no rich people, but every one looked quite content and happy. Some parts of the city reminded me of Holland and its waterways. These were my first impressions of Denmark and the Danes. They were the most primitive, faithful, honest, simple-minded people that ever lived and this opinion of them was endorsed many, many years later on.

We very soon visited the work my father had in hand for the Corporation, and it was evidently a very large contract with a very large number of men being employed. A number of them were the old English navvies my father had sent out, but how few of this class of workmen seem to be left.

At the time of writing this – 1908 – John Ward, a navvy who worked for my father's firm on the Suakin and Berber railway in Egypt, is now a MP; one of the now many Labour representatives elected in this present radical government. What is to be the end of it?

I found my younger brother Alex here, engaged on the work. He had left school and my father had sent him here to get a practical knowledge of this kind of work on the Continent. He was also a good linguist, with both French and German which was very useful here. My father also no doubt saw a great future on this side of the water.

My father, on his first visit to Copenhagen to start the work, saw there would be a difficulty in finding proper carmen with apparatus for carting out the number of iron pipes to be laid. He therefore decided to send for horses and trolleys from London for the purpose. These arrived by ship at the same time as our visit.

It was most amusing and interesting to see the astonishment of the natives when they started to work. They had never seen such great horses before (they were all splendid cart horses) and crowds of people followed them about. The large, ungainly-looking trolleys also bothered them, but delighted them at the same time to see how useful they were. The poor Danes somehow then seemed a long way behind.

During the time my father was engaged, I wandered about seeing what there was to be seen, not much but very interesting. The palaces of Christiansberg, Rosenburg and Charlottensburg had many beautiful specimens of the Danes' great sculptor Thorwaldsen's work. Poor Denmark! Some few years after my visit, the Germans went to war with them and pretty well knocked Copenhagen all to pieces.

Before returning home again, the idea occurred to my father to give a holiday and a picnic to all employees, both engineers' and contractors' people. Finding the idea appreciated, he carried it out in first rate style. He gave the proprietor of the hotel a *carte blanche* order to provide a sufficient number of charabancs for the party – some 60 or 70 I think – plus an ample and first-rate luncheon with wine, beer and waters, of course. The hotel also made the arrangements for tea later. I remember even now what great enthusiasm was shewn over this little affair, particularly by the Danes who were quite unaccustomed to such events.

My father's manager here in charge of this contract was James Eagle. A few years back the poor fellow had asked my father for a job on some pipelaying work and was picked up one day in the street. He was now a splendid specimen of an Englishman and one of the most practical men my father then had in his employ. Eagle took the idea of the party in hand at once and did everything to make it a success. Having some quoits, bowls and trapball made for games, he also got ready some little prizes for running races and tug of war. It happened, fortunately, to be a beautiful day - I remember, very hot.

After a drive of a couple of hours, we arrived at the Castle within an immense park and then prepared that our guests should thoroughly enjoy themselves, which they certainly did. Some few of the Danes, however, poor fellows, never having been accustomed to excitement of this kind or indulgence in cooling drinks or wine, succumbed after the lunch and wisely went to sleep under the trees in the shade. However, by the time some refreshing tea was concocted, they were quite all right again. The Chief Engineer of the work, Mr Colding, a typical Dane, was really like a boy and the soul of the little party, although so severe and stern on the work. Well, all's well that ends well, and that little outing was not forgotten for a long time after.

Returning home again, my holiday being over and glad to find all well, I returned to my work at Gwyder House. Here all seemed much the same. One or two of the inspectors were a little anxious as to their reports being completed in time, and so things went on. The Board, having taken what evidence they required regarding the water supply of London, had prepared their report. As several proofs of it were available before being finally printed and presented, I borrowed one of them and took it home to shew Mr Quick and my father whom, I knew, were very anxious to learn what they proposed.

Perusing it, Mr Quick was much disappointed to find there was no recommendation that the water company should be purchased and put in a trust with Mr Quick as Chief Engineer, as he had hoped. The recommendation was that pressure should at once be put on the water companies to abandon their present pumping stations and go higher up the river.

We had already heard from Mr Quick who represented three of the companies – the Southwark and Vauxhall, the Grand Junction and the West Middlesex – that they were already constructing new works at Hampton above Hampton Court. Mr James Simpson of the Chelsea and Lambeth Company was preparing to move to Surbiton just above Kingston on Thames, and Mr Wicksteed of the East London Company was moving from Old Ford Bow to Lea Bridge and also Walthamstow. The New River Company was moving higher up and the Kent Water Company was to sink more wells. All this must mean a lot of work to John Aird, particularly pipelaying.

Curiously, that very depressing feeling seemed to be coming over me again. When at home I seemed to prefer shutting myself up away from the family and refusing to join in any little pleasure going on. Again at this particular time I fancy I did not get on very well with my brother John. I remember well he had a very bad temper. Even now I have some recollection of a dispute or quarrel with him which almost knocked me over altogether, and being walked round the little bit of garden at the back by John Taylor, my father's old clerk who happened to be in the house at the time. Somehow or another he was always a good angel to our family.

I also remember how greatly troubled and anxious my dear father and mother were about me, and so tried to shake off this dreadful feeling, but it was no use. However, one day I asked my father to let me take a couple of rooms in London somewhere, to try and keep myself. Poor dear old father was terribly upset at this proposal and tried his best to make me alter my mind, but finding me obstinate, he gave his assent. I therefore looked about and took a couple of rooms in Chester Terrace on Westminster Bridge Road, not half a mile from our offices in Whitehall. To suit the state of my mind at that time, this terrace immediately fronted Bedlam, the lunatic asylum.

I think I must have remained in those rooms something like six months; occasionally going home to spend the Sunday or to accompany my then great friend, Mr Innes to his pretty little house at West Wickham. His wife had recently died, so he only had his niece Miss Mynott, or Minnie as she used to be called. He had succeeded Mr Munro as the Engineer of the old Phoenix Gas Company and every Saturday he drove from Bankside (the gas works) in his dogcart home to West Wickham. It had become an understood thing somehow between us that whenever the spirit moved me, I was quite at liberty to join him, which I did frequently and as naturally as possible. Furthermore, the quiet and delightful charm of the pretty place, and the country round me did me a wonder of good, for we were all kindred spirits and lovers of nature. How kind and good that man was to me in afterlife!

Any thought of my leading an irregular life by this arrangement of new quarters for myself must at once be abandoned. That was not my way for I never cared for, or liked anything in the way of fast life. I seldom, or ever, stepped outside the front door in the evening. At that time there were no such brilliant music halls as are in existence now. The only two I can remember being Canterbury Hall, a public house by Westminster Bridge, and a kind of musical restaurant under Hungerford Market, kept by some Swiss people, Gatti, who made a fortune.

What a strange place Hungerford Market was then. There was no Charing Cross railway station or hotel; only the market and the suspension bridge over the river. It will be remembered that this bridge, when taken down for the railway, was re-erected over the River Avon at Clifton. [In fact, according to Bamber Gascoigne, who researched this in about 1960, he found out that as the Charing Cross Bridge (also by Brunel) was pulled down, the Clifton Bridge was nearing completion so they used the old Charing Cross chains for the new Avon bridge].

Fish and shellfish were, I think, the principal items on sale. How well I remember when quite a little lad, long before Gwyder House was dreamt of, many, many times when passing by, buying a pint of shrimps or winkles, and eating them with a pin as I walked along.

I never, however, entered either of the two places I have named, and I even didn't much care for the theatre then, unless there was some well-known play written by our Secretary, Tom Taylor. He was always clever and worth seeing, particularly with Alfred Wigan acting. What a wonderful actor he was.

Now and then I went to the old Adelphi to see Wright and Paul Bedford in 'Paul Pry'; ' I believe you my boy' and 'I hope I don't intrude'. Or Mr and Mrs Keeley at the Lyceum with their charming daughters. Also Webster, Buckstone and Madame Celeste in the Green Bushes or Uncle Tom's Cabin. Melodrama was the order of the day then.

On Sunday, when not going home or spending that day at West Wickham, I always went to Kennington Church and liked doing so.

Now and then in the evening I would call on my father's very old friend, Mr Munro, who lived close by with his niece Jessie Christie and nephew Robert Munro Christie. I became very friendly with them but not, I am afraid to my advantage. There were two other families living close by, friends with whom I also became very friendly over many years. They were the Coplands and the Morrisons who were close to all the Aird family.

As a rule, when returning to the lodgings after my day's work, I had a chop or a little bit of fish, with a glass of sherry or beer, for I was very careful and saving. Now and then for a little treat (but very rarely) I would go to a little restaurant in the Strand opposite Exeter Hall called Simpson's. My father had once taken me there, but although reasonable enough in its charges, it was too expensive for me. Some few years later, however, I frequented it very often and became one of the Knights of the Round Table. Beside the side boxes with tables for four, as it used to be at Izants, there was a large oval table. As a rule this was only used by certain men, well-known actors and journalists – Ledger, the Editor of the Era paper, Alfred Mellon, Toole, Robson and many others. I was invited one day by one of them when I was better able to afford it, and in future I always sat there.

The great feature of this place, apart from the certainty of the fish or the joint being of the very finest quality (each joint up from the kitchen was wheeled to you on a little table) was that the carver (in a cool white linen suit) might cut you any particular part you wished. In those early days of Simpson and Dawes, both as well known to habituées as Paddy Green at Evans's, the two proprietors would walk about the room and chat with any well-known customers. I can see them both now after all these years. Simpson was a short, stout man and lame, but Dawes was very tall and thin. Port wine was then a very fashionable drink for those who could afford it and I am afraid I was extravagant enough sometimes to beckon to Mr Dawes and ask him to bring me a pint bottle of the 'Blue Seal'. It was very amusing to see him return from the cellar carrying this pint bottle so carefully and affectionately, covered with cobwebs, and draw the cork himself. The wine certainly deserved such respect.

Another feature about this place was that without going outside in the street again, there was a very large cigar shop where, after paying 1/-, you could select a first rate cigar from any number of boxes. After that, you then received a counter to go up to the next floor for a cup of splendid coffee in a large room called the Divan. This had lounges and chairs all about and tables with nearly every paper or magazine published, all to read with your coffee. There were also several tables for chess and always some very interesting games going on between great and well-known players.

When the time came, and when I could afford to frequent this place often, I frequently used to meet Mr Morris there, the Engineer of the Kent Water Works who was so kind to me about fishing when living at Greenwich.

Well, things now seemed to go on pretty much the same with me, month after month. But my father's business, from what I heard, seemed to grow more and more in interest and importance. He also seemed to be becoming well known to all the great engineering men in Westminster, who very frequently and kindly asked us to any little dance or festive occasion they had going on.

As a rule, in my father's office it was the practice, every Friday, to send a cheque to the manager in charge of the work going on in any town for wages and small accounts. As a check my father would sometimes send someone on the Saturday instead with cash and pay the men. When very busy and pressed he would ask me to go for him, which I always willingly and cheerfully did. I can remember Rugby being one of the many towns and Oxford another. Although it may create a smile from those who may read this, I can also remember a certain cold pigeon pie I afterwards enjoyed at the tavern or inn where I had paid the men.

It was a custom then in nearly all country towns to have a little kind of glass case or cupboard fixed inside the entrance door of the inns displaying some very appetising items of food. Cold sirloin of beef, cold fowl or pigeon pie were some of the items displayed and such fare, with a glass of home brewed ale, was always fully appreciated. Not caring to remain the night at the inn, I usually returned home again by the night mail in time for church on the Sunday.

I had now been two years at Gwyder House and in one more year I would be 21 years old. I little dreamt then, after this last two years what my future was to be. With most young men who start life, meaning to do their best, they select some single occupation and persevere through life with it, sometimes this is their father's business or profession, as the case may be. In my own case, I could never see any future, particularly with the knowledge that the past four best years of my life had been wasted because I had learnt nothing practical.

In my brother John's case, it was quite different. His future was distinctly secured. On leaving school, he was called in by my father to assist him and in a very short time he shewed how fit and able he was to do so. I believe it was marvellous the way he seemed to grasp all the many ins and outs of the different works. More marvellous still was his masterful power over men who were much older in authority than himself and in a much greater position in life. However, as I have said before, he always had the gifts of great force of character and firmly believed in himself; whatever might arise he stuck to that opinion.

Now his older brother Charles was quite different, he never believed in himself and in whatever position he might be placed, he always had the feeling that he had no business there. All the same, he can look back with pride to the fact that in whatever work he had to do, he always succeeded and pleased. He also never failed to make a lasting friendship with whom and for whom it was his fate to be associated.

After two years I think my salary was increased from three to four guineas a week. I certainly remember my holiday, as it enabled me to have another pleasant little journey with my father and some others to Berlin. (These were some financial men in the City who had formed an English company to supply Berlin with water, as in England).

It is curious how much foreigners of every nation disliked the idea of washing themselves too much. In any hotel one entered, there was no larger quantity of water supplied for your ablution than would fill a slop basin of about a pint. A bath was a thing unknown, and the dreadful smell of the WCs was indescribable.

The Engineers, being a well known firm of construction engineers at that time, consisted of Charles Fox (afterwards Sir Charles), Thomas

Henry Keeble Aird, son of Henry Aird (1849-1914),
sixth son of John Aird (1801-1876)

W. Crampton (better known as Tom Crampton), and a lame man. He only had one leg and walked with crutches, he was a most amusing man and one of the kindest-hearted I ever met; he was also a great favourite of my mother's. Some little time later on, as he was a bachelor, I frequently used to stay with him and also with Tom Crampton. He was, along with Mr Collins, a well-known Society man living in one of the great West End Squares. Mr Crampton had just made a great name for himself by taking out a patent for greatly increasing the speed of locomotives by enlarging the driving wheel. He was also a very handsome man (as was his eldest son Tom) and the most perfect gentleman. The daughter, I remember, used to sing divinely and married some great man, then an ambassador in Spain. All of that set, being very musical, had great musical parties when Tom Höhler and the Duchess of Newcastle used to sing together.

Despite the prospect of war with Russia, everyone was thinking and talking of the proposed first great International Exhibition in 1851, to be erected in Hyde Park. Sir Charles Fox had the contract to erect the building, conceived from a sketch or design made on a sheet of blotting paper by Mr Joseph (afterwards Sir Joseph) Paxton, the head gardener of the Duke of Devonshire at Chatsworth. It had also been arranged that John Aird should do all the work necessary to supply the vast building with water and gas.

I was much impressed with Berlin on our visit there, so different from any of the other cities I had yet seen – Paris, Copenhagen, Holland. (Poor Paris - at this time it was in a state of revolution and bloodshed; the great *coup d'etat* of Louis Napoleon was taking place and his election as President.).

There seemed to be a great strength and power about Berlin and its people, there was no frivolous gaiety, day and night as in Paris. At that time there was no united Germany, no Emperor, only the old King of Prussia, the grandfather of the present Emperor.

My father and I, on arrival, pitched our tent at the Hotel du Nord in the Unter den Linden. Our companions, the engineers of the company, stopped at the Hotel de Russie, leading out of the Unter den Linden which was built right through the city. It could not possibly be called a street as it was at least three times the width of any street in London and facing the

River Spree. We were together nearly the whole day. My brother Alex, I remember, was there to meet us, as was I think James Eagle, his work in Copenhagen nearing completion.

While my father and the engineers went about during the day, making arrangements for the construction of these new water works, I took advantage of any opportunity that offered to see what I could of the many interesting public buildings and places. I little thought how much more I would know of Berlin many years later on.

When evening came, we generally dined together at the one hotel or the other. These two hotels were the best in the city at that time; very quiet and most comfortable. There was not the least sign or attempt at grandeur, but with substantial plain food and very often veal. Their wines were superb, their lager beer came in great glass tankards of which one could never tire. In the evening, I remember we would sometimes all go to Crolls beautiful concert gardens, or the Tiergarten and listen to the splendid music and drink coffee or more lager beer. Every other man was in uniform, indeed it seemed to me the whole city was full of soldiers. Even if you went into an office, most of the clerks were seated at their desks in uniform.

The old king's palace was almost next door to the hotel and during our stay, at 10 o'clock in the morning I never once missed seeing one of the regiments on duty march down the Unter den Linden from their barracks outside, with their magnificent band. They would draw up outside the palace and play for one hour, then before the regiment left, the old king never failed to open the large casement window on the ground floor and give the military salute.

After a fortnight or so, matters regarding the new water works were settled, my father had the contract for the work, so we returned home.

Some little time after our return, a great grief and sorrow fell upon us all, our dear Dr Keeble died. This was indeed a terrible loss. I know I personally felt it keenly, for we had been such good friends and kindred spirits. But to dear Mrs Keeble, the blow was even more terrible, as they had no children and no near relations. But we one and all did our best to cheer and help her in her trouble. After a time, I remember my father took her and my mother to Scotland for a little while. Dr Keeble, when my brother Henry was born, expressed a wish to be his godfather, his name therefore was Henry Keeble Aird and Dr Keeble left him £500. Our dear doctor's successor was Dr Gay Shute, also of Greenwich. He was a clever man and always very kind and nice to us all, particularly to my dear mother. He was a very tall man and well built in proportion, but his wife was the smallest and shortest little lady you could find. To see them walk into a room together was very funny.

Returning to my work I found matters as usual. What a number of towns had been inspected and reported upon, and I think and hope with some amount of success. The Board of Health, not content with only the report on any town, followed this up by compelling the local authorities at once to remove the serious evils then existing. In nearly all cases, the evils in all the towns were much the same. The rivers were polluted; the wells for water lay alongside cesspools; where any system for drainage did exist, it was faulty with no proper fall and frequently in the wrong direction. There was no idea of disconnecting the pipes leading from indoor sinks by open air traps, and cisterns filled with stagnant water were used for water closets and drinking water at the same time. It really was wonderful how people continued to live under such circumstances.

In some towns, cholera was then raging and there was still continued agitation against the existing water supply to London in spite of our Board's report. Eventually a Water Bill was brought into the House of Commons to buy up all the water companies and give a better supply to London.

This Bill was not introduced by the Government, but by some private members, and there evidently would be a big fight in Committee, with a large array of counsel for and against. These committees on public matters always interested me greatly and whenever I had the chance of getting out I could always be found in one of the committee rooms.

How well I can remember all those well-known Parliamentary counsel during those days, seated in their wigs and robes before a committee of five Members. On a Bill like this there would be a leading counsel and two juniors, with the same number probably to represent each water company opposing. I can see them all now, although all dead and gone long ago – Sergeant Wrangham, Sergeant Ballatine, Hope Scott,

Edwin James Talbot, Phinn, Bellasis, Beckett Denison, Sir John Karslake, Bovill and others.

These great parliamentary fights would occur on gas as well as water, and it brings a well-known name to my memory – Baxter, solicitor of Baxter, Rose and Norton. I remember being present one day in a Committee Room on a London Gas Bill. It had arrived at a very critical stage when it was seen that no Counsel for the Bill were at that moment present. Mr Baxter (a nice-looking grey haired old gentleman) rose, and quietly asked the Chairman whether he might be allowed to speak for Counsel. The Chairman assented and a finer speech *a propos* the subject in hand was never heard. It had its effect, for the preamble of the Bill was passed.

There were other men beside the counsel and solicitors who had to get their living by parliamentary work and they were the parliamentary agents. They were supposed to pilot the several acts through and at that time the two principal firms were Pritt Venables and Wyatt and Metcalfe: how well I knew these firms and their people later on.

During the time I had been at Gwyder House, I had many opportunities of seeing many of the great ministers, statesmen and the well-known members of Parliament at that time; either in the House itself during debates or Committee Rooms, or by being pointed out to me. I was also at that time, much interested in collecting autographs of various notabilities.

One day, I remember having occasion to run down stairs to the chief messenger's room for some drawing paper and found him (Mr Lang) having a very warm dispute with the postman. I stood aside until it was ended, but while it was on, a little old gentleman came in the room and asked if he could see Lord Ashley. Lang, without even looking at him gave a surly reply - that he would attend to him presently. The old gentleman and I therefore amused ourselves by staring at one another until the dispute ended and Lang then offered to take the old gentleman's card upstairs. On his return, he confided to me that it was the Duke of Westminster, reputed to be the most wealthy man in England and grandfather of the present duke (1908).

It wanted now little more than a year to the opening of the Great

Exhibition of 1851 in Hyde Park and nothing else really seemed to be thought or spoken of at that time. I remember Punch that year was very amusing with its jokes and illustrations. To take up an old volume of that year now, one can very well realise the event.

I do not know whether the coming of this exhibition had anything to do with it, but to the great surprise of all the officials of the General Board of Health at Gwyder House, we were requested by the higher authorities to shift our quarters to No. 1 Richmond Terrace. Richmond Terrace was a row of very palatial buildings leading at right angles from Whitehall, or rather Parliament Street, being some few hundred yards from Gwyder House and nearer to Westminster Bridge. The entrance to this No. 1 was from Parliament Street.

Our new quarters had evidently been used for government offices for some time before and when we had moved and settled down, the rooms were quite as large as those we had quitted, in fact somewhat grander. Lang, the chief messenger and his men were left behind and we found a new staff ready to receive us.

The Chief was simply an ideal butler or major domo; tall, stately and important, but a real good fellow all the same. I cannot for the life of me remember the man's name. There was evidently great demand for sufficient accommodation for government officials. At Gwyder House I had a room to myself. Here to my great surprise, I found another located in the room who was evidently an Admiralty man. His work, so far as I could judge, was to plot on maps certain charts from recent surveys and gaugings of the flow of water on certain rivers.

In the next room to mine were a couple of men from the Charity Commissioners' office. One of them named Baker was really a gentleman, but the man in my room, whose name was Haile, was not. Anyone at a glance could see that he was in some way connected with the sea and looked more like a custom house officer than anything else. He was also double my age at well over 40, but there was no help for it.

Things went on much the same in these new offices so far as I was concerned, the same sort of work over and over again, learning nothing fresh and often wondering what the end of it all would be. My Indentures with Mr Austin for five years would soon be up, or had they been cancelled by my official appointment? It seemed a curious state of things, all left very much to take their chance, but looking at things in this light, I was unjust to my dear father, as I had reason to find out later on. For I did not know or sufficiently appreciate his great goodness to me after all these years.

Another little holiday during vacation time came round. I decided to take a holiday alone this time and journey up the Rhine as far as Heidelberg, and also go up the Moselle some distance, stopping a few days at some of the most interesting towns en route – Wiesbaden, Coblenz, Mayence and Cologne. At that time I cannot imagine a more interesting or beautiful trip. Travelling then was not as it is now. There were no vulgar tourists swarming about, but everybody was neat and clean, quiet and au naturel – as one would say. How well I can remember the steamer on a hot summer's day, gliding slowly along on the beautiful river with its romantic scenery and castles, enjoying my *dejeuner* on the deck under the awning, with a beautiful soft breeze blowing, at about 12 o'clock when one would be quite ready for it. Perhaps a trout, also a couple of cutlets as they can only cook them on the Continent, chipped potatoes, fruit and a pint bottle of hock. This is a beautiful wine, which I have always said can never be fully appreciated except in the country in which it is grown and made and then it is a dream, as it certainly was on this particular occasion with a good cigar to finish. We ought to be content with such days as these, but they don't last long.

This holiday had really been a very enjoyable one and I know I returned to work in somewhat better form; little thinking where my next holiday would be.

We are now in the year 1851 and I am 20 years of age. The Exhibition has opened with great ceremony and seems to give a lot of pleasure to the people. Of course I went whenever I had the opportunity, the building being erected in Hyde Park where the Albert Memorial now stands. There could be no pleasure grounds outside, but there was plenty inside to interest and amuse. The magnificent sculpture of the Amazon and Geoffrey de Bouillon and the Greek Slave interested me as much as anything. The Austrian Court was also very attractive with its beautiful leather work. I remember also that Mrs Sturmy, the wife of my father's solicitor had herself worked a very large and beautiful carpet for the occasion. This was exhibited in the English Court in the Gallery and always had a crowd round it. What a pompous old lady she always was in our little home society – bad enough before the days of the Exhibition, but after there was no holding her. The Exhbition opened on 1st May and closed on 1st October, by that time the people seemed to have had enough of it.

The constructors of the building – Sir C. Fox and others were bound to clear it all away at once. In the meantime they had formed a company to purchase it (I think for £70,000) and re-erect it on a much larger scale at Sydenham Hill, Norwood. Sir Joseph Paxton gave all his time to make it really a Crystal Palace. The lovely grounds outside were laid out to rival any others in this country and with terraces and fountains to rival even Versailles. Here my dear father comes in again. As he had laid all the mains both for gas and water at the '51 building, he had the contract to take them all up again and do all the work of that kind on the new building at Sydenham, a very large and important work.

In speaking of our removal to No. 1 Richmond Terrace, I mentioned that all the other houses were tenanted by several aristocratic families, one of them being the Earl and Countess of Harrington. The Earl was a very eccentric man; a great connoisseur of snuff: he consumed prodigious quantities. Not only that, but all the servants' liveries were of snuff colour, the carriages all so painted and the inside lining the same; the horses as near as possible. One could always recognise their carriage. The Countess was a well known and popular actress, I cannot at the moment remember her theatrical name. The reason I refer to these people was because Miss Robinson, the sister of Mr Charles Robinson, the Secretary of Mr Quick's water company at that time was lady companion to the Countess and I heard a good deal of them when meeting her.

Some more months have now passed by and we are now in the year 1852. At the end of it I shall be 21. I am sorry to say that the strange, well, unaccountable friction between the other occupant of my room, Mr Haile, seemed now to have got to fever heat. So much so that I am ashamed to say it one day boiled over and we actually came to blows, and there was no doubt from the way we went at one another we meant business.

The occupants of the next room – Baker and another of the Charity Commissioners' office – hearing such a noise, came in to see what was up and soon separated us before much damage was done. I can sincerely and honestly say that I never felt so thoroughly ashamed of myself in my life and I really think my partner in the foolish act felt the same. After being considerably chaffed by our neighbours, at all events we seemed better friends afterwards. The foolish man, being the elder in years, thought I could be bullied to obey his orders. He simply made a mistake.

6
Surveyor in Canada at 21 (1852)

And now comes a strange event in my career. One day there appeared in all the newspapers an announcement and the prospectus of a most important financial scheme. It was said the largest ever put before the British public and evidently supported and backed up by all the principal banks and financial houses. This scheme was to construct a railway in Canada 1200 miles long from Montreal to Toronto along the River St Lawrence and Lake Ontario. This was to be called the Grand Trunk Railway of Canada and to communicate with the sea coast at Portland. I think it was then also stated that Glyn Mills Currie were the principal bankers.

It was also stated in this same prospectus that a contract had been entered into with all the three largest of English contractors – Thomas Brassey, Sir Morton Peto and Betts and Jackson of Birkenhead to survey (or locate as the Americans say) and construct all this great work. This included the gigantic Victoria Bridge over the River St Lawrence at Montreal for one large sum of many millions of money. I remember at the time that the grandeur of the scheme almost took one's breath away.

I read all this in the papers, as no doubt many others did, merely thinking at the time it seemed a tremendous big job and thought no more about it. Some months after there was a further statement in all the papers, that the contractors previously named had nearly completed all their arrangements for an early commencement of this great work. They had appointed Mr Alexander Ross, a well-known railway engineer, to be Engineer in Chief of the whole work. He and the whole staff under him (about sixty) would be ready to leave England on a certain date, about two months hence. Some few days after this, when I had returned home in the evening, my father, to my great astonishment asked me whether I had read or noticed the statements of this great scheme, and whether I would like to join the staff going out and try my luck. If so, he thought he could manage it, as he knew Mr Ross, the Engineer very well. I well remember that this sudden proposition knocked me over more than the idea five years ago that I should be articled to Mr Austin. However, after a long and serious conversation with my dear father and mother (for I was older now) it was decided, if it was not too late to get an appointment on the staff, that I would go. My father therefore arranged that I should call at the company's offices in Great George Street and see Mr Ross, who was a stern, severe, hard-looking man of about 50. He asked very few questions and remarked that if I decided to go, I could probably make myself useful somewhere in the work. That I assured him, I would certainly try and do. I remember he also stated before we parted that as all the arrangements had been completed for himself and the whole staff to leave Liverpool on a certain near date, it would be impossible for me to be ready in time to accompany them. Therefore, I should arrange to follow on by the next Cunard steamer and he would expect me to report myself to him at Montreal on my arrival there.

Well, this was a sudden change indeed; almost more than I could grasp at the time for it must be remembered that I was still very young and North America a very long way off and very little known. There was no telegraph there and the most speedy communication was by the old-fashioned paddle wheel steamers of the Cunard company taking 15 days to get to New York. What little I myself knew was by having read Charles Dickens's book *Martin Chuzzlewit* and his *American Notes* which I believe gave great offence to that nation at the time.

Fate had so arranged it, however, and I had to try and pull myself together and make it a success. The first thing, of course, was to explain the state of affairs to Mr Austin and to send in my resignation to the Board of Health. Curiously, my Indentures with Mr Austin were just then expiring. I approached him, as I had always done, with some little nervousness. But although at first a little surprised at the news, he spoke very kindly and thought it very probable that the change might prove to be of great advantage to me.

This rather cheered me up. He wrote me a very kind sort of formal testimonial which I forwarded to Mr Ross before he left England, and another very kind and long private one which I kept with me for many years. I had also to explain to the inspectors the change of circumstances. Nearly all of them knew Mr Ross and indeed Mr Austin had frequently met him, he told me. I have often since thought it possible that Mr Ross may have asked them some few questions about me before leaving. However, I found it something of a wrench to say goodbye to my friends the clerks – Mr Walbridge and the others downstairs – and I could but observe the curious kind of admiring look at such a young fellow going so far away from home. Even the messengers and my old friend and companion Haile had a kind word to say. The latter giving me at parting two or three interesting autographs which I still have in my possession.

I had now only to say goodbye to the many kind and dear friends at home and start off on this new career. Some little dances were arranged that I might meet some of those many young lady friends for whom I had always had the greatest regard.

Before leaving home, I got to know how all this came about. A younger brother of Mr Ross, Mr Hugh Ross (also an engineer) was a great friend of my father's and I had met him at one or two dinner parties my father was then accustomed to give at Dartmouth Villas to his special friends. It was he who had suggested the idea of my going away.

The time had now come to start on this expedition. No one could say for how many years, or whether I was ever coming back. The day before leaving London for Liverpool, I went down to Sydenham to say goodbye to my brother John, as he was now in charge of all the mainlaying work at the new Crystal Palace being erected there.

I found him occupying two rooms above a chemist's shop at Norwood and he seemed to be enjoying his lunch or dinner (as it might be) of cold roast loin of pork. Unfortunately, poor fellow, he had sprained his ankle the previous day and was obliged to rest it. However, we parted wishing each other good luck.

The next day, accompanied by my father and mother, I went to Liverpool. We stayed the night at the Adelphi Hotel and then went on a little steam tug and embarked on the Cunard steamer the Cambria which had been lying opposite in the river. It seemed to me then a very large vessel, although it was not much more than 1500 tons, about the same size as the steam colliers are now, that bring coal to the London gas companies. It had the old-fashioned paddle wheel of course, as the screw propeller was not then dreamt of and it then took 15 days to get to New

York, even with fine weather. Having seen my cabin berth and the general accommodation of the ship with my mother and father, the time came to say goodbye and part.

It was not, I must admit, a very cheerful parting, for we none of us knew for how long it was to be. But although I was not much more than a boy, there was not the least nervousness or regret on my part and when alone and a start had been made, I soon began to look round the ship and take it all in. These Cunarders were the pioneers of a steam voyage to America – either to New York or Boston, and the Cambria was for the latter port. There were only some three or four vessels at that time, but others somewhat larger were being built.

The accommodation of this ship was none too much. The main cabin was below deck, with the sleeping berths surrounding it and two men in each – one upper and one lower. Mine was an upper and as ill luck would have it the man who occupied the berth under me was a Frenchman who was sick the whole voyage. As the cabin was very small, it was no joke.

There were more steerage passengers, poor creatures, with a lot of children. They were mostly emigrants. The people in the main cabin seemed to me to be mostly commercial travellers and only three or four lady passengers. There was certainly not a single tourist or anyone crossing for pleasure. That time had not yet come, for America then was almost unknown to the general public.

There was also no telegraph across the Atlantic; the only means of communication was by post meaning then a month's interval, sometimes more. How marvellously different now when writing these lines nearly sixty years later on. (In fact a cable was laid across the Atlantic to America six years later in 1858, after several failures, by Charles Bright - afterwards Sir Charles. He was only 21 years old and became very friendly with many members of our family).

The only other cabins on the main deck were some half a dozen single ones scattered here and there for the captain, doctor and other officers and no reading or music room. There was a kind of hut constructed a little in front of the engines alongside the 'caboose' where the cooking went on. This was called the Smoking Room where some of the passengers and the doctor played poker all day long and, as far as I could see, the doctor did pretty well. The roof of these small structures, having some few steps or stairs, made a very convenient little upper deck to sit quietly and read and I gladly availed myself of it. I had brought some few books with me on surveying and railway work, and I intended to try and pick up *en route* some small knowledge of the subjects before joining the others. It also enabled and amused me to watch the cook, his assistants and the butcher at their work preparing the meals (there seemed to be a lot of livestock on board).

I never tired of watching these men. The sea air simply made me ravenous and I had no sooner finished one large meal than I was on the look out for another. I was never a great smoker, but I found I could even enjoy a cigar just now and then. I need not say therefore how glad I was to find myself a fairly good sailor.

When about half way the voyage became a little monotonous and the weather most fearfully cold. It seemed impossible to get warm and the Captain said there were icebergs about, but none could be seen. It was I believe about the time that the ice on the River St Lawrence began to break up and a day or two afterwards several icebergs could be seen with the aid of the glass, and monsters they looked. After this little incident we found ourselves suddenly enveloped in a dense fog and, if possible, colder than ever.

We were supposed then to be off the Banks of Newfoundland. As the fog continued day after day, I could see the Captain was getting very anxious as he was unable to take his daily observation and with icebergs about, it seemed to many of us to be getting very serious.

I do not know anything more trying to one's nerves than to be in a fog at sea, one seems so helpless. However, both the fog horn and the siren were sounded day and night and a hideous noise it was to endure, but nothing more could be done and the vessel still steamed onward. One morning, feeling very anxious I got up before breakfast and went on deck. I think I was the only cabin passenger up. If possible, the fog was worse than ever and the cold too awful, shewing we must be getting nearer the icebergs, but nothing could be seen. I could not even see which officer was on the Bridge on duty. Suddenly, all in a moment the look out man on the forepart of the ship shouted or rather shrieked out "breakers ahead". At the same time the fog seemed to lift a little shewing, for a moment, right ahead rocks and the seas breaking over them with a great roaring noise. Almost deafening.

What a moment!! Never to be forgotten. I happened to be standing just outside the captain's cabin on the deck; he opened the door and came out having heard the cries, poor fellow. What an awful look on his face and so ghastly white. The First Officer was on watch on the bridge, at the time being over the engine room and the moment the shouts came from the lookout man, he directed the engines to be reversed full speed and in his excitement and agony, threw his glass, I remember, below.

The engines continued to be reversed, but now slowly, and soundings were taken until the Captain thought it safer to anchor and await help, if that were possible. The next thing was to load the two large cannon that were on the forepart of the deck and continue firing them off at intervals the whole day. I need not say that these cannon, added to the noise of the foghorn and siren were simply maddening.

When the alarm was first given, most of the passengers, both cabin and steerage, rushed up on deck, many in their night gear and in a frantic state. I noticed the officers had much difficulty in keeping them quiet. The sailors were also very soon at work, getting out the life boats and equipping them in case of emergency. Fortunately there was no wind or heavy weather, but there was a very heavy rolling swell, and the Captain being able to anchor the ship so easily, shewed how dangerously near to land we must be.

As for myself, during this anxious and critical time, I simply felt paralysed. The evening was approaching without any answer to our signals of distress and all of us were dreading the night, while the fog was as thick as ever, when as silent as death, with no warning or sound whatever, a great lugger sailing boat loomed up alongside our ship. Oh what joy. It was one of the large cod fishing boats working off the Banks of Newfoundland. The men had heard our signals, but there being no wind, they had to row most of the way. I do not know how many men they had on board, but three of them scrambled onto the steamer.

I can see them even now, looking more like amphibious animals than human beings, they were all so wrapped up in yellow oilskins from head to foot. However, the Captain soon got them into his own cabin, brought out his own chart or map (I heard afterwards) and asked them to point out the spot on it where they now were. They did so, and I believe their first remark was, in their own forcible language – "How the .... did you manage to get here?" However, all's well that ends well, but it was made quite clear to every one that had the steamship continued the course she was on when the fog so providentially lifted for a few moments to shew our danger, in another quarter of an hour the Cambria, and every soul on board, would in all probability never have been heard of again. We were just off the most dangerous part of the coast of Nova Scotia.

The fishermen were sent into the chief cabin to get a hearty good meal and I remember I went down to have a good look at them. They had put aside their oil skin coverings for the time being and one could see what very hard, strong, hardy men they all were; and in the meantime, the ship got up steam and the anchor. Although the fog and cold was as bad as ever, it was decided that in case the weather should become worse, these men should pilot us from the hole in which we had managed to get and put us on our proper course. And so they did this and I need not say how much it was to the gratification of all on board. We then parted, having all insisted on shaking hands with them and giving them a little sum that had been collected amongst the passengers. With this, and no doubt a further amount that would have to be paid them by the Cunard Company, they would probably remember the event as a good day's work – which they most richly earned.

Curiously, when put on our proper course again, the fog seemed to be disappearing and the sun made an effort to get out. As it was still three or four days yet before we could get to Boston, I found myself again on my little deck, watching the cooks at work preparing the different meals.

At last we arrived at Boston and having found the Custom House officers more annoying and disagreeable than any other country, I landed and went to the Tremont Hotel – quite American and wonderful to me. The next day, having gone over this beautiful city - so clean, so quiet and with so many beautiful churches (they were evidently a religious people), I went by railway to New York from Boston instead of the Long Island Steamer.

New York was simply astounding to me, its people even more so. I stayed at the Metropolitan for the night and shall never forget my first trial there of an American drink called Sherry Cobbler. I then went up the Hudson River to Albany – how beautiful – and then to Montreal at last by the beautiful Lake Champlain.

The first morning on my arrival, having made enquiries, I found Mr Ross, the Engineer in Chief of this great undertaking was still at his hotel. (The steamship he had come on with all his staff had been delayed en route). Having sent my name to him, I was received kindly, but with very few words as he was evidently greatly pressed with so much work and getting things in order. Having now reported myself, however, and given an account of my journey, he told his secretary who was in the room with him, to give me a letter to Mr George Tate who was to have charge of the last section of the line from Toronto to Kingston on Lake Ontario. I heard afterwards that Mr Ross divided the whole length of the line (1200 miles) into three sections of 400 miles each with a chief engineer for each length. Mr Tate had the most distant length, Mr Jackson the next and Mr P. Ogilvie had the Montreal length.

Mr Ross, having heard all I had to say, then told me that I had better get on to Toronto as quickly as I possibly could and see Mr Tate who was about making a start. I therefore left him, feeling a little bit bothered and troubled at having to make my own way on this further 800 mile journey, when it was in such a far away part of the world and I was so young. However, I pulled myself together and having left Mr Ross, I wandered about the city during the rest of the morning finding it very different to Boston. Montreal was much more busy and animated and very French, in fact hardly any other language was spoken. There was a part near Mr Ross's hotel called the Haymarket that interested me. I noticed also that they were already preparing to construct the great Victoria Bridge over the River St Lawrence. This was to be a square steel tubular bridge supported on massive granite piers with enormous projecting cutwaters to crush the floating ice, when breaking up, on the river.

Later in the day, I again called at Mr Ross's hotel, saw the secretary and received from him the letter of introduction to Mr Tate. He was also, I remember, very kind in putting me in the way to get information as to my further long journey. The next day I once more embarked on a steamer to go up the river to its entrance on Lake Ontario. We went through the Lachine Canal, to avoid the rapids; coming the other way, the steamers would descend them, a very exciting time.

Arriving at the lake, I think we embarked on a somewhat larger boat, as the water on these American lakes can sometimes be very rough. This steamer went straight on to Toronto, stopping on the way at the few little towns or settlements that then existed. I forget how long this journey took, but it must have been many days, and in spite of all the novelties of interest and the lovely scenery, particularly that of the Lake of the Thousand Isles where the river seemed to have widened out, I was awfully glad to get to the end.

The fact is, I never could get accustomed to the very disagreeable motion and smell of these American steamboats. Their curious, high beam engines, being so far above deck seemed to cause such a joggy, short, palpitating motion, with a nasty smell like bilge water, that I never could eat or enjoy the meals on board. I felt very sea sick which was so different to the ravenous appetite I always had at sea.

The boats on this lake were much the same as those on the Mississippi, perhaps a little smaller. They had a house of three or four stories built on the deck with a verandah or balcony built round each story, and with the pilot box sticking up on the top. The fuel used on these boats was wood, not coal – wood already chopped into what was called chord wood and stacked in large heaps on the lake shore for the steamers to stop and load up.

It was curious to find that there could be very dense fog on these freshwater lakes, as well as off the Banks of Newfoundland. It was very interesting to me to see the simple way of signalling they used. We all know for what a great distance sound will travel over water, therefore when in a fog the boatmen would throw a large long plank of wood down sharply on the deck. This would be heard by the men and a very long way off by the men on the little pier or town we were next to stop at. They would signal back in the same way in reply and continue to do so until the boat came slowly alongside.

At last we steamed into the beautiful Bay of Toronto early one morning. I remember the weather was getting very warm and how bright and beautiful everything looked, and yet somehow I could not help feeling nervous and dispirited. But once more pulling myself together, as the day was yet so early, I decided I would leave my luggage at the depot on the steamboat wharf while getting a check for it. (What a splendid system they seemed to have in this part of the world for dealing with travellers' luggage). I found my way to the Grand Trunk Railway's offices instead of going to the hotel first. These offices were in Front Street facing the bay and standing alone at the corner leading out of King Street. This was the principal street in the town (one could hardly call it a city then, as I believe the population was barely 20,000 people).

The building itself was a huge barracks built entirely of wood and had evidently been previously used for some military purposes. Entering the large open doorway and a general sort of office where several clerks were at work, I ascertained that Mr George Tate was in his room. I therefore sent in my name with Mr Alex Ross's letter, and in a short time I was taken to him. He was sitting at a large table on which were, I well remember seeing, a number of ordnance maps of that district. A man of about 30, or a little more perhaps, was standing by the table evidently examining these maps with him.

This man, I afterwards ascertained, was Vetruvius Wyatt, his right hand man and one I shall have to speak of, many years later. Mr Tate received me very kindly, but like Mr Ross was a man of few words and I should think had never smiled in his life, always being too occupied with his work. He was about 50 and so close shaven, that he looked very much an actor. Having asked me a few questions as to previous occupation, I could only tell him I had no actual knowledge of railway work, but that I had been considered a pretty fair draughtsman and perhaps rather good at figures.

Mr Tate then explained that he and Mr Ross were going over the country endeavouring to locate a line most suitable for this proposed railway. In a few days the two parties of surveyors would be sent out to make the necessary survey and take the levels. One party of men were to start from Toronto, the other from Kingston (the other end of Mr Tate's section) until the two met half way. He further mentioned that as the country at this end would in all probability be found to be very wild and uninhabited, it would be desirable that each Englishman should be provided with a rifle as a means of defence. This would also provide a means of obtaining food during the time we had to camp out. We therefore should prepare ourselves during this interval with such means to meet the journey. Some few experts, bred and born in the country, had also been engaged as pioneers and assistants.

Mr Tate also stated that as there was so much accommodation in the upper rooms of this building he had decided that a portion of it should be roughly furnished and that the engineers under his charge should reside there and have a general messroom – in fact a kind of club. He then rang the bell for the messenger and caretaker and desired him to put my name down and supply me with a room. Mr Wyatt then took me upstairs and introduced me to the several young fellows there. How well I can remember them. Thomas Mahony Quill, a Dublin University man; Horace Coombe, E Ingram, G Kent and others - altogether eight or ten men when at home at mess. The caretaker, Alex, and his wife were always very kind and obliging. When staying there, the food supplied was of the plainest and cheapest, but good and plentiful.

Walking through the town for the first time it looked to me like a town just born, so very new and primitive. There were then only two or three good streets in it – King Street and Front Street – every building on each side being built entirely of wood and nearly all shops and stores. At that time there was only one hotel which really looked like a private house and was no larger. The pavements throughout the town were simply of wooden planks and very rough to walk upon. The university, town hall, law courts and bank plus some churches, were solidly built of brick and stone.

When leaving England for North America I brought with me two letters of introduction, curiously both happened to be for Toronto. One was to the Manager of the Bank of British North America and the other to a Mr George Boyd, a wine merchant and chemist, who was a cousin of the brothers Smith, who were the architects at Greenwich engaged by my father and Mr Quick in their building operations.

I therefore took the first opportunity of calling on them and was received very kindly by both parties. At the bank I think a dinner, one or two calls and a couple of very enjoyable dances was the outcome of the introduction during my time in America and here I met Mr and Mrs Tate and many of the best people in Toronto. Even in those far away towns, at that time they knew how to dance and such music and singing was a wonder to me and as for pretty girls, as time went on, I had to give up counting them.

With the Boyds it was different. I soon became almost one of the family, they were so kind and friendly. His place of business was close by in King Street, and his private residence in Front Street, close by our barracks. Mrs Boyd was very nice and I think there were two or three little children. I remember the natives, and people generally, looked upon the grand trunkers as we were called, in a very favourable spirit.

After a time I felt certain that Mr Ross, Mr Tate and other principal engineers must have been over here the previous year and pretty well made up their minds as to the route of the line, before the scheme was made public.

In railway making the most important part of the business is to purchase the land required as cheaply as possible before any local speculators get hold of it first. Several agents came over with our engineers to work with us for that purpose.

We were now told, however, that our party at the Toronto end were to be fully equipped and ready to start work within the week. The party would consist of Thomas Mahony Quill "Chief", myself and two other young engineers, about a dozen backwoods men, some four or five porters and chain men including two Indians. The Indians had their own guns as they were engaged to supply us with food when obliged to camp out, carry tents and other odds and ends.

The backwoods men were splendid specimens of mankind – Canadians bred and born – whose usual work was lumbering (cutting down timber). We had to cut a way through the forests and the many horrible cedar swamps with such thick undergrowth as to make them almost impossible and full of goodness knows what. Having received this notice from Mr Tate, we at once set to work to get ready.

As the weather was becoming very hot, as far as dress was concerned

I simply followed that of the men. A light flannel shirt and collar, a leather belt with pockets, a rough, strong pair of trousers, a large, wide-brimmed felt hat and high water boots. Sometimes moccasins such as the Indians wore for the feet, as circumstances required.

The first thing was to get hold of a handy sort of rifle. None of us could afford to buy anything new, therefore we tried to pick up the best we could. As I had never had experience with firearms, I asked our Canadian guide Kent, a most useful man to the engineers and contractors, to help me. I therefore purchased, at a moderate price, an old-fashioned American rifle, a muzzle loader of course, and found it very useful in helping to stock our larder when camping out. But I am glad to say it was never used for any warlike purpose. It was tremendously heavy and very long. When out at work, Michael, one of the Indians, who for some reason had taken a great fancy to me, always carried it for me.

Having killed my first deer with it, for as we got more into the interior of the country they were very plentiful, I felt rather proud. When the time came for me to return to England again, I brought this rifle and the head, or rather the handsome antlers, of this same deer home with me. Mr Quill also purchased a rifle, a rather dangerous one, called, I think, a single trigger. The other two engineers of our party preferred to buy shot guns.

As Quill was quite ten years older than me, his father being a Colonel in the army, and Quill himself having shooting at home in Ireland, he was safe enough. But being out of practice, he suggested that before we started, we should get up early in the morning and practise; and this we did. As soon as daylight came, several of the other men with us, we went to a large tract of waste land outside the town and blazed away at a fixed mark. So much so, that the inhabitants nearby all got up and came to find out what it all meant. However, when our practising was over, the conclusion was come to, I remember, that I was the best shot of the two and this was curiously verified later on when out at work far away in the backwoods. I cannot help thinking Quill's eyeglass was the cause of failure.

Before starting on our expedition, Quill very kindly decided to give me the charge and responsibility of the surveying and chain work, and he would be responsible for the levelling.

On a railway of this magnitude, I need not say how necessary it was that there should be no errors or mistakes, as it would be very serious. I cannot tell you how anxious and nervous I was at first and how often at heart I used to thank my dear old friend Mr Fraser for giving me those little wrinkles and lessons when surveying the Wainwright Estate at Holloway; little thinking then where they would come in so useful. Well, Quill never had reason to regret his trust in me and I remained in the same position all the many months we were out. Our length was 200 miles and we were to meet the party who were surveying and levelling towards us from Kingston. On our length the country was fairly flat and level, with no tunnels, heavy cuttings or embankments.

Well, one morning we started work, (we were about 20 in number) at the spot where the terminus of the railway was to be constructed just outside the town and for some few weeks I continued to sleep at the barracks. The work was a little heavy and our progress slow until we got quite clear of the small suburbs of Toronto. I remember it was a little awkward at first, surveying the necessary crossing of the River Don, but I managed it all right.

The course we had religiously to follow was a distinct marked line on the ordnance maps of that part of the country. These maps, in the absence of any people or buildings about, were simply squared out in blocks and numbered, with concession or base lines and also side lines. When on the ground we would have to search for the particular trees the ordnance surveyors had blazed with numbers and pegs driven into the ground.

The tree line being thus found, every tree in our way, however, large, that interfered with our measurement or levels, would have to be cut down. This cutting down of timber was terrible hard work, for some of these trees were leviathans, many, many feet in circumference. It was marvellous to watch the men with their great axes almost as sharp as a razor and to see how quickly, neatly and exactly they were cut and how they fell in the very position required. Many of these trees to be cut down were maples full of sugar.

After a little time I tried my luck with an axe of lighter weight.

Having on several occasions nearly cut my legs off, or those of my companions nearby, I improved and often lent a hand, that we might push on with our work. Quill always preferred to look on. I can see him now with his eyeglass fixed (he always used one and I believe slept with it). He was a small man, perhaps somewhat insignificant looking, but how clever!! A Dublin University man, afterwards articled for five years to a civil engineer at Wrexham, a Mr Pilkington, a well-known man with a large practice in railway work who gave Quill a knowledge of this to his fingertips. How I used to envy him and to think how different it was with me and my five years of wasted life. As for books and reading generally, I used to think I had read a good deal, but I was nowhere. There seemed to be no book he could not quote from, and both of us loving books, it made our companionship in the evenings after work very pleasant.

We had now quite left Toronto and had plunged right into the wild part of the country, meeting just now and then some poor devil of a settler trying to create a little farm. The weather now was becoming very hot, and as for the mosquitoes and other venomous flies… It was too awful, the sight we presented to each other in the morning, our heads so swollen, it really was appalling; but painful as it was, no one could help giving shouts of laughter. With all these great lakes and rivers about, it seemed almost incredible that we should sometimes be almost dying with thirst and not even able to bathe or wash. This was awful work! When all right, and water plentiful, we all used to drink oatmeal and water from a tin pail alongside, or cold tea.

Quill and I, when we had the chance of getting a bottle of beer or two, would rejoice and toss for the last half pannikin of it. When obliged to camp, one of the men would do the cooking, generally of wild duck or goose, simply split open and baked in the earth. Sometimes, there would be a wild turkey or steak from a deer. The bread was the usual damper or some American cracker biscuits.

What a curious, strange sort of life this seemed to me, but I got accustomed to it and kept well in health. But how changeable was the country we had to go through. At one time, the most lovely scenery with pretty lakes and surroundings, beautiful little streams or rivers that could be paddled up for miles, and so silently. This brought to one's mind and thoughts the days of Fennimore Cooper's "*Last of the Mohicans*" or the "*Deerslayer*"; who has not read his books with delight? Another time, amidst the horrible cedar swamps, with their malarial smell, I was reminded forcibly of Charles Dickens and *Martin Chuzzlewit* in America.

We had got well into our work when a little excitement occurred caused by Quill and myself about the first deer. We had come alongside a small lake on which, close by, were two or three birch bark canoes belonging to some few Indians who were camping. As something in the way of meat was wanted, it was arranged, as deer always take to the water when hunted, that the couple of dogs we had with us for hunting should be taken some distance back to start the quarry and that Quill and I should each take one of the Indian canoes. Quill with one of the men at one end of the lake, myself and Michael the Indian at the other end, and wait. These Indian canoes are very frail and slight. You cannot sit in them, but kneel and balance the thing as you paddle. We took our stations and waited patiently, but it seemed a long time. However, we heard the dogs baying furiously some distance away. Hearing this, Michael and I just looked at one another, we were both much too excited to speak. I laid the rifle, already loaded, carefully by my side at the bottom of the canoe, and quickly took up my paddle ready. Michael, I could see was also ready with his. Just at that moment, a magnificent stag rushed from the forest and plunged into the lake, evidently with the intention of swimming to the other side.

It had taken to the water pretty fairly between us, but certainly nearer to Quill. I could see how desperately he and the man he had with him were paddling to get first shot at the poor, beautiful creature. I shall never forget as long as I live, how I and Michael paddled, but it was no use, Quill got fairly within shot first. He seemed to be taking very careful aim, but he missed. It was now my chance and before he could reload, Charles Aird took up his rifle, still carefully kneeling in the rickety, pointed canoe, took steady aim and cut the poor creature's throat as with a knife. In a moment, the water of the lake was red with the blood some distance round. I dropped my rifle and with Michael, paddled alongside the poor best just in time to keep it above water. It seemed quite dead, but Michael gave it a *coup de grace* with his knife and, twisting our paddles between its antlers, we towed its body to the shore. It took four or five of our strongest men to lift and carry it to camp. Quill was very kind and nice about it and insisted on my taking the head as a trophy which I did, and it is before me now when writing these words.

When we had done about 40 miles of our work, we suddenly came upon a little settlement, some three or four years old, called Newcastle, and it is probably now a very large town. This was the first we had come across and gave one an exact typical idea of what the beginning of these towns is like.

First, there was the tavern, or hotel as the owner preferred to call it. This was a long, low kind of building on two floors, built as roughly as possible with planks simply nailed together. As for cleanliness, that was an unknown quantity.

The man who owned this place was an American who would give one the idea that he was not quite to be trusted. He gloried in the name of Benjamin Hinds, or B. Hinds as our men used to call him. He seemed glad to hear that if he could give us (the engineers) decent accommodation, we might arrange to sleep and board there for some little time, instead of the camping out.

Our arrival caused considerable excitement and curiosity. Besides the hotel, there were several stores scattered about, forming a street and there was also a small church and meeting house. The tavern was simply built on the ground level and had wide open doors into the large front room where all customers and the storekeepers about would loaf and congregate, waiting for the gong to sound for the coming meal. The wooden walls of this room were plastered or pasted all over with gorgeous coloured bills giving notice of some travelling circus to arrive and exhibit there shortly. Where the audience was to come from, was a problem to me, but as all the many other taverns we met with on our route were advertised in a similar manner, it must have paid somehow.

There was a bar for drinks with its many coloured bottles on shelves behind. (It seemed the custom in this part of the world was to take the various drinks between meals and nothing but iced water at them). There was also a white cask of American cracker biscuits alongside and a very strong cheese *pro bono publico.* There were a number of spittoons about, but the condition of the floor from the expectoration was too dreadful to describe. Finally, there was a great iron stove fixed in the middle of the room.

Having arrived shortly before their usual midday meal, the gong sounded, a door opened behind and in a moment every man who had been waiting or loafing about rushed in. The room was very dirty with no other furniture than a long deal table with a long bench or form to sit on. On this table a dirty, horrible looking creature (who I afterwards found to be a brother of the landlord and an awful drunkard seldom free from DT) was dabbing down several dishes of food, mostly in tin dishes. There were pewter plates and black handled, two-pronged forks which might have been put in water since the last meal, but I cannot say.

Before we (our party) could sit down, half the food had gone. As for the food itself, I will not attempt to give a description of it, but at that time I was so used to rough life that I could eat anything cooked or uncooked, and like Oliver Twist, ready for more.

Our men and the Indians looked after themselves outside. The storekeepers and other loafers, whoever they were, having finished in about ten minutes, left the place until the next meal.

Before resuming our work that afternoon, we asked B. Hinds to let us see what he could do for us as regards sleeping accommodation. He called his wife (a decent, but poor sickly looking woman) who took us up a narrow wooden staircase on to a very narrow little passage, hardly wide enough to two persons to pass along. On each side of this passage were three little rooms, or berths, about the same size as the cabins on the Cambria. There was a little window and a wooden bench fastened to the wall with a rough mattress and blanket to sleep on; also a tin basin, a little water and a dirty coarse towel.

As in our instructions, near this place we had to take some few cross sections, which would oblige us to remain in the neighbourhood some little time, so we decided to stay. We would have an early breakfast, a late supper and do a good day's work. In those times, weather made no difference to us.

In following the line marked on the ordnance map, we sometimes found ourselves a long distance from Lake Ontario; in this instance we were about 20 miles. About twice a week Hinds's brother would drive what they called an 'omnibus' (a kind of van without any springs) to a little port on the Lake to meet the steamer and pick up any stray passengers or bring some provisions.

Sometimes we would be as much as 50 or 60 miles from the Lake, but when going on straight for another week, we would suddenly come on the lakeshore again. To follow the contour of the Lake would, of course, have been absurd. The object of the engineers in setting out the railway was to make it as short as possible, with easy gradients and on such land as would be most likely to be settled and built upon and also farmed. I would also like to take this opportunity of mentioning, if I have not done so before, that the railway when completed, was to have only a single line. However, every bridge, every culvert and work of that character, should be constructed of solid masonry or brickwork wide enough for a double line.

Well, here at this settlement we settled down for a time and found that all the cooking and housework was done entirely by Mrs Hinds and two daughters. The younger one Harriet, I remember was a very pretty girl. The family lived in a little room at the side, very poorly furnished with, of all things, a wretched, poor little piano with half the strings broken. Poor creatures, what a godsend it was, our coming and breaking up the monotony of their lives a little while. Now and then, at their invitation we would sit with them in the evening and have a game of cards, or start the poor old piano as I could then manage to strum a tune out of it, and so time passed.

One day, to our surprise, Mr Tate and Mr Wyatt came on the ground to look into some little question of deviation of the line. Mr Tate returned to Toronto, but Mr Wyatt remained a day or two and got very friendly with the girls and chaffed me a good deal, because I had got from Toronto a copy of D'Albert 's musical album for them. (I remember many years later when at home in England, every Xmas I used to give my dear sister Agnes a copy of this pretty album). He and I looked at such matters with very different eyes and I should be very, very sorry ever to be of the same disposition and character as that of Vetruvius Wyatt.

Having finished the little extra work required here, we moved on again having many, many more miles to accomplish before getting into anything like civilization again.

I have omitted to mention that when I landed at Boston, I wrote to my father and mother giving an account of the voyage and safe arrival, with a line also from Montreal and from Toronto. Just when leaving Newcastle I got a letter from them in return, forwarded from Toronto and I need not say how glad I was to get it and to hear that all was going on well.

Our work and life, on leaving this place was to be very much the same for the next three or four months. Now and then dropping upon some settler trying to grub some of the wild land into something like a farm, and at other times some little settlement like the one we had just left. I remember Port Britain, Windsor, Coburg and Port Hope.

Port Britain was an awful place, so marshy, so unhealthy and near the Lake. The poor wretched place called the hotel was occupied by a man and his wife only. I cannot remember their name, poor creatures and how they managed to exist I don't know. The wife was the most sickly looking creature, so thin and her bare arms and her face so covered with freckles and sores from insect bites. One would think now, that seeing her making what she called a pumpkin pie for our supper would take away all appetite, but no, it had no ill effect on mine then. However, the intense heat and the torture from the flies was almost more than we could stand.

I remember one day a little incident occurring to one of our dogs. A porcupine ran across our path, the dog made a grab at it, the result being that the poor brute's mouth and tongue were perforated with the quills it had shed. Afterwards, while two of the men held him, I had the disagreeable task of pulling them out with my bullet mould.

There were any number of smaller wild animals about and at night, when lying on our bed of fresh-cut hemlock boughs, we would often hear something larger prowling about. However they, whatever they were, didn't like the fire which was always blazing for there was always a man on guard and to keep the fire going. There were also any number of snakes, especially in the cedar swamps. Seeing a very large one one day running along, I cut at it with a little axe or tomahawk I generally carried in my belt, and cut it in two halves. It was very strange to see each half scudding away as quick as it could in opposite directions.

One of our chainmen, an Irishmen named King (how we picked him up at Toronto I do not know) was a broken-down gentleman. He was well educated and I believe of good family, but drink was his curse. He was past middle age and looked a sodden wreck. I mention this poor devil because as chainman it was his duty to go straight and head with it through anything – cedar swamps or not. But his intense horror of wading a swamp was most painful and there were times when he would almost go down on his knees to me, asking to be spared. The fact was, he had snakes so much on his brain when his paroxysms of DT were on, that it was all one thing to him. When he, and B. Hind's brother got together, it was awful work.

The next place or tavern where we had occasion to take up our quarters was different to the last, being on healthy ground. The usual tavern, and three or four log huts about the tavern, were kept by a very decent young fellow named O'Brien with a wife and little child. He had a couple of woodmen with him endeavouring to clear enough of the forest away to sow some wheat. One evening returning from our work for supper, we were surprised to find the place occupied by a gang of drunken rowdies. They were woodmen on the drink and evidently very far gone, for they were smashing the poor fellow's stock and drinking whatever they could get hold of. In fact they were playing the devil and the little wife and child were pretty well out of their senses. Of course, any thought of supper then was out of the question – what was to be done? The unfortunate owner of the tavern thought that if our party would join him and his two men, we should be strong enough to clear them out. We all agreed at once and no sooner said than done, for we went into it and after a stiffish bit of a fight and struggle, some us getting some rough knocks and bruises, we got them outside without anything more serious. But it was warm work just for the time.

When outside they talked and raved, some of them firing their revolvers in bravado, but much too drunk to aim straight and do mischief. Having broken the glass of the few windows, they would have tried to get in again, but as we had in the meantime barricaded all the doors and openings and they could see several rifles and guns ready to give them a volley if they attempted it, they thought better of it and during the night cleared off, much to our relief. I think the knowledge that our landlord had sent one of his men to the Sheriff of the district, letting him know what was going on had got to them and rather hastened their departure. So all's well that ends well. I heard afterwards the Sheriff was a settler and farmer between this place and Port Hope.

On and on we went with our work without any further such adventures I am glad to say. Indeed, it seemed to me that the country we were now passing through was less wild and somewhat more settled as we were nearing our goal – Port Hope. Here we were to meet, after all these many months, the other party working from Kingston, and after another month or so, we found ourselves drifting into that lovely period of weather called (in British North America) Indian Summer, or the Fall. It only lasts some few weeks, for as most people know, in Canada or North America there are only two seasons – winter and summer, no spring or autumn. At the period I speak of, it is really impossible to describe its charm.

Instead of the intense dry suffocating heat we had been enduring, it changed into a soft, silent, loveable, hazy feeling. The leaves of the vast forests were all changing to the most beautiful colours and tints – brilliant scarlet, yellow and every mixture that can be conceived. It is impossible to describe it all and with it, we found we could do double the work during the day.

I remember at that time when chatting with my henchman Michael the Indian, saying to him – what lovely weather it was for a hunting expedition back inland. He replied in his curious broken sort of language that if Mr Quill and I could get a week's leave next year at this time, he would so like to take us to the home of his tribe, who were living very far back near the Lakes Huron and Simcoe, and go hunting and fishing. (I remember he was of the Huron Indian tribe). We had no idea of what it was like, but we promised if all was well, to look him up when that time came. That time did come and that week or ten days leave with Michael I have always looked back upon as the acme of happiness of this poor old fellow's life. I certainly have never approached anything like it since.

At last, at last, after all these months of rough hard work, we arrived at Port Hope, the prettiest and cleanest little town or settlement we have met with. Of course it is only the beginning of it all. The tavern or hotel as usual was built entirely of wooden planks nailed together with only a plank partition to divide the few rooms. The voices being so distinctly heard, it was like living in one room. There were only a few wooden stores as yet, with the usual little church and chapel. The bar room of the tavern was, as usual, plastered with coloured advertisements. I remember very well that the landlord of this place was rather a superior sort of man (a widower), with two grown up daughters who were very good looking and much more refined than the two poor girls at Newcastle.

Hearing that it would yet be a couple of weeks before the other party could join us, we thought it would be better to work on towards them and save time. It was of the utmost importance that the junction should be made before the ice formed on the Lake and prevented our getting back to Toronto by the last steamer of the season. Fortunately we did so, for we were just able to accomplish it, meet, compare notes, catch the steamer and all of us journey back very thankfully to our headquarters where we should be shut up in winter quarters for the next five months.

Mr Tate seemed pleased and we at once all set to work to plot everything on those long rolls of canvas-backed drawing paper. As ill luck would have it, I came back with a very bad chill and obliged to remain in bed for a fortnight. However, shaking it off, I very gladly came down to the drawing office to do my share.

This room was a splendid room for the purpose, with two tables of great length to accommodate any number of draughtsmen. One and all, however, we found it very difficult and very, very painful to write the figures so carefully upon the section paper (on which so much depended) with our fingers frozen. I do not exaggerate when I say the thermometer in the room was sometimes 20 degrees below zero. It really was awful work and yet when the summer came round again, the heat in these rooms was so great we were obliged to have great pails of ice about.

During all these months we were very busy for ten or twelve hours during the day, transferring our work of 200 miles on to paper. The other party were doing the same – shewing the figures of every level taken and then enabling Mr Tate to decide upon the several gradients to be adopted and the quantities of cuttings and embankments to be taken out. Mr Wyatt had been at work some time with others preparing working drawings of the bridges, culverts, stations and other such matters. In fact, everyone was as hard at work as possible.

There were also people and clerks of the other department who were equally important – that is the book keeping and finance – for there must have been large amounts requiring to be accounted for at headquarters in London.

After my return to Toronto, I took the earliest opportunity of calling upon the family of the Manager of the Bank of North America and also upon Mr and Mrs Boyd. I also had several letters from home at this time, giving good news.

While the work in the office was thus going on, Mr Tate nearly every day would come into the drawing office, walk round and notice every man's work. I know I was doing my best, for it was work I took much interest in; anything in the shape of figures always fascinated me and I think Mr Tate noticed it.

Everything outside was now completely snowed up for several feet. The snow was hard and dry and although the thermometer was so very low, it was not felt to be so severely cold for that reason.

After office hours, every man residing at the barracks was his own master. After supper, some went out and some remained in. Some few of us had hired a second hand piano as one or two could play a little. There were generally eight or ten of us in the room, some playing cards or chess, but if one of those able to play the piano began a negro breakdown tune, the effect was marvellous. Every man present jumped up, chairs and tables were pushed aside and then we seemed to dance for our lives, singing or rather shouting the usual refrain and chorus until we had no more breath left. Mr Wyatt, who boarded in the barracks with us was, I think, perhaps the most noisy.

I remember another amusement of his was to watch a certain hole in the room and with the shovel in his hand, chop off the head of the poor mouse directly he popped it out. A strange amusement for such a man; but the place swarmed with mice.

There were many curiosities among our staff – little Horace Coombe *per example*, one of the draughtsmen. He was a very little fellow, not more than 5 feet high. He did not walk, but progressed with a funny little strut like a little bantam cock. He was very good-humoured and the butt of everyone in the office.

Usually eight or ten of us would sit down to supper, provided and cooked by the caretaker Alex and his wife. There was plenty of it, but it was roughly cooked. Generally, there was soup one day, fish another, wild turkey, goose or wild duck at one end of the table; a joint of beef, mutton or venison at the other. Kent the Canadian would always carve at one end, the result in his case was wonderful. With eight or ten hungry men to supply from a wild turkey before him, there always seemed enough left to supply them again. But it was always the same with everything that he produced. We rarely went in for sweets, sometimes a few of the wild fruits when in season. At our midday meal, we sometimes had a pudding of some sort, but pumpkin pie for ever – nothing to beat that.

I think I have mentioned that during the whole of my career I was never a smoker, as other men as a rule are. When so disposed, I could quickly smoke a pipe or cigar, but no more and never when reading – generally as a defence from the beastly flies.

At our meals in the barracks it was usually water or a glass of beer, sometimes coffee. Drink in the shape of intoxication never happened. I can positively say that during the three years I was at Toronto I never saw anyone of our party in any way near such a condition.

During the winter months, the little theatre was sometimes open with a poor struggling company to act. Some of us went now and then, but very seldom as it was much too cold. I remember well, however, a farce called 'Poor Pillicoddy' by a clever American actor being played there. I never laughed so much before and when back in England I never failed to go and see it again when advertised.

Now and then there were also some perfectly good concerts, but one did not require to go and hear professionals. The pretty daughters of all the best families in Toronto that I had the good fortune to meet at the few events to which I was invited were equally talented. I was simply astonished to hear them.

In describing my life and work when out in the backwoods, I have omitted to mention the enormous tracts of wild fruit we used to come across. These were particularly in many open spaces where no forest had been. It was extraordinary, there seemed to be no one about to gather any of it and the fruit simply rotted. We ate as much as we could when passing along – too much I am afraid sometimes. There seemed to be every fruit that could be named, but the smaller kinds were the most delicious – raspberries, strawberries, currants, gooseberries and cranberries. Curiously, the larger fruits such as peaches, pears, apples, and nectarines, grown and cultivated in gardens, had no taste or flavour whatever and were usually given to feed the pigs. In Toronto, one could buy a cartload of beautiful-looking peaches for less than sixpence with not a single one worth eating.

Also, at that time as regards fish, the small lakes and the pretty streams we met with *en route* were full of fish of many kinds, but as in other parts of the world, not always to be caught. Nevertheless, I know they sometime gave me a lot of sport and pleasure and we hardly had a meal without some. In the Great Lakes I believe they are of enormous size and weight, but too coarse for eating.

The only extravagance during the winter to which I must plead guilty was to hire from a livery stable keeper close by a span of grey cobs and a sleigh with smart harness and pretty silver bells. It was only on a Sunday, now and then, that I could venture to indulge in this, but what happiness it was!! The cobs were in perfect form and enjoyed the drive quite as much as I did, for the snow was several feet deep then, hard and dry and as level as a billiard table. Sometimes I had one of the stable men with me, or sometimes Quill or Coombe. In any case, considering I had never in my life driven a horse before it took some little courage on my part and, perhaps, some large amount of impudence. We generally drove out to some well-known place, get a bit of lunch and rest and return in the afternoon.

It was, of course, absolutely necessary to be well-clothed and wrapped up in furs and even then your attention was frequently called to your ears or nose being frozen, necessitating an immediate application of snow violently rubbed on the frozen part.

The only other Canadian past-time in which I indulged now and then in the winter was to have a sail on the frozen bay in an ice boat – that is a large sleigh rigged with masts, sails and rudder. On one of these with a good wind it was simply like flying. I never could skate, but the sight of the enormous bay with so many expert skaters in picturesque costume was very beautiful.

How different these Canadian winters are to our English ones, the latter being never alike. There was either a wet, miserable, cold damp winter with hardly any snow or frost; then the next one with perhaps some three or four weeks of sharp frost, but without snow – the thermometer even then not near zero. Sometimes there was a sudden great fall of snow, blocking many railways, but seldom lasting more than a few days, then a thaw and everybody and everything in supreme misery. Of course, as everyone knows there have been winters in England of exceptional severity, when the River Thames has been frozen over, but not lasting very long and the thaw again leaving the same wretched misery everywhere.

During the whole time that I was in Toronto I can hardly remember a single night without a fire, more or less serious, occurring. The fire bell was a most startling one – there always seemed to me at first something uncanny about it to wake up so suddenly and see a great glare of light coming through the windows. Some of us would get up if the fire looked serious. But one had to be careful, for the engines were all worked by hand and if a sufficient number of the fire brigade men were not present to work the engines, they would press men from the crowd to pump. Poor little Coombe was caught once in this way and had reason to remember it for sometime after.

Well, my first winter in Canada has now come to an end and we at once jump into summer again. I had understood, as most of the plotting work had now been finished, that it was Mr Tate's intention to send out parties again to set out the centre line of the railway and also the fencing on each side. I hoped that Quill and myself might be selected again, but I was greatly disappointed to learn such was not to be. Quill was to go with one of the other Englishmen. Mr Tate said that he believed that Mr Aird was too useful to him in the office now to work outside. Of course, it was a kind remark, but I certainly felt very sorry for the simple reason that although the work outside was so rough and hard, it seemed to suit my health so much better than the confinement in the office at a temperature that was at times almost tropical and only liveable in by great tubs of ice about.

For some time past, in addition to my usual work in the drawing office, I was called upon by Mr Tate to do a lot of work for himself, say for the Chief Officer, such as filling up the monthly returns and reports always sent to Mr Alex Ross, giving very detailed particulars of the progress made each month and other special reports which Mr Tate would sign.

Well, it was no use shewing any feeling in the matter, but simply to try and make the best of it. I only wondered whether he would back up his opinion of me by increasing my salary. This he did soon after and not only that, but said it would be much more convenient if I had a table or desk in his room to prepare these different papers. This was at once arranged and there I remained at such work until the day I left for old England again.

As far as health was concerned, I got through that summer fairly well, but there was never any real rest. Mr Tate was one of those men who never seemed to want rest; his work was the only thing he ever cared for. (How little I thought that years later, when master of work and men, I should have the same love for it and never have time to think of pleasure.) At this time there never seemed to be any office hours for me.

How well I remember after a long day's work, I had gone upstairs to supper with the others still living in the building, and afterwards taking up a book to read, or going on with a little water colour drawing which I had lately been trying to amuse myself with again, when an ominous bell would ring and in a few minutes Alex the caretaker would make his appearance and say that Mr Tate was in his room and wished to see Mr Aird. There was no help for it, down I went. There I found Mr Tate wailing "Oh, Mr A. I have this evening just received some letters from Mr Ross requiring certain additional particulars which he wishes me to have prepared and take to him on Friday next to Montreal. I hope you can see your way clear to manage it for me". A week's good, hard work to get all this information prepared and ready for Friday!!

I could only reply "I will do my best, Sir" and went straight up to my one little room. It would have been impossible to have done any of this work in the general office where so much was always going on. It is curious that I should think on that particular occasion of the poor fellow Carton, the lawyer's devil in Charles Dickens's story *A Tale of Two Cities*, who did his work with wet towels wrapped round his head. Not that

I required wet towels, but it was very exacting work. However, having collected from the several offices downstairs the information required, I put it all into some sort of shape and on the Friday morning at about 8 o'clock when Mr Tate drove up in his carriage and pair on his way to the steamer for Montreal, I had the satisfaction of being ready for him.

So things went on, month after month, until that lovely time came again – the Indian Summer. Remembering my promise to Michael the Indian, I told Mr Tate that I was then really feeling very unwell, which was the fact, as for some little time past I had been spitting blood simply from weakness, and asked for a week's leave. This he very kindly granted. As Quill was away on his work I asked Kent to try and go with me, which he did and as he was considered to be an expert in the far backwoods, it added much to the enjoyment and pleasure of the little holiday. In that short time there was so much learn for such a lover of nature as I was; so many interesting incidents with the Indians and that lovely, beautiful foliage of the thick woods. But the time was much too short for much hunting. Such a happy time never came to me again.

On my return we were plunging into another long severe winter, when very little work could be done out of doors. Quill had come back and everyone in the office was now busy preparing the contract drawings and specification so that the contracts might be let and the actual construction of the railway at our end might be commenced directly the winter ended. In preparing for these contracts, with which I had a good deal to do, for the life of me I cannot recollect whether the fencing on each side of the line was included in those contracts, or whether it was dealt with separately. I remember I often used to think what a nice little contract it would be to get some of those miles of fencing in partnership with one or two of the smaller contractors whom I knew and liked. I could not, however, summon up sufficient courage to make the venture, for the reason I mentioned when starting life, that I never could feel sufficient confidence or belief in myself. How many good chances have I lost during life, from that unhappy failing and weakness in my character.

The fencing on both sides of the railway was what is called snake fencing – that is trees split up and cut into lengths, I think of 9 feet, and about six inches thick in the rough. They are not planed or trimmed in any way, then built up, zigzag, one on or over the other to about 3 ft 6 inches high, with a couple of stakes driven in the ground to steady them. We had cut down all the trees on our first survey and these came in useful.

It was a very busy time this winter for us all with these preparations. Many well-known contractors called to see Mr Tate about the work, and during his absence (which was very seldom) I had to do my best to explain things. I cannot now remember their names, but I think when the time came, a contractor who was well-known and doing some large works both in Canada and the United States named Zaronski or some name like it, had the biggest length. Some two or three smaller contractors had shorter lengths. All these contracts when signed by Mr Tate and the contractors, also bear the signature of Charles Aird as witness; an insignificant matter, but I was young then and could not help feeling a little proud of it.

Having thus gone through two winters and now begun my third summer here, I have lived to see an active and good start made with the actual construction of this large railway.

At about this time I found out that my family at home had been in some sort of communication with Mr Tate. I am afraid I had been a little neglectful in writing home and as the Secretary of the Grand Trunk Railway Company, a Mr Christian, who was also the private secretary to Sir Morton Peto (one of the contractors) had come to live at Blackheath near my people, my father had asked him to relieve their anxiety by enquiring of Mr Tate, when he next had occasion to write to him, if all was well as regards myself. I was therefore somewhat surprised, one fine morning at work in his room, at his mentioning their anxiety; of course I at once wrote home to put matters right.

Up to this time, I had no idea of returning to England, certainly not yet. But when I received the next letter I could not help thinking there was a hidden wish that I should see my way to do so, and I decided I would think about it. Work went on as usual, however. After the contracts were all let, Quill and I were sent to Port Hope again to make a survey of some additional land for sidings and workshops.

I was greatly surprised to see how much the place had grown since I last saw it two years ago. How often in my later life have I wished to run across the Atlantic once more and see it again, but never had the chance.

I think it was in June of this my third summer, that Quill, who was then not feeling very well, asked me if I would join him and Horace Coombe to go over and see the Falls of Niagara, if we could get a week's holiday from Mr Tate. Of course I assented and permission was given, as the work just then was a little easier. But I never knew a man like Mr Tate who so hated anything in the shape of a holiday that he certainly never took one himself, all the time I knew him.

We started, however, in great spirits. It was not a great journey, taking the steamer to London at the end of the Lake, and then rail and car, getting there in the evening. Miles before we got there, we seemed to hear the booming noise of the Falls and feel the wet and damp of the mist. At last we arrived at that palatial building, the Clifton House, and "pitched our tent". In the summer season the place was very full, as so many of the wealthy planters from the southern states came north to get away from the malaria there. As there were generally 300 or 400 people in the house, we found on arrival the place already pretty full.

The Clifton House stood alone on the Canadian side, giving a splendid view of the Falls themselves; together with Goat Island and the marvellous suspension railway bridge over the river looking like wire stretched across the latter, its enormous height above the river and the precipitous banks on each side. At the back of the house there was a very pretty garden and built around it were some neat little cottages of four or five rooms most simply furnished, which were used *en supplement* to the hotel. We decided to occupy one of these cottages, as they were so much cooler than the main building, but we generally had our meals at the hotel.

It really was wonderful that we should have been alive after all these heavy meals. At breakfast, beginning with three or four eggs almost raw, emptied in a large soda water tumbler and stirred up with a lot of pepper and salt – this was only a beginning. Then some oysters almost as large as a plate, cooked in a variety of ways; then sausages, kidneys and chops; also fish, hot rolls, Indian corn, corncakes molasses and every variety of preserves. The other meals were as heavy and plenteous, with occasional visits to the gorgeous Bar room to see the small barkeeper in his exquisite lawn shirt sleeves and splendid real diamond studs in his shirt front, prepare a champagne cocktail, mint julep or sherry cobbler for you. He had an artistic way of tossing the fascinating drink from one great tumbler to another with the ice in it without spilling a drop. I remember being told that these men get large salaries. It was the universal custom in America then to drink only iced water at meals.

What a happy, enjoyable week this was, exploring everything to be seen. Going over to the American side and having a look at the International Hotel there; having a little trip in the tiny little steamer, the 'Maid of the Mist' that runs about at the bottom of the Falls; putting on oilskins and going under the Falls, then the awful-looking whirlpool on the river below the Falls, the rapids and many other interesting points round about. But as a lover of pure Nature in itself, I could but feel sad and disappointed to see all beauty and interest destroyed by the hideous and vulgar advertisements pasted about all over the place, together with the number of itinerant photographers about with their travelling caravans. I found this miserable practice to exist in every part of the United States that I travelled through later on. Whatever ugly eyesores were met with of this kind, they never seemed to hurt the feelings of Americans. They had not a shadow of refinement about them – dollars were their only thought.

Three evenings in the week a 'ball' was given in the dining hall by the management with a really first class orchestra – the Germania (I have never heard anything better). I need not say that we did not miss one of those dances. Both Quill and myself had a real good time, but poor little Coombe did not happen to possess a dress coat and therefore was out of it. But ever if he had one, with his curious little strut, he could not have danced with any pleasure either to himself or a partner, so he contented himself with looking on and making impertinent observations on ourselves (of course in joke).

What a pleasant, happy time that was. What a number of pretty women there were and, having once been introduced by the Manager, there was no difficulty about partners. I remember a certain young lady dressed in blue, who was most anxious that I should be introduced to her father and mother, and very kind and nice people they were. They came from Kentucky and having understood that it was my intention to make a little tour through the States before returning to England, were most anxious I should pay them a visit.

Time being up, there was no help for it but to go back to work again after a most happy and enjoyable week. Having returned, my work was much the same as before in Mr Tate's room. Correspondence, reports and estimates; occasionally going out on the Works to measure up a sub-contractor's monthly certificate.

7
A trip through the United States, and then home (1855-1856)

The heat this summer was fearful – far greater than the two previous summers I had gone through here and so exhausting at times as to make one almost collapse. Lately also I had felt very unsettled and worried about the letters from home. Although there was no expressed desire in them that I should return, I could very well read between the lines how pleased they would all be if I decided to do so. It was a serious matter and made me very anxious to do that which was best. I could not but think that if I remained until the railway was completed (another three years as all the other men intended to do) I should then be so much older and it would be difficult for me to get work in England.

Until I heard that Mr Christian had been writing to Mr Tate about me, I had no thought of returning home, as I seemed to be content as things were. However, as I continued very restless and unsettled about it all, I determined to write plainly to my father and ask his candid opinion. After the usual interval then of a month, it came – to the effect that if I saw no distinct opportunity of making progress in the future, it would be better that I should return home and try to find work again in the old country.

I therefore decided not to worry any more about it, but let Mr Tate know at once what my feelings were. I did so. He evidently was much surprised and bothered about it, but considering what a hard, driving man he always was with work, he really was very kind and listened quietly to all I wished to say. I told him that before returning to England I would much like to spend about three months going through the United States, arriving home about Christmas, and therefore it would be sometime in September that I would propose to leave, and so it was arranged.

Of course it soon became known in the office and created some little excitement, as I think I was a little bit of a favourite. But it was soon forgotten about as all such things are.

Well, this was about August. By the middle of September I had made all my arrangements to start; said Good bye to all, even to Michael the Indian, whom I asked to come and see me, much to his delight and gave him a lot of odds and ends in the way of clothing and other things which seemed to please him very much. Mr Tate gave a most kind letter to take with me and so did Mr Wyatt. It was a bit of a wrench, this parting for I had been very happy with them all during the three years I had been in Canada.

Before starting on my long journey through the United States, I would like to jump back to my boyhood days to explain that when my father was starting his career as contractor in his little premises in Emerson Street, Southwark, he then commenced a banking account with Currie's Bank – a small but much respected bank in Cornhill. It consisted of a room on the ground floor with some three or four clerks and a very small room at the back for the private office. I had frequently to go to this bank for my father to pay in money, get cheques cashed for wages, or leave a pass book, and would always see old Mr Currie strolling about the room noting what was going on and sometimes asking me some little question. But there was one always there, who for some reason or other I always particular noticed and disliked. This was Bertram Currie (the youngest son); the two eldest sons were attachés in the Foreign Office. Bertram Currie's manner, whenever he asked a question was so objectionable and I thought offensive; so different to his old father and I little thought then how much I should have to do with him years later on.

Now, I will proceed with my journey from Toronto. It may be thought I was somewhat bold in attempting this journey through the States, but I felt I should never have another chance and although alone and so young, I had no fear or nervousness then about myself. As regards ways and means, having made a careful estimate of the cost, I found I could very well undertake it with a clear conscience for I had managed to save a good bit of money. I therefore arranged for the bank to let me have sufficient to take with me and send the balance on, as I did not care to travel with more than I required. America than was a somewhat rough country up West, I was told.

After an affectionate parting with my old comrades and also having said goodbye to Mr and Mrs Boyd and to the manager at the bank, both of whom had been very kind and had introduced me to many of the best people in Toronto, I went down to the end of the wharf on the Lake to meet the steamer that would take me to Hamilton. When the steamer arrived alongside, those passengers who desired to land did so before I embarked. How great was my surprise to see Mr Bertram Currie walk down the gangway. I do not know whether he recognised me. I should think not, but his face was one never to be forgotten. However, having embarked, the steamer went on to Hamilton – a small place where I stayed the night. The next morning I journeyed by Stage (as there was no railway then) to Detroit, which was a long, tiresome, tedious journey. When on the Stage (a great clumsy affair) I found myself seated on the top of it with an elderly spinster who, having ascertained that I was an Englishman, stuck to me like a leech and pumped me quite dry of any news regarding Queen Victoria.

At last we arrived at Detroit, then a small town but now I believe a large city. But as ill luck would have it, I was taken very ill on my arrival at the hotel, having evidently picked up a chill with fever which developed into a bad attack of bronchitis. Had it not been for the very great kindness of the landlord in nursing me, I should certainly have come to grief. If ever there was a good Samaritan in this world, that man was one.

This was a bad start, but after a fortnight, I got all right again and, nothing daunted, determined to stick to the programme I had made and not give up the journey as some might have done. It was my intention to cross the river here and go on by railway to Chicago and afterwards straight on west to St Louis on the Mississippi. This was a long, rough journey over the prairies. There were not many railways at that time in America, and what there were were rough and dangerously constructed with wooden bridges and most dangerous curves and gradients. Fortunately they then did not travel at any great speed.

The cars on these rough western lines were all of one class and mostly for emigrants with not the slightest attempt at luxury. Each car was of great length with a corridor passage running through the centre and seats for two persons on each side the whole length. If you wanted to stretch your legs, you could open your car door at the end and go into the next. There being perhaps four or five cars, it would make a good walk. Between each car there was a little platform where, with a tip to the conductor, I could sit for a time and get a little quiet. Railways in America, like everything else there, are on a gigantic scale. A thousand miles long is thought nothing.

It used to amuse me when the train pulled up by the side of some small river or creek to allow the giant locomotive to take in water, and wood cut and stacked for the purpose. The men in charge of the train got down and with long scoops or pails filled up with not a sign of any people about. Then further on the train would stop to take on board a couple of the inhabitants waving a great big cotton umbrella; railway stations or depots were few and far between on these lines. The interior of these particular cars was extraordinary. Crowded with passengers, evidently emigrants from all nations, they seemed to have all their belongings of every description with them packed anywhere. Rifles, guns and even chamber utensils were made use of quite openly by the children. In a corner of the car there was a little stall for the sale of fruit, cakes, books and papers, iced water etc. Boys would also now and then pass along selling such things.

Taking it altogether, I was not sorry when this part of my journey came to an end for the weather was fearfully hot I remember and the mosquitoes and other venomous flies the very devil.

When I left Toronto, I made up my mind to keep a sort of diary and did so throughout, but many years afterwards I took it into my head to destroy this plus every letter, testimonial, photo and private paper I had and have sincerely regretted doing so ever since. Many little incidents during this journey were recorded in it, which I cannot now remember. I know I was awfully glad when we arrived at Chicago and was impressed beyond measure at the high pressure the inhabitants seemed to be working under. Dollars, dollars seemed to be the cry everywhere. Also the enormous buildings erected for stores and offices; this was a go-ahead city if ever there was one on this earth. But it spoilt everything to see the vast, vulgar advertisements pasted about wherever there was a chance. The hotel I stopped at was immense, but fairly comfortable and not expensive. One thing with these gigantic hotels, you know exactly the cost as they are all on the boarding system. It seemed to be common practice in this part of the world for the train, on leaving the depot to pass along the principal streets and take up passengers from the hotels. There was no other protection for the general public than the noisy, clanging of a bell on the giant locomotive and a large notice board put up on it saying "Look out when the bell rings".

What a change from that time to this! The misery and discomfort of railway travel in those days in America, now there are de luxe trains everywhere, splendidly equipped with sleeping accommodation and dining cars. Since the completion of the Grand Trunk in Canada (the pioneer of this great railway system) to Toronto, others have been added to it striking northward to the vast grain-growing district by Winnipeg, Lake Superior and the Hudson Bay Territory; then over the Rocky Mountains to Vancouver in British Columbia on the Pacific and now known as the Canadian Pacific Railway.

The other Railway on the American side, called the Union Pacific, goes through Salt Lake City and Mormon country to California. There can be no other countries in the world to approach these constructions.

How many time during my journey through the States have I seen with great sadness gangs of Negro slaves, nearly nude and chained together, pass along the railway platform and be put into open trucks like cattle, either coming from some plantation sale or going to one. There was no thought then of that fearful bloody civil war between North and South on the question of slavery, which occurred some six years later.

I think I remained quite a week here in Chicago greatly interested and taking two or three voyages on Lake Michigan, which at times was more like a sea than a fresh water lake, but the steamboats were large and safe. However, the disagreeable, jerky motion of their engines produced the same feeling of nausea as on Lake Ontario.

My next journey to St Louis on the Mississippi was longer and I think even more appalling and rough than the last one – Detroit to Chicago – the cars were again crowded with emigrants. However, as I had taken this journey to see what America was like at that time, it was not for me to complain. These vast, level plains with great herds of buffalo and Indians swarming wild on them must have been very exciting and interesting. Some few years, six or seven after the time of my visit, I came across a very interesting little book written by two travellers – Viscount Milton, son of Earl Fitzwilliam, and his doctor, Dr Cheadle. They journeyed from Chicago north by Lake Superior (the largest of the Canadian lakes) and the vast districts of the Hudson Bay Company, over the Rocky Mountains to the Pacific Ocean. At that time they found Winnipeg and all that part almost uninhabited, and now at this present time it is densely populated by fortunate settlers and farmers. This little book of which I have a copy is well worth reading. Also another book on the same subject is the '*Queen's Highway*' by Stuart Cumberland.

Having at last reached St Louis all right, I remained some time greatly interested with the place and its movements on this great Mississippi River. It is considered next in importance to New Orleans and full of Hudson Bay hunters and Indians. Indeed, I was so much interested that I conceived the idea of going north up the river to St Paul and the Falls of St Anthony – wild and uncivilized enough there. But thinking it over quietly, strong as my desire was to accomplish this, I could not but feel that as it would take a so much longer time going up stream, the distance being about 1000 miles, it would perhaps be unwise for me to get so far away from home and friends.

I therefore determined when I had seen everything, to go down south to New Orleans 1200 miles away and took my passage on one of the swell, fast steamers. It was a great nuisance with these steamers as they seldom left at the time named, waiting until fully loaded. However, after some few hours delay in this instance we made a start.

On the second day after leaving St Louis, a town that had interested me more, I think, than any other I had yet seen, chatting with the captain I happened to mention that I was going right through to New Orleans. He said probably I was not aware that yellow fever was raging there and indeed all along the southern coast, Florida, Charleston and South Carolina. He thought I should be incurring a great risk by attempting it unacclimatized as I was. This was a terrible disappointment, for had I known of it before leaving St Louis I should in all probability have gone north instead. However, I did not care to go back and finding the captain willing to make an allowance in the fare I had paid to New Orleans, I decided to disembark at Cairo.

At Cairo three states can be seen at one time. To the right Kentucky, in the centre Illinois and to the left Missouri. Here the splendid river Ohio joins the Mississippi and goes on to Cincinnati, some 700 miles. This river was very interesting and at times very pretty, but like the Mississippi, very dangerous, particularly at night - numerous gigantic lumber rafts of timber slowly creeping along, the danger of getting stuck on a 'Snag' (a large, uprooted tree floating about wild), also when a couple of the steamers get racing. But all this increased the interest and excitement of the journey. At night it was very weird and ghost-like with the flashes of light in the darkness when the nigger stokers were firing up at the boilers. The whole of the crew were niggers and now and then would amuse themselves by singing plantation songs and very well sung too. Some of them could play the concertina very well and when they sometimes did the cakewalk dance on the deck, it so reminded me of my old companions at Toronto that I was almost inclined to join in. They always seemed to me a merry lot, these nigger crews and there was always something or other going on on these large boats to amuse everyone. The river from St Louis to Cairo was very wide, but the land in each side was very flat. At Cairo I had to change to another steamer, quite as large and also with a nigger crew.

The scenery of the River Ohio in some places was very striking, the river passing between precipitous high banks. The whole of the journey to Cincinnati (700 miles) was very interesting, there being so much to see on both sides of these great rivers. There was also the amusing life on the steamer. En route we stopped at two important towns in Kentucky – Raleigh and Louisville – with sufficient time to get a very good idea of them.

Arriving at the city of Cincinnati, I was astounded at the immensity of this city; even more so than when at Chicago. Everyone seemed to be rushing about and working in the same state of high pressure – dollars, dollars the everlasting cry.

During my journey through the States, I took the greatest possible interest in the Negro life and its many phases, little knowing then, as I have previously said, of the fearful Civil War that was to come a few years later and cripple and half ruin this part of the Union for a long time. When that sad time occurred, I remembered how very interesting it was to me then to be able to follow the scenes of the many sanguinary conflicts.

The field hands among the different plantations of cotton, sugar and tobacco, were really treated and looked upon more as cattle than human beings. No idea of education whatever, and yet they seemed to have some sort of idea of religion, frequently singing hymns with much feeling.

I once had the curiosity to go into one of their Methodist chapels. But a very short stay was enough, for the antics of the sermon of the black clergyman were too funny and the dreadful smell too much - poor creatures. How fond they were too of bright and gaudy colours with their dress.

Indoor servants in hotels, private houses, stores and so on were quite a different class, but yet with all the characteristics of the Negro. How well I remember the time when, on my return home to England, the book 'Uncle Tom's Cabin' was published and every mortal soul in Great Britain reading it. It was most curious to see nearly every one in your train with the book and the extraordinary run it had when dramatized on the stage – the old Adelphi Theatre with Alfred Wigan and Nellie Farren.

The steamer that brought me to Cincinnati continued its journey another 500 miles up the River Ohio to Pittsburgh. However, I had made up my mind to work for my goal (New York) in a zig zag sort of way, seeing everything I could in the several states I could venture to visit. With all the many vast, wonderful rivers and waterways spread like a net over every state in the Union, who can wonder at the enormous and rapid piling up of wealth and prosperity in this great country, as they call it? Take the lumber timber trade for an instance. How many millionaires have been made by the use of these same waterways, many of them more than 1000 miles long? Take the six great Canadian fresh water lakes, but really seas by their immense extent and the tonnage of the ship sailing and trading on them. At the time I was in that country between 50 and 60 years ago, I remember it then seemed to me as immense; what would it be now?

After remaining at Cincinnati some time, seeing the vast hog killing and pork packing business and the many other manufacturies, all evidently coining dollars, I turned into Kentucky for Lexington and Louisville again. I wanted to see more of them and then go on to visit and explore that marvellous creation of nature – the Mammoth Cave – indeed marvellous. At the house I was staying at, they arranged for two Negro guides (slaves of course) to accompany me carrying some provisions and also a good supply of torches and candle lamps to view and realise the extent of the several vast chambers. Of course they were all in darkness being underground.

I must confess I remember feeling a little bit serious after entering the cave, a dark hole about 10 feet high in the side of a hill. On starting, a deadly chill of cold air seemed to creep over me, but soon the violent exertion of crawling on my hands and knees through short tunnels and passages, and then climbing over great mounds of stalactites soon removed any feeling of chilliness. I cannot really attempt to give any description of the wonders of this extraordinary place, but shortly to summarise it.

The cave was 18 miles long and before the extreme end there was another short branch which made it something like a two-pronged fork in shape. After struggling along for about three quarters of a mile, we came to two vast halls or chambers called the Rotunda and the Church, said to be large enough to seat 5000 persons. Then ascending a flight of steps for some 20 feet, we entered what was called the Gothic Avenue and Chapel where there were a number of splendid stalactite pillars as if supporting the roof, and beautifully polished as if by hand - whereas they were quite naturally formed by the constant drip of lime water for many years. The bottomless pit was an awful-looking place.

About 3 miles within the cave we came to a large river called the River Styx and the Dead Sea. It was full of fish, but without eyes. There was a boat in which I travelled several miles, but it goes much further. Two miles from the river was an immense place called Martha's Vineyard, the tops and sides being covered with lime formation in bunches resembling grapes. When lighted up with some torches it had a most beautiful appearance. Another place further on had walls and roof covered with sulphate of lime as white as snow and almost as brilliant as diamonds. The place called the Snowball Ravine was one of the most beautiful places imaginable, but there were any number more, the beauties of which exceed description.

It was awful hard work moving from place to place, climbing up and climbing down many feet and all so wild and rough. Glancing back over these few lines, I find I have omitted to mention a little incident that perhaps interested me more than anything. In the first or second large chamber we came to after entering the cave, I observed a kind of soft brown lining completely covering the roof and walls. Approaching it with a lamp, to my astonishment I found it, when touching it, quite soft and warm and that it was a thick layer of bats at least three in depth. There must have been millions of them.

At last we returned and got back to the house I was staying at late in the evening, having left very early in the morning. I can safely say that was the hardest day's work of manual labour I ever did in my life. During that 12 hours underground, we had not met a soul; the Negroes said that at that time of the year travellers seldom visited it.

The following morning I was so bruised and sore and stiff that I could not move and it was several days before I could get about again. Before leaving I arranged that a few of the most beautiful of the stalactites should be collected and packed. Also a couple of bottles of the eyeless fish from the River Styx and sent to my other luggage waiting for me at New York.

When making this exploration of the cave I had with me a very stout, thick walking stick which I brought from England, thinking it might come in handy in case of a row (a dog's head had been very cleverly carved on the handle). The Negro who took charge of me and acted as guide used to hold the stick frequently that I might have the full use of my hands and arms for climbing or crawling. He seemed so proud and pleased with being trusted, and had behaved so well, that I when I left the place and said Goodbye, I made him a present of it. Poor fellow, how delighted he was.

Well, goodbye to old Kentucky. I liked the State and the people in it and shall always remember it, in spite of their beastly habit of chewing tobacco and expectorating everywhere. Tobacco seemed to be grown extensively with plantations everywhere. I wonder too, why Americans cannot sit and rest as other nations do, but must always tilt their chairs back at a dangerous angle and put their heels up, higher than their head.

They were also very rough and coarse in their manner and ways – a good deal of the cowboy element about them. Sometimes there were scenes in the bar rooms of the hotels that had better not be referred to. A fight between two rough Yankees is the most brutal and disgusting scene that can be imagined.

From Kentucky I went further south to Nashville in Tennessee and also Knoxville. Wherever possible I always preferred to make use of the rivers or stage to railways for getting about, as I could see so much more of the people and the country.

Having so foolishly destroyed the little diary I had made during this journey, I find it rather a trial for my memory to remember many incidents and impressions of the several towns visited. One thing I can certainly remember was that the summer of that year – 1855 – was a very hot one. It was so when I left Toronto and continued all through my journey so far.

Leaving Nashville, a most interesting city on the Cumberland River, I went further south into Alabama and to Florence. At once I perceived a most sudden and strange change both in the country climate and the people. Although now the end of October, the heat here was really almost tropical. Palms, pepper trees and many other southern plants and trees were growing luxuriantly. The plantations now were nearly all growing cotton or sugar with Negro slaves swarming everywhere.

But the change in the people themselves seemed so sudden. French now was so much spoken, the dress and the manners of the people also seemed so French and so refined. Indeed one could sometimes almost believe we were in France. Most of the gentry were wearing thin white cotton suits, and the ladies most picturesque costumes. The same disgusting habit of chewing tobacco existed, but it did not seem so general and most of the ladies smoked cigarettes. Altogether it was indeed a change coming south. Still, there were great and beautiful rivers everywhere.

From Florence, I worked my way through Alabama to Montgomery where all the country and the people were still of the same character. From there I went on to Columbus, Milledgeville, Atlanta and Augusta in Georgia – all still very southern in every respect and again most interesting.

Augusta is on the Savannah River. On arriving at the railway depot there I remember jumping into the omnibus going to the hotel and presently three or four others entered it, but all wearing little nosebags of camphor. I need not say this was a bit of a shock to me, particularly as I noticed driving along that several of the houses had red cross on the door denoting that fatal cases had occurred. I must admit I was somewhat disconcerted, but not alarmed. Curiously only a few days previously I had put a little packet of camphor in my breast pocket. Barring New Orleans, Charleston and South Carolina, no danger was apprehended, but Augusta being on the Savannah river and on the skirts of South Carolina, I must have come a little too near. However, no evil occurred, thank goodness. I soon left and journeyed northwards by Chattanooga to Raleigh in North Carolina.

Milledgeville, although a small and interesting town of only 3000 inhabitants, is the capital of Georgia and has its town hall. It struck me, I remember, as curious that every town or city, large or small, possesses a town hall and proud they may of it, for all those I had the opportunity of seeing were very handsome.

During my journey through the States I did not meet many European travellers, but plenty of the natives, and found them a very great nuisance as a rule. The men were bad enough, but with the womenfolk, it was most irritating. I had taken a comfortable corner seat in a car and perhaps reading a newspaper, when there would be the usual Yankee drawl – "Say, stranger, guess I would like to sit there". There was no alternative but to get up at once, lift my hat and retire. If when seated the seat next to me should be unoccupied, someone of the female sex would sit down in it, no doubt having formed some sort of opinion of that young stranger, and commence an interview. It was always the same, whether in car, or river or stage. The latter the worse as there were no means of getting away.

Passing through these southern states I could not help noticing the many very handsome young girls about (Creoles) who were half black, half white with some Spanish blood. I thought they were still slaves, but later understood they were mostly engaged as housekeepers, companions or maids as they had been fairly well educated.

Raleigh was an important looking town, although somewhat small and seemingly not a great population. The town hall, or State House as they preferred to call it was really a splendid building. I remember it was said to be built after the Parthenon at Athens and was of granite.

From Raleigh I went straight away into Virginia to Richmond, the principal town. It was long tedious journey, mostly by stage. *En route,* I remember we passed by an immense morass, or cypress swamp. It was called the Dismal Swamp and it certainly deserved its name. Although so many, many years ago, I remember being told it was 45 miles long and 25 wide. It was a vast quagmire of decaying wood and plants and of course with abundance of that strange, evil creeper, Spanish moss, about. No doubt it was the hiding place of many a runaway slave.

At last I arrived at Richmond, Virginia, after a long and somewhat rough journey. I had been looking forward for a long time to seeing the State of 'Ole Virginey'. I don't know why, unless in my boyhood days I had read some exciting books referring to it. Anyhow, there was certainly not the least feeling of disappointment on my arrival, for it seemed the most quaint, quiet, restful city I had yet met. Busy, but not that feverish rushing about at high pressure as with some of the other large cities I have passed through.

It was beautifully situated on the falls of the James River. There seemed to be any number of busy factories - cotton, woollen and iron and also a very large number of tobacco factories. In fact as far as I can remember the growth and manufacture of tobacco seemed to be the principal trade. Any number of Negroes were employed at the works and on the plantations and I really think I got a better knowledge of what Negro life was like here than at other places. The people were very interesting, very southern in their ways and it was a hot bed of slavery.

As usual there was a very fine town hall and in it a very fine marble statue of George Washington. (How often I have wished I had never been such a fool as to destroy the diary I kept during this long journey, it would have helped me now). There were many interesting excursions to be easily made round about, among them the celebrated Virginia Springs – these are to the Southerners what those of Saratoga are to the North, but at the time I was there, the season was over. However, as I went to see some of the beautiful scenery and not the beautiful women, it did not much matter.

In most of the Southern towns I had passed through, when seeing a little red flag up outside some building (denoting a sale of some sort), I had sometimes strolled in from curiosity, if it happened to be a sale of slaves. One day when strolling about in Richmond, seeing the usual flag up, an incident occurred which became indelibly fixed on my memory. Anyone reading these reminiscences would perhaps smile at my memory being able to remember after all these many years such a detail as the few steps that led up to the entrance; nevertheless it is before me now like a photograph.

Going up these few steps I found myself in a large hall and found a large number of slaves (men, women and children) squatting down together round the walls, with a number ticketed on each one. There seemed to be several families and in fact it was the entire stock of some large plantation being sold off. At one end of the room the auctioneer was seated on his rostrum; a real clever Yankee and quite as convincing in his remarks as Sam Slick the clockmaker ever was. Just in front of him was a little stage upon which the poor creature had to stand half nude and be subject to examination of every kind – feeling, hard punching and other tests – by the crown of men round. It really was a distressing and disgusting scene. I noticed that the auctioneer made an effort now and then to sell a family all together in one lot, but if he failed and it was objected to, they were sold singly and that meant separation which of course was most sad and painful for the poor wretches.

While standing outside the circle of those looking on, I felt someone pulling at my coat and looking down, saw a little nigger boy at it. Enquiring what he wanted and stooping down to him, I ascertained with much difficulty from his queer language that 'Nanny' or some such female name wanted to speak to me. Somewhat flabbergasted, I allowed the child to lead me to a group of slaves nearby waiting their turn to be sold and found that a very good-looking young woman had noticed me among the bystanders. Possibly observing on my countenance the feelings of sympathy and pity I felt, she wished so much that I would purchase her when her turn came and anxiously told me what kind of work she had been accustomed to. I could only explain, however, that I was a stranger, not an American, therefore only looking and not a purchaser. After a few more words of sympathy, I gave the poor creature a dollar or two and went my way, not having the courage to wait the result. This was the last slave sale I ever attended.

As time was now getting on and I had much ground yet to get over I decided to go straight on from Richmond and its beautiful surrounding scenery to Washington City. I found the journey was a very patchy sort of one at that time – rail, river and stage. I remember feeling very irritated as I wanted to get on, but probably I saw more. No doubt by this time it is very different.

Arriving at Washington, I was greatly struck by its admirable position and the many handsome buildings about, particularly the President's residence, the White House and the Capitol, superbly situated in the middle of the city on a very high hill which can be seen for miles round. From it there is a magnificent view.

The several large and wide avenues leading from the city to the capitol also seemed to add very much to the character of the place. Anyone walking up one of these avenues, ascending the whole time, and then continuing up the immense number of wide marble steps to the Capitol itself, would be awfully glad of a rest when they got there. The building itself was most beautiful and interesting, both inside and out and it would be difficult to imagine in any other country a senate or parliament house to excel it.

Although my memory is wonderfully good as a rule, I cannot now remember the name of the President in office at that time, but I think it must have been Lincoln.

The river on which the city of Washington is situated adds greatly to its attraction, the scenery round about being very beautiful. I can remember being told when there, that George Washington himself selected the site of this presidential city. The hotel I stayed at was, I think, called the Windsor and I have some idea in my mind of seeing a visitor at that hotel pulling out a revolver from his hip pocket and shooting one of the Negro waiters. I believe such did occur, but as I cannot quite fix it as a fact in my mind, I will not further refer to it. In those times, there were many sad and brutal scenes, for every man (except the blacks) had such a weapon about him.

Having seen all that there was to be seen in Washington with much interest, I went on to Baltimore. This was only a short distance, some fifty miles, and on arrival I found it a very different place to Washington. A very large city, even at that time, very busy and noisy and, I thought in the trading part, very dirty; but it was evidently a money-making place. It was situated on one of their large rivers entering Chesapeake Bay, by which steamers and sailing ships of moderate tonnage could come into the docks to load and unload. However, as the mercantile part did not so much interest me, I made the most of my time by taking some one or two excursions to the magnificent scenery further back – of Harper's Ferry and the Allegheny Mountains. Such scenery would recompense anyone for crossing the Atlantic to see and enjoy.

Pittsburgh was not then of the importance it is now, the manufacture of steel being in its infancy. As it was 400 miles from Baltimore and my means and time were getting limited, I did not visit it, but I would like to take this opportunity of mentioning a little matter that occurs to my mind while writing these lines. Of course, at this present time some 50 or 60 years since I journeyed through America, everyone throughout the world would know the name of Andrew Carnegie. At the very time I was at Baltimore, this young man, about my own age, must have been in Pittsburgh making and piling up the many and solid millions of money that he ultimately made. As a lad he emigrated to America from Scotland with, I believe only the proverbial two and six in his pocket. It was only a few days ago that I observed in the newspapers that since he had retired from the steel business and taken a beautiful estate in Scotland, he had distributed about 25 millions of money amongst certain educational institutes and that at the present time, he still possessed 45 millions capital and from two to three millions income. Wonders will never cease in this world.

From Baltimore to Philadelphia was only about 100 miles and on my arrival I unhesitatingly gave the palm to that city as being the most handsome and cleanest of any I had seen. New York was then, I believe, considered the largest; Philadelphia the next.

It was beautifully situated on the Delaware river with, I think, a frontage of something like five or six miles and with accommodation for large steamships and shipping. I well remembered how delighted I was with the place. There was such a charm and elegance about it. The many handsome buildings were built of a lovely, brown stone, beautifully carved, and with bronze light ironwork introduced. There can be no question that the architects in America are very clever, artistic men, but what struck me more than anything here was the admirable supply of water to the city from the Fairmount Water Works, which could be seen perched up high at the back of the city. The water was pumped from another large river there, which made a great feature in the view.

The principal street was Chestnut Street, where the State House and the best hotels were situated. The hotel I was staying at was in this street and I well remember how interested I used to be going out in the early morning before breakfast and seeing the Negro servants outside every building with a hose sluicing and washing the front down. It reminded me so much of Holland and its wonderful cleanliness. I should mention that throughout this trip, I always made a practice of getting up very early; one then seems to look at things so freshly and cheerfully.

There was I remember a very pretty excursion up this river by steamer, but the name of the river I cannot now remember. Well, I shall always think of Philadelphia.

I cannot help thinking how wonderfully lucky I was during this journey to have had such good weather. Even now, although nearing the end of November, when as a rule it should be winter, it was still pleasant and enjoyable.

New York at last; more noisy and hideous I think than when I was here just within three years ago. Yet there were still many interesting places to visit in it. The very first thing I did on arrival was to go to the office of the Cunard Company and secure a berth on the next boat for Liverpool which I ascertained was to leave in three days. I also looked up the luggage I had sent on from Toronto, together with my few curiosities from the Mammoth Cave in Kentucky. I was glad also to find at the post office some two or three letters from home saying all was well, and among the little odds and ends of news, that my father was having a very nice house built for him in the Lee Road about half a mile from Dartmouth Villas.

This time I stopped at the St Nicholas Hotel, not the Metropolitan as before. It was built entirely of white marble, I remember.

The organisation and management of these immense palatial hotels is something wonderful, and the servants are all Negroes. So many young married couples then made a practice of making it their home for a start, where they seemed to live quite quietly and even moderately.

During the short time I had before the steamer left, I amused myself looking about and crossing by the steam ferry boats to Staten Island and Brooklyn (there was no wonderful bridge erected then). Having been so much interested in the Fairmount Water Works at Philadelphia, I went and had a look at the great Croton Aqueduct supplying New York with water, a fine piece of engineering.

At last the day came to say Goodbye to America and start for home. I left, I remember, in quite good form by the steamship *Asia* under Captain John Lott. The steamer was a newer and larger ship than the poor old *Cambria*, in which I came out, and I was fortunate this time in having a cabin to myself, for there were not many crossing at that time of the year. The voyage this time was quite uneventful. There was no ice about (at least we met with none) and very little fog. Our Captain was a queer sort of character. He seldom or ever spoke to any one of the passengers and was, I remember, considered by all a very surly sort of man. He was a good sailor without doubt, but he made not the least effort to brighten up things during that tedious fourteen days. I head afterwards that that was always his manner.

Fortunately, I made friends with a fellow passenger, a director of the Surrey Dock Company at Rotherhithe in London; he was very nice and knew my father very well. But for the life of me, I cannot now remember his name. However, I often met this gentleman in London afterwards. As a rule my memory is wonderful, but sometimes, very curiously, it plays tricks. If I am standing together with two men with whom I may be conferring, the name of one comes to me at once, the other I cannot remember.

After a voyage of, I think, 14 days the *Asia* arrived at Liverpool. As it was in the afternoon, I decided to go on to London by the night mail, after clearing the Customs and having a good, plain old English dinner at the hotel.

I remember well, thinking it all over afterwards, how much I had gone through in the last three years; how much ground I had travelled over and how grateful I ought to feel that, barring the unfortunate illness I tumbled into at Detroit soon after leaving Toronto, my health had been good in spite of the yellow fever scare. Also, the weather had continued to be so very favourable and further that I had been able to keep within the estimate of cost I had prepared before leaving Toronto.

Arriving in the early morning at Euston, I crossed over to the hotel and after a good wash, clean and breakfast, I went to Charing Cross station for a train to Lewisham. How strange it all seemed again. When the cab drove up with my unworthy self and luggage to Dartmouth Villas, the front door was already wide open and I can remember a very pretty young girl, with a smiling face, was standing there to receive me. This was my dear sister Jessie, whom I certainly should not have recognised had I seen her elsewhere. I had already given them notice of about the time of my arrival. My father and my brother John had left for the London office, but I need not say with what pleasure and delight my dear mother and sisters received me home again. My younger brothers I think were still at school, except Alex who was in Berlin. The first real bit of news I heard on my return was that my brother John was engaged to Sarah, the youngest daughter of old Mr Ben Smith the wharfinger at Deptford Creek and they were, I think, to be married the next year. How strange it seems to write this now when only some three or four years ago, both of them celebrated their Golden Wedding.

In the evening when my dear father and brother came home and Mr and Mrs Quick came round from No. 6 to see me, there was much excitement and talk, and the opening of the packages with the few little presents I had managed to bring home. There was the Indians' wonderful clever work I had bought from Michael's tribe when out with them on Lake Simcoe, such as moccasins, belts, snowshoes and different skins, furs and rugs. I somewhat astonished them all, for the articles were wonderfully clever and some of them really beautiful. Then there were the beautiful stalactites and the eyeless fish from the great Mammoth Cave in Kentucky and not forgetting the large head of the deer I first shot. For Mr and Mrs Quick I remember bringing home a very striking large picture of the Falls of Niagara.

I was much rejoiced to find all our family so well, but in that respect, unhappily it did not last very long. A little brother, George, had been born during the last year [in 1855], but sad to say, only a few days after my return home, the poor little child had an attack of croup and died in my arms. This sorrow of course upset everything and everybody – particularly my dear mother - and made our Xmas a very sad one.

I can remember this Xmas also being a very severe one with a great fall of snow. There was also at this time great excitement and outcry throughout England at the shameful, indeed awful condition of our soldiers, poor fellows, in the Crimean War, without food and proper winter clothing.

Having taken an opportunity of calling on my very dear old friend Mrs Keeble and some few others, I began to think very seriously about myself and my future. It was perfectly clear that I must begin again and made another fresh start in life at 25 years of age. I had had quite enough of travelling and holiday and the sooner I made some sort of beginning, the better. It seemed clear that it must be something in the engineering way. I had long talks I remember with my father and Mr Quick. I had only some three or four hundred pounds, not enough for any partnership that might offer; therefore it must be in some official capacity. We therefore began to look about and see what might turn up.

It was curious, but I don't think any member of our family were at first very proud of me and my appearance when I returned. I remember wearing a suit of ready-made clothes and a cowboy's felt hat, all invested in on the other side of the water. I certainly was not a fop then and my beard and moustache had been allowed to grow wild. But what seemed to distress them more than anything was the peculiarity of my manners and speech. I had managed to pick up thoroughly the Yankee drawl and sayings. I had only to chew tobacco and expectorate and I should have been perfect. The curious thing was that until the family called my attention to the vulgar habits I had fallen into – "say", "I guess", "Sir-ee" and "So long" - I was myself unaware of it. However, it soon became a nine day's wonder and a razor soon settled the beard and moustache difficulty as the moustache was not then grown in England. My father also very kindly told me to go to George Flaxman, his tailor in the Borough, and get a fresh rig-out.

In the meantime I devoted myself entirely to the society of my two sisters, Agnes and Jessie. Never in this world was there a brother so proud of his sisters as I was. Although only what is termed of the middle class, they dressed always with such perfect taste and always looked such perfect ladies. They had evidently been well looked after during the time I had been away. Agnes had had many lessons on the piano from Benedict, the great pianist and composer at that time.

Their society I noticed was much sought for. Mr and Mrs Penn seldom gave a dance without them. Since I had been away, a wealthy old gentleman, Mr Brandman, a City man, who had been living for many years at a very pretty place in the Lee Road opposite St Margaret's Church called 'The Cedars', died. Mr Penn bought it and was now residing there, Mr Bazley White having taken The Priory off him. I was taken round of course, to see the new house that my father was having built for us, just by The Cedars on a piece of ground Mr Penn had leased to him. It was nearing completion and a more perfect residence could not be imagined. The brothers Smith were the architects.

I was also taken to see the new Crystal Palace which had been completed at Sydenham. I was simply astounded at its vastness and beauty as it was then. I had of course seen the Exhibition of '51; but although very beautiful and attractive, it could not approach this in any way. When strolling about the building, with what pride I used to notice how many people would turn round and notice my two sisters.

The fountains and water displays in the building, and outside in the grounds, were also splendid and supposed to equal those at Versailles in France. But they were found too costly to keep up and were afterwards unhappily abandoned. Indeed the poor palace now seems to be a kind of white elephant.

My father, now thinking I looked rather more respectable, suggested one day that I should go up to town with him to see the new offices in Belvedere Road, and also my old friends Taylor and Ingleton. What a change indeed from the little, insignificant yard and place in Emerson Street! In Belvedere Road, Lambeth between Waterloo and Hungerford Bridges there stood a large, important-looking building set back by itself from the road and not at all unlike Gwyder House in appearance and very nearly as large in accommodation.

A large gateway led to the entrance up several very wide steps into the building, then there was a large space for a hall with a staircase in front. On the left on entering, there was a waiting room. The next room behind this was a very large room for use by my father and brother John with a large window at the end of it looking out on to a leaded platform, the latter being a roof to the basement below, about 50 or 60 feet square. From this platform were three or four steps leading down to a yard, the entrance to which was by another very large gateway at the side. This yard contained a long range of offices for some half a dozen store keepers under my old friend John Taylor; a large range of stabling for horses and carts for the two broughams for the London work of my father and brother. There were large workshops for carpenters, smiths and others and for stores. At the end of the yard was a crane for loading and unloading barges of plant and various materials from the River Thames.

Returning to the inside of the building. On the right on entering was a room for doorkeeper and messenger, then behind this was a large room corresponding to my father's room on the other side. Here I found my old friend Ingleton, with some three or four other clerks, having charge and preparation of the wages sheets every Saturday. At the end of the hall or passage, a door led out of the leaded platform and a staircase down to the basement.

Ascending the stairs to the first floor, on the landing were four rooms and a w.c. On the right was one occupied by the chief cashier and book-keeper, then a most pleasant and efficient young fellow, a Mr Neilson, a Dane, who my father had taken a fancy to and engaged after the completion of the Copenhagen Water Works. The poor fellow, however, died some few years after from consumption, to everyone's great sorrow and regret. The room at the back was occupied by two assistant book-keepers. On the left side of the landing was the drawing office, occupied by three or four men including Mr Jones, the chief draughtsman.

One of the clerks in Mr Ingleton's room downstairs was a Mr Wallis, a stranger to me then and a rather comical-looking little man in appearance and very like a clerk, but a most useful and faithful man. Some few years later he rendered much kind help to my father when those sad troubles occurred with my younger brothers. Mr Wallis had been permitted to have his two sons (quite young lads) in the office to learn and be useful.

My father finding the elder one – George – to be a shrewd, clever, young fellow, although young, sent him into Staffordshire as a timekeeper on some work he was doing there for Mr McClean, the great and well-known engineer of that time. When the work was complete Mr McClean asked my father to lend George Wallis to him for a time, as he was about to carry out a small water works at Eastbourne for the Duke of Devonshire. My father of course assented and George Wallis went there, did what was required of him satisfactorily and pleased the Duke and also Mr McClean. He was shrewd and clever enough to see the opportunity before him, elected to remain in Eastbourne and strike out for himself. He never returned to John Aird and Son and after a time sent for his brother William to join him.

This was Belvedere Road, and I was simply astounded at the progress my father had made with his contracting business during my three year's absence. But after all it was not to be wondered at, for he was a man of such great capacity and ability for that class of work; he had it all to himself then with no one to compete with him. There were plenty of builders of houses and public buildings, but no other man who understood ground work and its many difficulties. Consequently his assistance was sought by nearly all engineers and companies.

Gas, as I have before mentioned, was quite in its infancy. Before my father became a contractor, engineers when dealing with gasholders and tanks for storage were, as a rule, greatly afraid of water if discovered. Instead of sinking down into the ground, they would build up a cast iron tank out of the ground with large cast iron plates carefully bolted together and, of course, at great expense. It was for John Aird to show them the folly of this and that there was no danger or impossibility of sinking down any depth, using suitable pumps. I have often wondered, during his lifetime, how many gasholder tanks he must have constructed, many of enormous size, and also depth, and under very difficult and exceptional circumstances - many, many hundreds and very possibly thousands, both in London and the provinces and foreign towns also. The great danger in this work was dealing with the air in the ground afterwards, which only he seemed to understand.

Being just now like a frozen-out gardener and with no work to do, I took the opportunity of going round some of the works with my father and brother John. I felt greatly interested, and I remember on these occasions my father telling me of some interesting works he had carried out during my absence in America. Shortly after I left England, he was asked to do some work at Eastbourne, a small seaside place on the Sussex coast where he had never been before. It was to collect some springs of water near the Downs into a small reservoir and convey it by a main to an iron tank erected near the seashore to supply fresh water to some dozen houses that had been built in that area. He said they were called 'Sea houses and let-in apartments'. The only town that then existed was about a mile and a quarter inland and away from the sea and very, very old. My father said he was so pleased with the beautiful air and quaintness of the place, he decided to send my younger brother Joseph to school there. It was the only one that seemed to exist, called The Gables, opposite the very old church and kept by a Mr Stroud.

The three water works at Hampton I found nearing completion and the great trunk mains from them into London, already laid. There were several other works going on of a similar nature at Old Ford Bow, for the East London; at Barnes for the West Middlesex; at Battersea for the Southwark and at Kew Bridge for the Grand Junction, which latter brought to my remembrance a little adventure when I was but ten years old.

As for the main laying, it now seemed to have become an immense business and seemed entirely in the charge of my brother. It was a great responsibility, for nearly all the gas and water companies in London, in all the provinces and other places on the Continent required attention; but he was equal to it all. I was astonished to see how thoroughly he had made himself master of the work.

I can remember when going round the different gangs, how happy it made me to come across men again whom I had known in the trench, when I was but a lad and one of my father's timekeepers. Some of these men were now advanced to the position of jointer or ganger and came up with a broad grin all over their faces and a "Very glad to see you again, Master Charles". It always was astonishing how men of every class remained in my father's service nearly all their lifetime.

When my father made a start as contractor, his first wages sheets would probably have been for 50 men; now they would be for at least 500.

8
An appointment with the Grand Junction Water Company

After a fortnight of my enforced idleness in this way, one evening Mr Quick came round to mention that at the board meeting of the Grand Junction Water Company (of which company he was consulting engineer) it had been suggested that as the district of the company was now becoming so much built over, the engineer Mr Hughes required assistance in the outdoor work. This was approved of and it was arranged that steps should be taken to find a suitable person.

I became that suitable person. After a little talk about it, it was arranged that I should send an application for the position. I remember when Mr Quick, my father and myself were talking the subject over, Mr Quick remarked that the collectors realised as much as £500 to £600 a year by their commission, but I thought I was worthy of a better position than that, irrespective of remuneration.

A short time afterwards, I received a notice from the Secretary of the company to attend at the office on a certain day to meet the Chairman, Sir W. Clay MP. This, of course, I did, and rather nervously found myself in the presence of the great man in the Secretary's room at the offices in Brook Street. I found him awfully stiff and formal at first, but having answered several questions and stated what my occupation had been, he seemed much interested about the Grand Trunk Railway and at my having been engaged on it. He finally became quite pleasant, but expressed no opinion about my application, so I therefore bowed and withdrew. But it was all right, for in a few days I received another letter from the Secretary, Mr E. O. Coe that I was appointed at £300 a year. Well it was not very grand, but it was something to do.

I did not lose any time I remember, entering upon my duties, seeing the engineer Mr Hughes and having a talk with him. He was a breezy, rough sort of man, but pleasant and kind.

The offices were at the corner of South Molton Street in Brook Street. This was rather a large building as the Secretary, Mr E.O. Coe, and his family lived there in the rooms above. They were very nice people in good society. There was Mrs Coe and four daughters - two married a little later - and our family became very friendly with them.

Mr Coe was a very clever little man, very much given to theatricals. He had written two or three successful plays and was very good in private theatricals. I remember some being given by his family at home that were very well done. One of the players was a very nice and clever young fellow with whom later on I became very friendly. Arthur Blunt was a son of Mr Blunt, a well-known solicitor representing most of the water companies.

As my work was almost entirely out of doors, I was only in the general office for a short time, first thing in the morning. It was a large room with three clerks, Hazelden the chief, Horsey and another. There were also six collectors; two attended each day to receive rates. I can well remember their names – Barnes, Ferdinand Ball, Frank Carter, Billy Raymond, Alfred Tenniel (brother of John Tenniel the artist) and George Strother. The engineer Mr Hughes had his own room and there was a small yard at the back and a room for the turncock's use as they had to see me and report every morning.

Having seen the turncocks and extracted from the large office books the several entries requiring my attention. I would start out to work upon them until the next morning, and so it went on day after day.

The company's district was of great extent and covered about the best part of the West End; starting from Piccadilly Circus, both sides of Regent Street to Oxford Circus; all the houses north of Piccadilly including the fashionable squares – Grosvenor, Berkeley, Portman – the whole of Paddington, Bayswater, Westbourne Terrace, Notting Hill, Campden Hill and Maida Hill.

All the ground at that time north west from Notting Hill Gate, I remember, was simply wild ground; fields covered with rubbish, refuse, broken china and any number of pigsties. It was an awful part of London then. I remember a large public house called the Elgin Arms standing alone in the middle of the oasis, but it was all very soon covered over with new buildings as far as Kensal Green. Westbourne Grove where Whiteley's palatial establishment now stands - then a very poor little shop; no one looking at it then could have imagined what its future was to be. It must be remembered also that there were no underground railways or Thames Embankment, therefore in my getting about for my work it was only by the old-fashioned omnibus.

My duties were simple enough. The Secretary would receive notice that certain houses had been built and completed, with a request for water to be laid on to them. Before this was done, it was my duty to inspect them and satisfy myself that the service pipes were sufficiently large and the cisterns also. I very well remember my first day's work was to inspect Cleveland Square, Leinster Square and Gardens and many other very large, handsome houses just finished. In addition to this work, there were also complaints from consumers of insufficient supply, or leakage, or rated too high, or some grumble or another. This kept me pretty busy.

Again, when the collectors attended at the office, they generally brought and entered in the office book, complaints they had met with of some sort or other. Such complaints were nearly always in the old parts of the district and the investigation into these was very dirty work. Disgusting sometimes in certain places, climbing outside bedroom windows to examine cisterns and the fittings.

Such complaints as a rule would come from very shady parts such as round about Golden Square and Regent Street which were always filthily dirty. It was a pity the Aliens' Act was not in operation then. However, in a very short time I found it would be better for Abbott the foreman turncock to do the climbing part of the business, for nine times out of ten, the so-called defect was found to be local and a plumber only wanted. The occupation of these householders could very well be judged, I think.

The complaints of no supply of water to the cisterns upstairs were numerous and I found they were distinctly caused by neglect on the part of some of the turncocks. Each man possessed a copy of the engineers' regulations as to the proper time for opening and shutting certain water cocks in his district. This was to ensure sufficient pressure to fill all the upper cisterns and prevent complaint. Of course, if all the cocks were left wide open, that object would not be effected.

I very soon put an end to this. One morning I told Abbott, the foreman, to bring a set of turncock's irons or tools and accompany me, and having attested many of the water cocks, found it to be as I had imagined. The cocks were wide open and the men inside a public house drinking. This being a very serious matter, I made a special report and sent it to Mr Hughes. Having talked the matter over, I suggested they should have one more chance and that I would speak to them and caution them. As he assented, I did so and they promised to reform, but I never trusted them again and during the time I remained with the company, I remember two of them had to be dismissed. Some time after, I was very pleased to hear that it had been mentioned at the Board that since Mr Aird had been with the company, the outdoor service had been much more satisfactory.

I find I have omitted to mention that when I joined the company, I was told that it was imperative that I should reside in the company's district. I therefore took a couple of rooms in Cambridge Street, Hyde Park – No. 42 or No. 44 – on the ground floor. This was considered necessary in case any very serious fire occurred, or any of the large trunk mains burst, so that I might be within call to deal at once with the trouble whatever it might be.

How well I remember the awful dullness and misery when occupying those rooms. I can see myself staring out of the window looking at nothing, with one of those infernal old box organs opposite playing hymns. There were no piano organs then.

I had only my breakfast there; getting a snack of lunch anywhere on my work and dinner at some moderate eating house. There were no such restaurants and grill rooms as there are now. I remember patronising a place in Piccadilly, somewhere where Prince's rooms now stand, I think. A place, if it existed now, I should be ashamed to be seen in. As for the food and the greasy way it was served up, it makes me shudder to think of it. If beer was wanted, Henry the waiter would go to a public house next door and come back with half a dozen tankards.

As a luxury I would sometimes go to Simpson's in the Strand, or sometimes when feeling very tired or unwell, I would have a chop or something with tea at my lodgings. Mentioning Simpson's reminds me of a then well known place opposite, Exeter Hall, where the several oratorios were so splendidly performed under Michael Costa. Mr Quick's eldest son

Joseph, who had now grown up and worked in his father's drawing office in Summer Street would sometimes accompany me, although this young man had his peculiarities. We were almost constant companions through life and like myself, passionately fond of music. I very well remember that a few year's later on, he was greatly in love or fancied he was, with that great songstress Adelina Patti. He persuaded me to accompany him to Paris to hear her sing again, and acted very foolishly.

After a while I found that several of my colleagues in the office were very musical; Hazelden the Chief Clerk was a clever pianist; Frank Carter played the flute (or thought he did). He had rooms near to mine in Connaught Terrace and whenever I called, always found him puffing and blowing away and making an unearthly noise. I remember my brother John attempted to play it and had one of Rudolf Carte's best instruments.

It has always been a mystery to me how any human being could take to the flute for music. To me, even expert professionals fail to produce real melody and nothing but a most painful wail, like from a wounded hare or rabbit.

This reminds me of an incident that occurred some twenty years later on and I will make a little jump just for the moment to that time. I was sitting in my morning room, a pretty little room with French windows looking out into an Italian garden on rising ground leading to the beautiful woods round the place. I think it was on a Sunday and I was quietly reading, when suddenly I heard a painful wail or cry outside like the flute. But I knew. My gun happened to be standing in the corner of the room. I jumped up, got hold of it, slipped in a cartridge, went to the open window, aimed and fired. I ran up the lawn and picked up what I expected to find – a poor, wretched rabbit trying to escape from a weasel on its back sucking its life's blood. They were both dead. Now I will return to the time I was writing.

After a time, I therefore thought I would have a little musical party. My colleagues all had good voices and could sing some glees fairly well. As my rooms would not hold more than half a dozen, five with myself, it had to be limited to Ferdinand Ball, Alfred Tenniel, H. Hazelden, Frank Carter and Horsey. I remember some difficulty with my landlady about the matter, as it was necessary to clear my bedroom for the supper, but I got over that. The difficulty remaining now was providing the supper. I had hired a piano for the evening.

I was about to order the supper from a confectioner close by when I received a message from that dear mother of mine, who had heard in some way or other of my intention, that she would feel very much hurt if I did not let her provide it. This being just one of those things that always pleased and delighted her to carry out for any one of us. I at once assented. What a dear good mother she then was to us all.

The day came and in the afternoon arrived her cook or parlourmaid, I forget which, with a supper for a dozen, even wine. With such a supper, success was assured. After a pleasant time with some few glees and a little music, the supper was fully appreciated and I fancy those men remembered it for a long time afterwards.

Alfred Tenniel, like myself, was very fond of music and we often would go to a first night at the opera at Covent Garden. I remember we did so at the first night of Verdi's *Il Trovatore*, and how delighted we were with it. We could then only afford seats in the gallery. When the opera season was over, the house was let on a short lease to a man named Anderson, a conjuror called the Wizard of the North, for a pantomime. This turning out a success, he finished up with a very rowdy fancy dress ball, and the place was burnt down. It was only just outside the Grand Junction Water Company district. I remember that Lucas Brothers, the great builders, afterwards had the contract to rebuild it.

There was another event of some importance occurring at this time. Peace was proclaimed with Russia and the disastrous and disgraceful Crimean War ended. It was celebrated by firework displays all over the kingdom and in the London parks. They were thought very grand at that time.

I strolled about Hyde Park looking on. The crowd was enormous when suddenly I came face to face with a young lady with whom I had had many an enjoyable dance and pleasant time – Miss Harriet Glover, a younger sister of Mrs Quick. She was in company with some gentleman friend, but as I had not met her since my return from America and saw not the least disposition that I was in the way, I remained chatting for some time and had rather a good time. I am afraid, however, her *cher ami* wished me at Jericho.

It was now about time that I looked them up at home at Lewisham to thank my mother personally for her kindness regarding my supper. I had heard in the meantime that my father had been very busy with his work for the many gas and water works with whom he was now engaged, but he had also been very busy at the same time in arranging for other great events about to take place. My brother John was about to be married to Sarah, the youngest daughter of Mr B Smith our neighbour. John then became my father's partner in the firm John Aird and Son. The newly married couple then occupied a little house somewhere nearby. John's family tree appears on the next page.

I recollect a little time after their wedding, my brother invited some few of his old bachelor friends to dinner (or supper). There were some half dozen including his wife's two brothers, Henry and Tom, and myself. On that occasion I grievously hurt Sarah's feelings by spoiling one of her new silver spoons when hammering with it on the table in my excitement when their health and long life were proposed. However, fifty years afterwards, when they came to their golden wedding under very different circumstances, I remembered the incident of the battered spoon. I think there was a very pleasant little dance on the wedding occasion.

I went round to my father's new house, The Grove, to see how it was getting on and could see it would not be very long before the family moved into it.

I continued on with my work at the Grand Junction Water Works. It was not very interesting, but I stuck to it, determined to do my best to please. Sometimes on a Board day, if I happened to be in the office, the Chairman Sir W. Clay would send for me before the Board met, to ask some few questions in the Secretary's room. These would be about the new district that was opening up and so on, but I think this was only to show his authority. I never seemed nervous or afraid of him and although he was always very stiff and formal, he was also affable. I know he always had the highest opinion of my father.

Curiously I remember about this time that he moved the Second Reading of the Bill for the Abolition of Church Rates in the House of

FAMILY TREE OF JOHN AIRD (second) 1833 -1911

John (1833-1911) Baronet 1901
married 1855
Sarah Smith (d. 1909)
dau. of Benjamin Smith, Lewisham

Sarah	Jessie	Jo	Kate	Ada Eleanor	Vida	Malcolm	Gertrude	Dorothy
(1858-1938)	(1860-1932)	(1861-1934)	(1863-1905)	(1867-1947)	(1869-1947)	(1872-1934)	(1873-1942)	(1876-1931)
married 1878	married 1883	married 1886	married 1884	married 1889	married 1892	married 1898		married 1899
Basil Ellis	G. Abernethy	Ellen Renton	Alfred Bean	Louis Mieville	Edward Anderson	Nellie Dummett		Spyridion Mavrojani
(d. 1907)	(d. 1923)	(1856-1951)		(1867-1944)	(Rev)	(d. 1954)		(d. 1930)
4 children	3 children	5 children		4 children	7 children	5 children		6 children

Commons. My brother John and myself have always been great politicians throughout our lives and never missed political news in the papers. I well remember the interest I took at that time in Disraeli.

After a time I found that Frank Carter, one of the Collectors, was fond of fishing and after several chats on the subject in our rooms in the evening, we arranged to go to Twickenham the following Sunday and try our luck in the River Thames. The first fine Sunday that turned up, we started on our expedition early with some tackle we had looked up for the occasion.

On arriving at Twickenham we went down to the ferry by the old church in the town and engaged one of the fishermen and his punt for the day. We embarked and went to what he called the best pitch on that part of the river. Although nothing sensational occurred, we had a very good day's sport with the roach and dace.

I have heard many people say that it is only fools who can sit in a punt all day long, bottom fishing. I cannot agree with them, as given a fine day, to my mind such fishing is the acme of rest, peace and laziness; or as an old friend of days later on (Sir Myles Fenton) used to say *"Dolce far niente"*. As the day on this occasion happened to be a very lovely one, we thoroughly enjoyed it and out pitch was very nearly opposite that very pretty and well known residence, Pope's villa. How well I remember when chatting with Carter in the punt, remarking what a lovely place it was and what tremendous luck to own such a place to live in, for indeed it was enough to make anyone envious. The beautiful lawn sloping down to the river like an exquisite piece of green velvet; the quaint picturesque Chinese pagoda style of building; the large boathouse with a pretty summer house above it and the beautiful garden with roses seemingly everywhere. But Frank Carter was always very matter of fact, not an ounce of sentiment or romance with him. So long as he had enough to eat, a bed to sleep in, with his flute and his pipe, he was quite content. He therefore merely remarked in reply "Oh yes, I dare say, but as such luck is not likely to fall in our way, we need not fret about it". The fisherman I remember at the time remarked that the place then belonged to some old gentleman in the City; that the original grottoes of Pope's time were very pretty and interesting and went under the road to the beautiful garden on the other side. Also that he had heard that many Americans, when visiting this country, had called and asked permission to look at the grottoes. However, we put all this romance aside and went on with our fishing.

At about 1 o'clock we felt as if a little bit of something to eat would be useful and, getting a passing boat to land us, leaving the punt, we found a little public house close by. On opening the doors (it being Sunday) and on going down a step into it, we found there was a tiny passage with a bar and two tiny rooms right and left. Asking if they could cook some chops for us, they said they could, but there was a nice joint of roast beef with Yorkshire pudding just being dished up for their own dinner and we were welcome to have the first cut of it if we liked. Of course we liked and thanked them, and I need not say that with a tankard of good ale and a bit of cheese to follow, we made a good and hearty meal. How many times did I pass that little pub afterwards and remember that dinner. We resumed our sport afterwards for an hour or two and then returned homewards. I think during the remainder of my time with the Water Company we made another similar expedition further up the river.

In these reminiscences I have written of a College at Putney for the education of young civil engineers. It was fated that at about that time I should become acquainted with very many of them and see many of them from time to time in later life. First of all there was Robert Christie, and a great friend of his (and also of mine) H. P. Stephenson. Then J. W. Gray, J. Vaudray, Henry Gill and Alfred Williams. At this time I used to meet Christie very often and sometimes stayed the night at his house off the Bow Road, for he had been appointed Engineer to the Commercial Gas Company at Stepney. This was a new company which seems to have prospered better than any other gas company in London; no doubt by working on smaller capital. At that time there was a good deal of uneasiness among the existing companies as a Mr A. Angus Croll started in opposition to the old City Gas Company. He was a well known engineer in the City and was able to form a strong Directorate. He constructed a very large works at Bow Common called the Grand Central. They had a great fight to get their bill through Parliament, there being a strong feeling against the pulling up of the streets. But that was overcome, I remember, by the promise that all the mains should be laid during the night and not in the day, my father having the contract for the work. This gentleman, Mr Angus Croll, became a great and influential authority on all gas subjects and formed and leased many other companies. On one occasion he was also Sheriff of London.

Curiously, at the same time, another opposition company started at Kensal Green called the Western Gas Company. This was no doubt caused by the immense amount of building that had been going on for some time in that district – as I had reason to know, representing the water works. My father also did most of the work for this company. The engineers appointed were, I remember, Mr Robert Harris to the Great Central and R. Pritchard to the Western Gas and it was interesting that I should know them both well a few years later on.

I have already mentioned the interest I used to take in the Committee rooms of the House of Commons whenever there was a good fight on. As these two new companies had to get their Acts of Parliament through, it was a splendid opportunity for the existing companies who were bitterly opposed to them, and what an array of Counsel there was!

Going back to my old friend Robert Monro Christie, better known as Bob Christie, I remember that on one occasion when staying with him, he gave a bachelor's dance. It was rather a tight fit as the rooms were so small. However, he managed it somehow. Mrs Morrison, old Mr Munro's friend at Kennington was there, acting as chaperone and with her two large daughters. Some of the young Coplands were there, and also Christie's sister Jessie Christie. All of these I knew well. Among the few strangers to me was a Miss Gertrude Platt, a daughter of the then Judge Platt and also a Mr Crawford, at that time MP for the Ayr Boroughs, a sort of kindred spirit with Christie.

The curious part of this affair so far as I was concerned was Mrs Platt. For some little time before this Christie had always been talking to me of a young lady, very handsome with money who, he was sure, would make me a good wife. Whenever we met, he was always on the same subject and my reply always was that I had no idea whatever of getting married. In fact, whatever the charms and merits of the young lady might be, it was quite out of the question. However, the young lady, Miss Gertrude Platt, was at this dance and certainly a very handsome girl.

She was about my own age and very dark. In fact, she reminded me very much of some of those Creole girls I used to meet in South America. I danced with her several times and Christie I remember made himself very ridiculous about it. Certainly there was no response on my part to meet his views and, about a month afterwards, to everyone's surprise (certainly to mine) he married the young lady himself. I always had an idea at the time that if he did marry, it would be to Mr Innes's niece, Miss Mynott.

It was at this little dance I first met his great friend H. P. Stephenson, who also became a great friend of mine.

Soon after this, Christie's sister Jessie married a Mr MacBey, a factor, surveyor and estate agent in Elgin, Scotland. Mr father was very friendly with this gentleman, and often when he could get away, he would go and stay with them for a little while. Indeed on one occasion, I remember he took my mother and Mrs Keeble with him. When my brother William left school (I think Tonbridge) he arranged with Mr MacBey to take charge of him for a year or two, and give him some notion of his work. My dear father always had a strong wish that one of his boys might strike out in that direction – poor father!

Still at work for the Water Company, the same thing over and over again, day after day, but always trying to do my best to keep things in order, I should mention one of the Collectors, Raymond, commonly called Billy Raymond, got into some trouble and was obliged to leave and a Mr Terry was appointed. He was the father of Kate and Ellen Terry.

One day, passing along Pall Mall and always being fond of pictures, I popped into the Institute of Water Colours for a few minutes one early morning. I had a look round and was leaving when at the same moment the door opened and the Prince Consort came in with one of his equerries. There was hardly any one in the room and without seemingly take any notice of him I had a good opportunity to see what he was like.

Having heard that my father had at last been able to move the family from Dartmouth Villas to the new home – The Grove – I took an opportunity of going there to congratulate them.

I really was astounded to see the result. The arrangements were most complete – the acme of comfort and an ideal house for any family such as our own. The great feature of the house to my mind was the large hall which went right through, and was quite large enough to dance in in addition to the drawing room. As for the furniture for any little decorative work, it was simply carried throughout with most perfect taste – nothing *outrè* or vulgar, not a sign of it. Of course, I had not been in the house many minutes before my dear mother took me in hand to show me everything; even the pretty bedroom over the vestibule that was to be my bedroom. I recollect with what pride the silk curtains in the drawing room, which were of a lovely amber colour, were shown. But like the Dutch housewives, she always insisted on their being covered up, unless on some great occasion – a dance or a dinner.

I remember being greatly struck with the hall. The walls were covered with a very suitable marble paper and some few large and choice proof engravings were hanging; my father did not then possess many pictures. My dear father did not say much looking round with him, but I could see how pleased and proud he was. I remember also my dear old friend Mrs Keeble was there at the time of my visit and how pleased we both were to meet again.

The old home at Dartmouth Villas when vacated was occupied by Mr Boustred, the Magistrates' Clerk at Greenwich whose family were for many years friends with the Aird family.

Some little time after this, a dance was given, a kind of house warming. This dance was a larger matter than any we had ventured on before, although goodness knows, the former were happy and joyous enough. On this occasion, however, there seemed so many new faces of neighbours and many from London and, of course, several new and pretty faces. One I remember well made a great impression on me at the time. I will not mention names, but I afterwards often thought what a wife that girl would make to any man, who had no objection to married life and who would be glad of a partner to help him through it. It was not only a pleasing face, but there was also a quiet, clever intelligence about her and no doubt some money. However, as I had not the least idea of anything of the kind then and with nothing to keep a wife, such ideas died a natural death.

Somewhere about this time a similar event took place at Mr Quick's; his eldest daughter Lizzie was married to Edward Wilson, son of his next door neighbour. There was a dance on that occasion and they made me sing I remember, as the older people usually did, and my father was generally the ringleader. I can recollect even now singing two songs for them – *Molly Bawn* and *The Bloom is on the Rye*. On such occasions I usually managed to break down and look very foolish, but this time I suppose I was in better form and voice, for the old gentleman, my father, came up to me and with tears in his eyes made himself very ridiculous, I thought. Nearly all Mr and Mrs Quick's friends seemed to come from London or elsewhere, never many neighbours. I can remember a Mr Young, a very wealthy man, and his daughter Nellie - a young lady I always feared to approach - yet she was then a kind of confidential friend of Mr Quick's other daughter Polly. There were also many prominent City people including Lord Mayor Causton and his family. One of this individual's sons I used to meet then is now the MP for Southwark, R. K. Causton and holding some office in the present Liberal government; he did not look like it then.

There was a Baptist chapel by the Blackheath railway station conducted by a well-known and very eloquent preacher, Mr Sherman, who had come from a place in London known as Rowland Hill's Chapel in Blackfriars Road. I remember my old Scotch grandmother going there. Mr Quick and his family always attended Mr Sherman's Chapel and I frequently went with them, when at home.

How well I recollect walking across the Heath those Sunday mornings with young Joe Quick and his sister Lizzie. Polly Quick then was not much more than a child. What happy times those seemed to be. Young Jo was the most humorous, amusing young fellow. It was impossible to be in his company very long before feeling quite ill and exhausted from laughing. I can remember poor Lizzie sometimes on these occasions, when walking home, being obliged to stop and lean against a wall to recover herself. Joe was always so, and the most amusing companion I ever travelled with; but he had his faults – well who has not? I remember him, something of an authority on those matters, taking me to some celebrated shirtmaker in Conduit Street to get a set of embroidered shirts.

There was another dance a little later on I will refer to. I have already I think spoken of a family named Robinson, living near The

Grove. The son Charles was the Secretary of Mr Quick's water company, the Southwark and Vauxhall Company. There was another son, Henry, who was a Commissary General in the army, and a daughter who was then acting as a lady companion to the Countess of Harrington and considered very fascinating.

I can't remember exactly how it came about, but they were invited to a ball given by a Miss Richardson at a large ladies' school in Blackheath and I was invited to accompany them. I suppose they were asked to bring one or two dancing men. When the evening came I went to their house to receive instructions and found Charles Robinson quite ready and he introduced me at once to a very pretty girl, a Miss Martin, to whom he had recently become engaged and who was several years younger than himself. Then Miss Robinson came into the room looking lovely and I felt bound to tell her so. But it was to be a day of surprises to me, for the door again opened and in rushed a gentleman, who I knew well by sight about the Committee Rooms of the House of Commons, as he was the principal clerk of Pritt Venables the Parliamentary Agents. However, I had not known his name or himself personally. He hurried into the room looking somewhat anxious, carrying a very large and beautiful bouquet which he presented to Miss Robinson, whispering some few words. This to me told and meant a great deal. I was not wrong. He was introduced to me as Mr Phillips. Some other young fellow had been asked to make up the party, but at the last moment apologised. We left therefore for the ball. Phillips and Miss Robinson in one carriage, Charles Robinson, Miss Martin and myself in another and a very pleasant and enjoyable evening I spent. Charles Robinson evidently was not much of a dancing man and I therefore got as many dances as I could from Miss Martin for she was a delightful dancer. Whether he felt huffed or annoyed, I do not know, but soon after arriving, without saying anything, he put on his hat and coat and walked home leaving me to take charge of his *fiancée*, which I did with much care and pleasure. Poor girl, she died a year or two after their marriage and he married again, a Miss Besley, sister of Mr Besley, a well known Counsel and, afterwards, judge. The brother, Henry Robinson, I have referred to, I thought was paying my dear sister Agnes a good deal of attention and I did not like it. He was a widower with nothing but his army pay. I never heard whether he proposed or not, but it seemed however, to die a natural death and my father and mother never referred to it.

John Arwell Phillips was the handsomest, best-looking and most gentlemanly-looking man I ever met. He was tall, dark with pretty curly hair; well set up like a soldier and always most carefully dressed and a most pleasant and amiable man. Among my many other close friends, during my life I can only remember one man who could in any way approach him, both in personal appearance and friendship, and that was John Sydney Hargrove, but he was very fair. Curiously both became well-known and intimate with one another through life and from the evening of this dance commenced a close friendship between Phillips and myself. He was not able to get married just then, but some two years later he succeeded in obtaining the appointment of Secretary to the old Chartered Gas Company in the Horseferry Road, Westminster, at a large and suitable salary. I rather think my father had a finger in this pie. Phillips then married Miss Robinson and I am sure never regretted it. They took a pretty little cottage at Weybridge and lived there all their lives.

There was another well-known firm of parliamentary agents in Westminster at that time called Wyatt and Metcalfe. Wyatt became a very prominent man as years went on and was much liked by everyone. I knew him well and my brother John knew him better, but his partner Metcalfe came to grief.

9

A lease of the Kingston Gas Works (1857)

I had been with the Grand Junction Water Company very nearly two years and one morning, I think at the end of the year 1857, I received a note from my father asking me to call and see him at his office in Belvedere Road. I took an early opportunity of doing so, and a future that I never dreamt of was put before me thus.

A gentleman whom he had known some time, a Mr Robert Paulson Spice, who was a well-known consulting gas engineer, and who had appeared before Committees in the House of Commons and arbitration cases, also leased and worked some three or four small gas works at Richmond, Hampton Wick, Staines and Kingston upon Thames, the last of which he had only recently obtained. His object in calling on my father was to ascertain whether he could be disposed to take the Kingston on Thames lease off his hands as he was just then rather pressed financially. Knowing my father had been in early life much mixed up and interested in the gas work, he was willing to part with this lease for 14 years for £2000. Having asked several questions, my father's reply was that he had given up gas making himself and really had as much work otherwise as he could possibly undertake, but his eldest son might like to accept such an opportunity and he would at once communicate with him. Mr Spice remarked that the matter pressed with him and he would be glad of a reply as soon as possible. He had not yet approached the Directors of the company as to his wish to transfer the lease. They were all businessmen, the Chairman being Mr James Simpson, a well-known civil engineer in Great George Street and President of the Institution at that time, and they might object to my father's son unless my father, who was so well-known, became surety and responsible. Thus the matter was left until my father had seen me.

I was simply dumbfounded at such a prospect or possibility for my future, and remember it took some little time to pull myself together, before I was able to say a few words to my father for his kindness in thinking of me and thinking me able and fit to undertake such a position.

The only little knowledge I possessed of gas making was during the time he was at Greenwich and I was only a boy trying to help him. All the same I had no doubt whatever of succeeding if I only had the chance and if the Kingston Gas Company and himself would have faith in me, he should not lose his £2,000.

In further talking the matter over with my father, we came to the conclusion that it would be better if he saw Mr Simpson and mentioned to him what Mr Spice had proposed. He should ask if he thought there would be any objection on the part of the company, as it would be a serious thing for me if I resigned my present situation with the Water Company and found myself left in the cold.

My father therefore called on Mr Simpson at his office at once, knowing him well, having carried out a good deal of work for him – satisfactorily – in Copenhagen and other places. They met. Mr Simpson was somewhat surprised, I believe, and would not say much, but seemed anxious or curious to know if my father's son was a quiet and steady young fellow (I had never met him) and whether he really thought that I was quite fit and able to undertake such a responsibility. He had reason to know, living himself at Kingston, that there were many young men there who were not desirable acquaintances. My father, I believe, assured him smilingly that there would be no danger of that kind. Mr Simpson then said that under the circumstances, the best plan would be for Mr Spice to arrange a meeting at Kingston with the Directors, my father and myself and come to some decision.

Shortly, let me say, this arrangement was carried out. My father and myself attended a meeting called by the Secretary of the company with Mr Spice; the meeting was held at the Secretary's house.

No one knows and no one will ever know what my feelings were on that day. Anyone who has read the previous pages of this book will, I think, admit that I have not been wanting in either pluck or courage in facing the many changes that have seemed to cross my path hitherto, but the ordeal on this occasion seemed almost too much for me. The awful fear and dread of failing to satisfy these six Directors – all shrewd businessmen – that I was able and qualified to represent them.

I do think they were pleased with my manner and appearance,

perhaps looking a quiet, steady, gentlemanly young fellow. At all events there must have been something about me that impressed them, or they would never, after some quiet talk and discussion and many questions to myself, have unanimously agreed and resolved that if my father would be willing to be surety for me, the lease should be transferred to Charles Aird. Let me say now that during the following years, the remaining length of the lease, they never once had occasion to regret it. Never once at our many future meetings on business matters, was there ever an angry word. Their confidence in me in the face of many great trials and difficulties was simply unbounded and I need not say how much this cheered and helped me.

I was quite a stranger to Mr Simpson, the Chairman, but he seemed to take a liking to me at once and during all those years was most kind. My dear father also, was very pleased, proud and gratified at the result. I could not help thinking afterwards that the Directors, in coming to a decision, reasoned that it would be better to have a young man to represent their interests. Moreover, one who would be always on the ground, rather than Mr Spice who would only occasionally visit the works and even then had too many irons in the fire. I feel sure the Chairman was of that opinion.

As it only wanted four or five weeks to the 1st January of the next year (1858) and the lawyers had to do whatever was necessary, it was thought better that Mr Spice should keep the works going on till then, when I would take possession and make a start. In the meantime, I would look about, find some lodgings and study the works and the district generally.

Now, the awkward task of resigning my situation at the Grand Junction Water Works had to be faced and carried out, as it was feared Sir William Clay might take umbrage at my leaving so soon. However, Mr Quick very kindly took him in hand, having always had great influence over him and my resignation was accepted in a very kind manner. I said goodbye to all my colleagues and friends in the office; even poor old Abbott and his men, the turncocks, seemed sorry. What a faithful, straightforward, reliable man he was. So ended that phase of my curious career. A nephew of Mr Quick, Joseph Hadley, was afterwards appointed in my place.

During the interval, having got rid of my rooms in Cambridge Street, I went to The Grove for a few days and received many kind congratulations. I found my father had commenced building a comfortable little house a little lower down the road, to be called Moray Villa which was intended for my brother John and his family. There were some stables also adjoining for the use of my mother to have a carriage and pair. Poor mother, I very much feared it would prove to be a very white elephant to her, for she never had any inclination to be the grand lady. The only friends she had were within a walking distance of her home and I do not think my sisters ever made much use of it when it arrived. I remember some years later on, once driving up to London with her on some particular occasion (I can't remember what). A more uninteresting drive cannot be conceived; through New Cross and along the Kent Road.

I recollect during this visit home I walked to the Crystal Palace to see that very extraordinary man Blondin perform there on the tightrope. He had been performing in America when I was there, walking across the River Niagara on a rope at a stupendous height just below the Falls, but I never had the opportunity to see him. As a rule I have never cared to witness daring feats of this kind, but curiosity I suppose in this instance got the better of me.

The rope, an ordinary ship's cable, was fixed high up in the roof, as high as it could possibly go, from one end of the Grand Transept to the other. There was no netting or protection of any kind should he slip or fall – the man's nerve must have been marvellous. I shall never forget my feelings when he made his appearance high up at one end of the rope with a great long heavy pole in his hands. An awful feeling of sickness came over me; indeed I was as near as possible fainting away. However, I saw him walk from one side to the other on the rope, doing some of his tricks when midway. That sufficed for me and I left. Nothing in the world would have induced me to remain the witness the remainder of the programme such as crossing the rope with a man on his back and wheeling another in a barrow. I believe he also pretended to slip and fall, clutching at the rope to save himself. That would have settled me.

During this visit to The Grove, my father mentioned that he was going over to Berlin again in about a week's time to see how the work was going on there, and if I liked to go with him, he would be very pleased. Of course I assented at once for Berlin then had many attractions for me.

As I was not to take possession of the Gas Works at Kingston for another month, it would suit me very well. However, as I was anxious to have everything in readiness and be quite prepared when I did make a start, I had in the meantime written to Mr Spice's manager to oblige by looking up some suitable lodgings for me near the works. I also wrote to Mr Spice mentioning my intended visit to Berlin with my father and that I would be glad if he could spare a day to meet me at the works and go round them. At the same time he could give me some idea of the office books required that I might order a new set to be ready for my own use. Also when writing I called his attention to the fact that there was no office of any kind at the works. I asked his permission to take possession at once of one of the sheds at present used as a storeroom for meters, service pipes and other things and I would have it turned into a room fit for a human being to sit in all day. He at once assented and I took an early opportunity of getting an estimate for this little job from a local builder who promised to have it all completed by Xmas. This man, James Macrostie, proved a very useful man to me for several years doing little odds and ends of jobs.

It will thus be seen that it was not all clover for me, there being so much to think of and arrange before taking possession. Mr Spice having replied to my letter and named a day to meet me at the works, I returned to Kingston and found the manager, a great big Scotchman named Strachan prepared with a list of some lodgings. Having seen some, I decided on taking a couple of rooms in the Fairfield, a row of small, neat-looking houses with a large open space in front and near to the gas works. This particular house was tenanted by a Quakeress, a spinster lady and one domestic, Rachel, also a spinster and of the same community as her mistress and her very image. Mr Simpson and the other Directors I am sure must have felt quite at ease when they heard I was in such good hands. I remained here twelve months very comfortable and satisfied. The rooms were afterwards taken by a very quiet young fellow, a Mr Penfold, a surveyor in London, whom I had met at the residence of one of the Directors, Mr Mercer, and became very friendly with him. A few years later he was a great authority in many large arbitration cases and ultimately President of the Surveyors' Institute. In appearance he was the most simple minded young fellow one could meet and would often in the evening come in for a chat.

Mr Spice, having arrived at the gas works on the day appointed, we went round them together and made certain comments upon their capacity and condition. The £2000 that had been paid him as a premium I must confess rather stuck in my throat, as everything about was of such cheap and flimsy character, but of course there must be a beginning to everything. He was kind enough to bring all his account books and also the carbonizing books and forms with him that I might take note of what would be required. I must admit they were very complete and of a thorough book-keeping character. Curiously, I recognised several as being the same forms as were used in my father's office when at Greenwich. I took note of them all and at once gave a local stationer (Phillipson) the order and to have a complete set ready without fail by the end of the month.

Having accomplished this, Mr Spice and I went over the district included in the company's Act of Parliament. I was much surprised at its extent and felt I should indeed have my work cut out. I asked him some few questions about his manager, Strachan, and as he had nothing to say against him, I thought perhaps it would be best to go slow and try him for the first year. Mr Spice was then living at Richmond and after our day's work, I remember he very kindly asked me to go home and dine with him.

Before going to Berlin with my father, I will try and give a little idea or sketch of this new field of action on my part. Kingston on Thames at this time was a picturesque little town or rather borough, with its coronation stone by the Market Place, and looked, when I came, as if it had just woken up after many years sleep. Being on the river and so near to many pretty places, Hampton Court, Richmond Park and others, made it very attractive for residents.

About two years before this, Mr Spice, then having a lease of the Richmond, Hampton Wick and the Staines gas works, thought it would be a good speculation to build a gas works here and get a lease of it if possible. This he did and got a lease of it also. He formed a company and an Act of Parliament was obtained. The Directors of the company were: James Simpson Esq. CB – Chairman; W Mercer Esq. – Banker; Dr Cox – Surgeon; S Raymond Esq. – Tallow Merchant; S Gray – Coal Merchant; M Wells – Maltster; J Chalk Esq. – Quaker; W A Fricker – Secretary; Messrs Shrubsole – Bankers. These were the men with whom I would be closely associated during the next 12 years.

Going over the district with Mr Spice, I was astonished at its extent, including New Malden and even as far as that beautiful village, Esher. This was where my brother John and I were sent to an infant school when only six years old, but how changed it was! A new church had been built on the pretty village green, the peacocks had all left the manor house wall and the people who had charge of us had also disappeared. As regards Kingston itself, no gas mains had been laid and no gas yet supplied beyond the town itself, and then only thinly, but I made up my mind to alter all this presently.

At the back of the shops in the Market Place on the north side, the pretty river runs along. The little bit of garden at the back of the two principle inns - The Griffin and The Sun - enabled boating men to land and this continued for about 100 yards to pass along behind the shops in the little High Street. Then all the shops on that side disappeared and the river opens on that side right away, along what was called the Portsmouth Road. A beautiful promenade existed between the river and this road as far as Seething Wells.

After leaving the shops in the High Street, the river side was occupied by a coal wharf, with barges to unload and some malthouses belonging to two of the Directors – Gray and Wells – and a small inn called The Anglers. Here another road turned off to the left from the Portsmouth Road, leading to Surbiton station on the main line of the South Western Railway. As there was no railway then direct to Kingston, a clumsy, dirty old omnibus travelled to and fro. Another road from this one branched off a little further to the left and went on to Surbiton Hill and towards Hook, which promised to open up into a residential district.

Where this road left the Portsmouth Road, a large house stood with large grounds belonging to Christy, the hatter in London, and at the angle of the roads facing the river, there was a small cottage then occupied by Millais the artist. [Much later Millais taught Edith Aird, Charles's daughter, while she was residing in Paris]. After that came the Roman Catholic chapel much frequented by the French royal family from Orleans House at Twickenham. From this point as far as Westfield, where I was told Mr James Simpson resided (a large place with a beautiful garden) there was a large, open estate which had recently been purchased by a speculating London builder, a Mr W Woods. It was a unique site for good, large residences and well laid out for that purpose being quite open to the Portsmouth Road and the river. He had made a spacious wide road in the centre called Palace Road, as Hampton Court Palace could be seen looking across the river with the Home Park and the Long Pond – a really pretty scene. Two other roads of less width were also laid out parallel to the Palace Road called the Anglesea and the Uxbridge Roads. There was another at the back called Maple Road where they had already commenced building. I remember making a note to see this gentleman (W Woods) as soon as possible with reference to a supply of gas for them.

Just beyond Mr Simpson's residence, some few small cottages existed on the Portsmouth Road as far as the Brighton Road. There was a road striking off at right angles, along which on each side, there were several small shops and cottages as far as the South Western Railway bridge. A few yards on there was a road to the left with a few more little shops and then Surbiton railway station.

Where the Brighton Road leaves the Portsmouth Road, there was then a small public house called the Brighton Arms and this particular part was known as Seething Wells. Immediately opposite to Westfield there was an ait or island in the middle of the river upon which a large boathouse had been built and a well-known waterman, Parker, took charge of gentlemen's boats and let out others. A little later on Phil Parker and I became acquainted. A few yards beyond this ait the Promenade by the river terminated and the Portsmouth Road went right on, but with land on both sides and on the river side, it was somewhat narrow as far as Styles Coal Wharf. All this land on both sides for some time past has been in the possession of two of the large water companies in London – the Chelsea and the Lambeth.

It will be remembered that the Southwark and Vauxhall, the

West Middlesex and the Grand Junction Water Companies, yielding to the pressure of the new Board of Health government office, had constructed great works at Hampton beyond Hampton Court for the new intake, my father having the contract. The two great companies now mentioned, of which Mr Simpson was the Engineer, and I believe almost the largest shareholder, decided to follow suit and move higher up the river to Seething Wells. These large works were then in progress, the two companies carrying out the work without a contractor and Mr Simpson's eldest son James, being the Resident Engineer. Thames Ditton village was a little beyond these works and as far as the Kingston Gas Company which then supplied gas in that direction.

These water works in their operations finding, I assume, some difficulty in getting rid of the earth excavated for their filter beds, had deposited it upon a piece of their ground alongside the river and fronting the few cottages before coming to the Brighton Road. This earth formed an enormous and hideous mound which was an eyesore to that pretty part of the river for several years. I little thought when I moved from my lodgings in the Fairfield after a year's tenancy to one of these cottages, that several years later I should be the agent to effect the removal of this very unsightly mound.

At Surbiton Hill building was going on. Also on the London Road to Norbiton, Kingston Hill towards the beautiful district of Coombe.

Gas works are always placed on the lowest and most out of the way situations for, although most certainly not unhealthy, the smell and the fumes, from quenching the hot coke, are very disagreeable to any private houses nearby. I remember children being brought to the works to be cured of whooping cough.

Practically, when I took possession there were very few private consumers of gas, but nearly all shops and other places in the town itself, with a certain number of public houses, were connected.

The Kingston Gas Works were situated at the very opposite side of the town to the Water Works buildings at Seething Wells and about a quarter of a mile below Kingston Bridge. At the foot of the bridge there were two coal wharves; one occupied by James Smith, a lighterman who then had a contract with Mr Spice to lighter the coal from the London

Pool and deliver it into the gas works by cart from his wharf at 4/6 per ton. The other wharf was occupied by Robert Sivers, also a lighterman. The road from the town and these wharves to the works was called the Lower Richmond Road. By the gas works, a narrow strip of meadow (nearly always under water) separated it from the towing path and also from the river. The towing path, I should mention, ended at Kingston Bridge, then crossed over and continued on the Middlesex side as far as Hampton Court Bridge, when it crossed over again.

On the short length of road from the town to the gas works on the gas works side of it, there were two houses. The first was a small cottage in which a man named Knapp then lived. He was the printer, and I think also Editor, of a very well managed local newspaper, the Surrey Comet. The other house, quite close to the works, was a large one walled all round with a high wall and only an ordinary wicket gate off the road to enter it. This place was occupied by a Mr Gibbons who was one of the Clerks of the House of Commons – and a most odd place for such a man to live. During the lease, he was the curse of my life, threatening to bring actions for no end of imaginary complaints, but never did. I remember his son was a splendid oarsman and stroke for several summers in the Kingston Eight.

This lower road went on past the gas works a short distance and then curved round in a lane and joined the Upper Richmond Road. This, turning to the left, would pass along Ham Common, Ham House, through Petersham and into Richmond Park, or along the same road at the foot of the hill. How often I used to take this beautiful walk on a summer evening and listen to the song of the nightingales which then seemed to swarm about there.

The day arrived for my going to Berlin with my father. I remember it was a very cold journey and Berlin itself is always very cold in winter. However, he seemed very anxious to go for some reason. The water works were then nearing completion and I think he had some views regarding the future of my brother Alex.

Germany was then not much more than a third rate power. Its ruler, the King of Prussia, was a nice old gentleman, very fond of champagne. The people all seemed very poor, but thrifty. I remember at that time being much surprised to learn what very poor salaries all officials lived on. However, now the new water works were near completion, it really did look as if they thought seriously of thoroughly washing and realising the good effect and use of a bath in a house.

My father's idea evidently was that a good opportunity existed at this time to create a large business in Berlin. Fitting up houses and indeed even the royal buildings with the new water supply, he proposed establishing in some prominent quarter of the city a place for offices. These would have a room set apart as a show room with a few up-to-date fittings such as baths, lavatories, WCs and some few good English plumbers on hand.

Mr Henry Gill, the resident engineer of the new water company, quite approved of the idea, but for very proper reasons could not be connected with it. (This gentleman, a most agreeable, pleasant man was one of those at Putney College with Christie, H P Stephenson, Gray and others). My father's desire was to give my brother Alex a start and make a fortune, if he would, for it was there to be made. He had been in Berlin during the construction of the water works, had made many friends among the German officials and he also had the experience of being at the Copenhagen Water Works contract, and the name Aird seemed to be then becoming known in most parts of Germany.

My brother Alex was then very young, only about 23. I know my father at that time used to think him the clever one of the family, and so he was. He was a scholar in every sense of the word. Give him a subject and he could write volumes on it in the most perfect prose and he could hold his own with many well known literary men. He was a linguist and could speak both French and German perfectly. He was also an artist and had it been then arranged for him to go in for that alone and study, he would in all probability have been a great success, but it is not the possession of such gifts that always ensures success in this world.

I remember very well at the time thinking what a splendid chance he had in a city like Berlin, compared to the little country town like Kingston on Thames. However, my brother seemed very pleased at the prospect put before him, it was finally arranged and we returned home. My father on his return deputing my old friend John Taylor to look out for a couple of reliable men as plumbers, take them over to Berlin, arrange matters for a start and hope for the best.

I remember spending that Xmas very happily with my father and mother and the others of the family then at home. On that Xmas morning, a real, old-fashioned winter's day, my father with three of his sons, including myself, walked from The Grove to the Crystal Palace and back before our Xmas dinner. On previous occasions my favourite walk on that day was along the Lee Road, past the Tiger's Head inn, to Eltham and back. I was always very fond of a good, long walk.

Xmas over, I went back to Kingston, having arranged to meet Mr Spice at the works on the last day of the year to measure up together what coal and coke was left and prepare a list of all meters, service tubing and fittings that I should be required to take over. This was done and I do not think I met Mr Spice but once again and that was at some dinner.

The first day of the new year arrived and I was in possession and absolutely my own master. I did not lose a moment in getting to work. How well I remember first sitting down in my little office, realising the great responsibility I had undertaken. However, it did not take me long to pull myself together and get things into shape.

The man Macrostie had made a good job of the little office. A fireplace had been fixed in the little recess. On one side there was a cupboard to receive all the various account and carbonizing books and there was a small safe for cash from the sale of coke and gas receipts. There was an ordinary desk in the middle of the room and a common deal bureau facing me for plans and papers. A window and door looked and opened into the yard and as this door was constantly being opened all day long, it was none too warm there during the winter months. A pane of glass in the window was made to work up and down for the issuing of tickets for coke and tar; this I could do when sitting at my desk at work. It may not seem very grand to be thus occupied, but there must be a beginning to all things and I was not proud. Indeed, I had plenty to occupy my mind on first starting with so many things to think of and arrange.

I had already seen the local bankers, Shrubsole and Co., and arranged to keep my bank account with them, and as their London agents were Glyn Mills & Co., it was very convenient. I had also had a long talk with the Secretary of the gas company relative to the *modus operandi* to be gone through each half year between the company and myself as to relative accounts and also as to the half yearly meeting of shareholders.

I had a little circular printed and sent by post to the most important residents, notifying that I had become the lessee and would be glad at any time to give information as to the supply of gas. I had also remembered that, when a boy at the gas works at Greenwich where my father was then manager, seeing a lot of pretty little green paper-covered books or pamphlets of not more than a dozen leaves in his office. This was headed *The Advantages of Gas Lighting* by J O Rutter, Engineer of the Brighton Gas Co. Remembering this, and hearing that the old gentleman was still alive and in harness, I wrote to him asking me to send me a hundred copies. When they arrived I sent them out also among the many non-consumers of gas in the town. It was an attractive little work and would impress a good many people. There was a diagram in it enabling any consumer to read his meter as easily as he would his watch or clock and at any time see what quantity of gas he was consuming.

I had also to see Smith the lighterman who had barged the coal for Mr Spice and make a fresh arrangement with him. I was in hope he would have reduced his price of 4/6 a ton, but could make no impression on the old fellow; he probably thought I was too young. But being anxious to avoid anything like ill feeling with any of the natives, at first I accepted his terms. I remember this was for the first six months until I could look round a bit.

Smith had hardly left my office when another of the natives came in, a Mr Jackson, with whom it was fated I should have considerable dealings for many years in connection with the most important of my coke residuals. Mr Jackson had a very large nursery just outside the town, both for plants and fruit. He had a very large shop in the Market Place for their sale and also had the management of some portion of the Hampton Court gardens, including the great Vine and Pinery, and therefore required a considerable quantity of coke. I was glad to find we soon came to terms. He had a certain cart which, when measured, contained exactly one chaldron. His man would therefore come with the cart, get a ticket from my office, load up and away. This went on for weeks and years with the same man and cart. How well I remember them.

Of course the very first thing I did on taking possession was to say a few words to Strachan, the Foreman, that he might quite understand his duties and also to the men generally who had collected in the yard for the purpose, about a dozen perhaps. I remember that, of the stokers, six were of the name of Wright. Two brothers and two sons of each brother and, barring one of the sons who sometimes was on the drink, they did their work satisfactorily.

It was curious to observe how keenly the travellers of the several and numerous firms and companies who traded in matters and material relative to gas manufacture, took up the scent when any change, such as at Kingston, occurred. In the first two weeks I am sure at least a dozen men must have called representing coal, iron retorts, retort fittings, purifiers, iron coke barrows, stoking tools, iron pipes, tubing and gas meters, wet and dry and also fire clay goods. There was one young fellow among them, a Mr Charles Horsley, who represented a well-known firm, James Oakes and Co. of the Alfreton Iron Works, whom I seemed to take a fancy to. He was about my own age and I remember he went away with the first order I then gave.

There were several applicants, also, for the removal of our tar and ammoniacal liquor, which was really almost a drug at that time and one was almost glad to get it taken away for nothing. But as we got older, we got a little wiser. I remember two firms of manufacturing chemists who came to see me. One was Frank Hills of Deptford, the other was Flockton and Bunning of Rotherhithe. I think I then agreed with Hills to remove the liquor for some fractional amount per gallon. At that time I could not understand how it could possibly pay them to send one of their tank barges all the way to Kingston and fill it with either tar or liquor. If the works had possessed a wharf on the river, it would have been simply pumped in, but as it was, it had to be pumped into barrels and carted to Smith's wharf and then emptied into their barge. This practice went on for some time until it was found out one day that these gentlemen had been coining money by making a most valuable agricultural product out of nothing. I will, if all is well, give an instance of this later on. It is quite certain there are very few men in this world more useful and who make larger fortunes than manufacturing chemists.

After a few days I found it would be desirable to increase my stock of coal. When my friend Mr Innes, the engineer, then of the old Phoenix Gas Co, heard that I had entered into the gas world to get my living, he seemed very pleased. When chatting over the subject of coal, I remembered he had said that the firm of Dixon and Harris on the Coal Exchange were the Phoenix Company's agents or factors. There was also a Mr Stephenson Clarke on the Exchange who did a very large business and whom I should no doubt come in contact with. I had in fact already done so, as one of his travellers made his appearance the first day I entered my office, but I then declined to give him an order.

As I did not wish to follow Mr Spice's haphazard practice of simply writing to a merchant to send in a couple of cargoes, irrespective of price, I determined to make my appearance on the Coal Exchange and see how it worked. Therefore, on the next market day on the Exchange, after as usual going round the works very early and looking over the night shift work report, I then gave the outside men their instructions for the day, booked their previous days work and saw all was in order. I left the man Strachan in charge, ran into the market place, jumped into the old omnibus waiting outside the Griffin Hotel and then to Surbiton station to catch the early train to Waterloo. It was then only a short walk over London Bridge to the Exchange, just behind the Monument.

This was a large building with a fine circular hall and a large gallery round above with offices, a staircase from the arena below leading to them. In the arena there were certain tall stands or desks with the factor or merchant standing alongside waiting for a customer. There were also a number of lightermen and captains of the colliers about, both inside and outside the building, waiting for orders. (There were no large screw steam colliers then, bringing coals to the market, as now). Seeing these old Geordies again hanging about in their shabby old top hats (they all wore them) reminded me at once of a time at Greenwich. I was with my old friend Captain Griggs, with whom I now and then used to lunch on board his ship, and I remembered the sailors' fare of salt junk and hard ships' biscuit.

Having found Dixon and Harris's stand, I introduced myself. Dixon was a very pleasant little man, but I remember I did not take to his partner as he was somewhat stand-offish. I mentioned that Mr Innes of the Phoenix Gas Company would tell them all about me if they desired to make any enquiries. After a little chat, I found they could supply me with a couple of cargoes of New Pelton, now lying in the Pool, at a moderate price. I promised to give them a reply in half an hour and not seeing any other desk representing Stephenson Clarke, I asked Smith the lighterman where he was to be found. Being told his office was in the lane just outside, I told Smith to wait for instructions and ran round to them. I found myself in a place not unlike my old quarters in Walbrook, quite as dirty, only very much larger with a constant stream of ships' captains going up and down stairs. The clerk's office was the front room on the first floor with Mr S Clarke's private room leading from it at the back. Having asked to see Mr Clarke and given my name, I was shown into his room at once and found myself face to face with the individual I was to know so well later on. He was seated at a desk in the middle of the room, smoking a cigar, the aroma of which, and noting the brand on the box, which was lying open on his desk, proved that he indulged in the very choicest.

He was a sharp, clever, active-looking little man, not more than two or three and thirty with black piercing eyes, but his manner was most extraordinary. I had hardly entered the room, the door of the clerk's room being still open, when he shouted and roared at me as if I had committed some crime. I cannot now remember the exact words, but it seemed to me very offensive, intimating that I had acted very meanly in sending his representative back without giving him the order as he expected, in the same manner as Mr Spice did. I do not know how I looked at this sort of reception, but I remember very well now I felt. However, I managed to keep my temper and simply replied that I felt sure his representative would acquit me of any discourtesy when he called. I noticed that the latter was sitting at a desk in the same room at the time and knew him well afterwards as Johnny Goldesham. I had thanked him and said, this being my first transaction with regard to coal, I preferred to go to headquarters, look round and try and feel my way. "You must understand, Mr Clarke, that with the help of my kind father, the lease of these works has been placed in my hands to get my living and do the best I can for myself and which I hope to do. I cannot therefore afford to purchase my coal as Mr

Spice seems to have done." When I had spoken somewhat thus, he seemed to think he had made a mistake and suggested that I might try a glass of sherry and a cigar. I at once declined and remarked that I was anxious to catch the 2 o'clock train back if he could kindly offer a price for a couple of cargoes. He shouted out to the clerks in the out office for information and then stated one, but as it was above that of Dixon and Harris I was obliged to decline it and left, running back to the Exchange. Although it seems to occupy a long time to write all this, I really had not been away more than twenty minutes.

I accepted Dixon and Harris's offer for two cargoes of New Pelton, gave the order for them to Smith who had been waiting and hastened back to Waterloo. I found I could spare a few moments *en route* to run into Pimm's old place in Cheapside and swallow at the counter half a dozen oysters and a small glass of stout at a cost then of one halfpenny each. They were not like the tasteless sickly looking little oysters we get now at a cost of two pence each.

I caught the 2 o'clock train and was back in my little office by 3.00. Then I started Strachan out to look after the service layers and see that all was going on well with them. I was indeed glad to find that the little book I had sent out – *The Advantages of Gas* – was doing wonders and the applications for gas to be laid on were numerous. I was even obliged to take on two more men to keep up with the demand. But what gratified me more than anything was to find from the carbonizing returns that we were doing so much better per ton of coal than my predecessor had. So much, I thought for being always on the spot.

With all this to encourage me at starting, I nevertheless had an anxiety on my mind that was worrying me a good deal. When walking round the works one day just before taking possession, I noticed a split or crack in the large chimney shaft of the retort house about half way up. When measuring up the coal stores I had to take over with Mr Spice, I called his attention to it, but he seemed to think it of no importance and felt sure it would not get worse. However, when actually in charge, I had three little screeds of strong cement placed across the crack to see if it was to extend. To my sorrow, about a fortnight after I found they had all broken.

When Strachan had left to look after the service layers, I remained to issue the tickets for coke and continued my book keeping which I determined should always be in perfect order. In fact for many years afterwards it got larger and larger: I took great pride in it.

At about 5 o'clock I started the lamplighters, with their little ladders and lanterns as it was dark then at about half past. After going on into the retort house to see the stokers draw the 6 o'clock charge, I went home to my quiet Quaker lodgings in the Fairfield and my little bit of roast mutton, milk pudding and glass of beer and cigar. I then went to bed feeling I had done a pretty fair day's work.

Thus the days went on; always at the works as soon as daylight and home to my rooms when dark. By the by, my landlady did not like gas and I had great difficulty in getting her consent to my having a burner in my sitting room. This was a little bit of a check on the works, for by that time, by looking at the burner, I could very well judge whether the gas was of proper quality and power.

When Shrove Tuesday came along, I recollect I was astonished to find every shop and house with its shutters up and where there were no shutters, all windows or openings were boarded over. In the middle of the day a great crowd of roughs assembled in the Market Place and the mayor of the borough, in accordance with the very ancient custom, gave the first kick to the football. The town was then entirely given up to the rough horseplay of these men for the rest of the day. Of course all quiet people were *non est*. This outrageous scene went on year after year all my time, and then was put an end to.

During the several weeks since I had taken possession of the works, I could but feel grateful to the directors of the company for keeping away and not coming to worry me. It gave me so much more confidence in myself and showed also, I thought, that they were satisfied. Sometimes I would meet one or two in the town and just shake hands in a friendly way, but after a couple of months had passed, one morning when at my desk writing, a gentleman passed by the window, tapped on the door with his walking stick and came in. It was Mr Mercer, the Deputy Chairman of the company who was a tall, fine old man with white hair and a most perfect gentleman.

I, of course, immediately got up and placed a chair for him, saying a few appropriate words. When seated, he remarked that he had been a little curious to know what kind of a start I had made in my new venture, but had thought it better to give me a little time to settle down before calling. He also said that as he had become an idle man, having recently sold the large flour mill he had here, since he left Maidstone, to a Mr Leonard and his son (very nice and upright men, whom he hoped I should become acquainted with) he always took a little walk round about in the morning. This morning, it being in this direction, he thought he might venture to look in. I reciprocated his kindness at once and felt justified in being quite open with him, mentioning the little books I had sent out and the very satisfactory response so far. Also, our greatly increased make of gas, which looked well for the future. I remember mentioning also my present anxiety at the condition of the retort house chimney shaft.

While chatting on these and other matters, I could see he was much interested and amused to see me, while talking, at the same time reply to the taps at the little window made by hawkers and others for coke tickets, write them out and put the cash in the safe. Smilingly, he remarked that I seemed to do all the work myself. Yes, I said, practically all that is possible for me to do. I have never been afraid of work and indeed it is my happiness. How well I remember the old gentleman saying it pleased him to hear me say that, as they were precisely his own feelings when a young man. He did not stay very long, but left I think very pleased with his visit and thus commenced a long and sincere friendship between two men for something like ten or twelve years. When he left, I walked outside with him to the entrance gates and pointed out the trouble in the chimney shaft and explained what I proposed to do should it get worse.

It would be impossible to find any gas works in the kingdom free from complaints from consumers. Generally, that they did not get enough gas or not sufficient pressure. From the first, I immediately took these in hand myself to investigate and quite as I expected, found the cause to be purely local. The fittings were very unfit for the purpose intended and in some cases a 5 light meter had been supplied for twelve burners, evidently the work of some local ironmonger.

Having at last, after much argument and a wonderful amount of patience, satisfied these people that the fault was entirely with themselves, and there could be no improvement until the defects, which I had carefully pointed out, were altered, I at once made up my mind that something must be done about this in the future. Things would never prosper with me in this way and I must exercise some supervision over the pipes and the fittings of the new buildings being erected before any gas was laid on. This was precisely the same way, as it was my duty to do, when with the Grand Junction Water Company.

Curiously, the worst instance among these complaints was at one of our largest consumers at Shrubsoles, attached to their bank in the Market Place. They had a large linen drapery business, a large rambling place with any number of work rooms and show rooms added on from time to time.

When at the bank one day, John Shrubsole mentioned the matter, suggesting as usual there was not sufficient pressure and would I kindly see what could be done. Yes, certainly was my reply and the very first thing is to prove to you that it is not as you suppose from want of pressure, but from inside.

The next morning by appointment, I met him again with one of my men and a pressure gauge and very soon satisfied him on that point. On going through the many rooms, taking the pressure in each one, he could but see the real cause of the trouble, and I could but say to him as I had said to the others, there can be no improvement until it is altered. He was then very anxious that I would take the matter in hand and do what was necessary. However, as I was so busy and with so much to do getting things into shape, I tried to persuade him to get a gas fitter from London, as there was no one here who could tackle such a job. But it was no use, I had to give way to his wishes and what job it proved to be!!

For several weeks the place was pulled to pieces, but the rooms were full of workwomen all the time and the business went on all the same. Mr Shrubsole and myself, all this miserable time, were like two sulky bears who had their fur ruffled up the wrong way. There was also a Mrs John Shrubsole in the background who was considered the *grande dame* of Kingston on Thames, which did not tend to smooth matters over. However, the work at last came to an end, as everything will do with patience, and they were evidently satisfied. I had previously told Mr Shrubsole that under the circumstances I should ask him to pay the exact cost of the work only and not one farthing for profit.

Unfortunately, after having so much trouble and so much worry in inducing these consumers to mend their ways, or rather their fittings and they having then expressed their satisfaction, when the next quarter's gas account was sent in as a matter of course after curing the defects, it was found to be much increased. I myself was much abused; such is human nature, but I put my feelings aside and still plodded on in spite of it all.

Anyone reading, or trying to read these lines would no doubt consider them very prosy and uninteresting. However, to the writer it is a pleasure to note down and try and express what his feelings really were on those matters and at that time, more than half a century ago. Also, now and then, to try and take his attention away from his current physical suffering.

Having, thank goodness, finished the work for Mr Shrubsole, I was much gratified shortly afterwards in receiving an application from the Board of Guardians for gas to be laid on to the workhouse, a large building on the high ground at Norbiton. At that time, the gas main did not extend so far, there being no other buildings to supply en route. This was certainly a matter to be considered and I therefore replied suggesting that I should meet some of the Board or their clerk, on the ground.

An appointment was made and I met a committee with the clerk and went over the building with them, noting at the time the number of lights they proposed which I was glad to find would justify the main being extended. I was also asked whether it would answer my purpose to carry out all the necessary work of the internal fittings and if so, to submit an estimate for this. I expressed my willingness to do so, for this would be a very different sort of job to that of the Shrubsoles. Here I should be my own master and it was new work with the possibility of making some profit out of it. I would not be making good mistakes and defects under the most adverse circumstances.

I went again to the workhouse and took the necessary measurements for the estimate and also the length of new gas main required. The committee in the meantime selected from an illustrated pattern book that I submitted to them of suitable fittings for such a building. The estimate took me some time to prepare, for I did it very carefully and to my great gratification, it was accepted; it was indeed a big order. Under the circumstances, I thought it would be better to make the Directors acquainted with the matter, particularly as it would be necessary to spend some further capital by extending the main. I came to the conclusion, at the same time, that it would be well to bring to their notice the condition of the chimney shaft.

I therefore wrote to the Secretary of the company, Mr Fricker, enclosing a short report upon both matters, and saying that I would be very pleased to attend any Board meeting he might think it desirable to call. As regards the chimney shaft, I proposed to make it temporarily secure by placing (the shaft being of octagonal form) three stout hard wood planks round the fracture with three strong, iron clips round them, securely bolted, to keep them in place. I cannot now remember the amount I estimated as the cost, but it was not very large.

The Secretary had thought it desirable to call a meeting and I was invited to attend, which I did with some little nervousness as it was our first, but no meeting could have been more pleasant. I remember being asked to give my impressions as to the future, which I did fully, saying that if I succeeded in increasing the number of consumers at the rate I was now doing, the rental would be doubled at the end of the year. Both my little estimates were approved, Mr Simpson saying he would drive round in the morning and have a look at the shaft before going up to town. This he did, simply remarking that nothing better could be done.

After all this, and realising that gas always goes to the highest ground (the workhouse was situated at that point in our district), I thought it would be desirable to take the pressure at various points throughout the night on the mains. I would regulate it, or check it to see if it disposed to rob the lower part. Therefore, for several nights Strachan, a lamplighter with his ladder and lantern, and myself, could be seen at all hours of the night prowling about. The result was that when the additional main was laid to the workhouse, it was found necessary to put in a valve on the lower ground. All this survey proved of great use later on.

Although this outside work required a good deal of my attention,

I still managed to watch and study many subjects on the works regarding the carbonization. Later on therefore, I found myself being spoken of in London as likely to be a useful man in gas and in my being invited a few years later to become a Director of one or two other gas companies. I need not say at that time that these proved very useful in the way of income.

Having thus suddenly jumped into a kind of gasfitting business, I found that although it seemed to increase my work so much, I seemed to revel in it. My great desire at that time was to get things in some sort of shape and organisation, increase the consumption of gas and of all things to prevent any recurrence of faulty fittings and consequently complaints. What a number of interviews I had with Woods, the speculating builder, about this, and in my anxiety I was fool enough to do some of the work for him; I remember well the difficulty I had to get payment. However, having engaged another man, a first rate gas fitter, and started him at the workhouse, I had already laid the additional length of main. With regard to the chimney shaft of the retort house, I had accepted the estimate of a firm of ironfounders close by (Priest and Woolnough) for its temporary repair.

I could now, therefore, give my attention to a matter of some importance. The first quarter of my first year (Lady Day) had now arrived. Up to this time it had been all money going out and none coming in, except by the retail sale of coke at the window. But now it was necessary for the index of all the consumers' meters to be read and the quarter's accounts to be made out and delivered. I remember so well how curious and anxious I felt as to what it might amount to. As I knew it was a very dirty job, groping about in filthy cellars, I decided to leave this duty to Strachan and I would afterwards prepare and write out the accounts and assist him to deliver them and collect.

There was really now so much for me to do and look after at that time, that when I look back, it is a marvel how on earth I got through it all. The general book keeping in itself was quite enough for any one man. The gas accounts I made out at my lodgings, and sat up half the nights to do them, so that I could also give some attention to the gas fitting then in hand and other pressing matters on the works. Somehow or other, I always seemed to be trying to cram 24 hours work into 12, but, with it all, I loved my Sundays.

Although still winter, every Sunday, weather permitting, I would walk alongside the river to the service in the Chapel at Hampton Court Palace. It is curious the great affection I always had for that place and its neighbourhood from days of childhood. It was a beautiful walk, even in winter, but in the spring and summer there are no words for it. The Chapel was a narrow building with a solemn quiet beauty about it in the way of its rich decoration, as one would imagine it to be in Cardinal Wolsey's time. On each side were two rows of very solid, richly carved, raised oaken benches or pews for the use of the lady pensioners living there. Among them I can remember very many handsome aristocratic-looking women. On the floor, between these side benches were about a dozen rows of plain, oaken forms for the use of the domestics. When they were seated, the Verger would, if there was then room, give a seat on these forms only to any stranger standing waiting in the entrance hall. As the beautiful organ began to play, a squadron of cavalry soldiers would march in, their swords rattling on the stone floor; there they were, seated in the entrance hall. A cavalry regiment was always stationed at the barracks outside. The morning service being beautifully read by the Chaplain, the Revd. James Bailey, and a short sermon given, the little congregation left and passed through the several quaint courts and cloisters into the lovely gardens and grounds. Sometimes I would wander through the rooms above and again and again admire the very beautiful wood carving of Grinling Gibbons.

If winter, I would walk home on the Middlesex side of the river by the Hampton Court Road and over Kingston Bridge, quite ready for my cold dinner at 2 o'clock. But when spring or summer came, I would wander about the lovely gardens and look at the goldfish. When disposed for a good, long walk, (for I could manage any distance in those days) I would cross the road from the chief entrance to the palace and grounds by the well-known maze to the entrance into Bushey Park, walk through the park to the other end in Teddington and then home. After a few hours rest and a cup of tea, I would go for another stroll in the other direction over Ham Common to Petersham and come back through Richmond Park, it being then quite evening. The song of the nightingales was almost too much. This was my Sunday's rest and happy enjoyment for many years to come.

Having thus made out and delivered all the accounts, money thank goodness began to come in and to my great delight the total amount of that quarter was simply splendid. There were also other accounts to be made out: for the public lamps against the corporation; for the lamps on Kingston Bridge; against the Bridge Estates for all the public lamps at Surbiton and against the Surbiton commissioners. There were also accounts to be made out for the quarter for coke, tar and ammoniacal liquor. All these, when added to the retail sale of coke each day, mounted up and, I remember, made me look very cheerful as to the future.

When all the accounts had been delivered, I sometimes relieved Strachan in the collecting. As I have said before, I was not proud and indeed it was an opportunity to introduce myself to many consumers where I thought it might be useful.

Many preferred to come to the office and pay the rate and curiously very many of these were ladies from the Surbiton district. Surbiton Hill and all that part was being largely built over and the houses were immediately occupied by newly-married couples. Many were young stockbrokers, the boating on the river being a great attraction to them. So many of these ladies came to my office to pay their gas rate that I got a little bit bothered. If they left, having obtained a receipt, all well and good, but they seemed to forget how much they were interfering, by their many questions, with other work I had to do. However, after all it really was my own fault, for when I became the lessee I had fixed on a little table in the room a very pretty glass lamp I had seen at Osler's in Oxford Street, combining a very pretty light and also heat. Of course this, and some one or two other things, attracted attention and rather trespassed on my time. Nevertheless, I had to get a number of them and have them fitted.

Even my old friend Mr Mercer speculated in one. Every two or three weeks I would hear his friendly tap on the door with his stick, but always he avoided trespassing on my time if busy. On one occasion I remember he called to ask me to dinner, as Mrs Mercer wished to meet me and he would at the same time ask his friend Mr Penfold.

This winter quarter being got through, I now had to turn my attention to the very important question of wear and tear. This is to renew the setting of the cast iron retorts (which only last one year's work) and the resetting of the furnaces; this meaning the purchase of new iron retorts and a quantity of fire clay material for I had decided that we were not yet making enough gas to justify the large outlay of new capital, necessary in the use of clay retorts.

I had not been quite satisfied this past quarter with the results from the furnace settings and decided to engage another retort setter and see if we could not do better. There were several expert men who advertised themselves in the then well-known journal of gas lighting. I had the good luck to meet with one, who seemed to meet and agree with my views on the subject, for it was indeed an important one. I remember at the time having before me a large drawing of a retort I had made for my father when I was a boy at Greenwich, only 14 or 15 years old. I have it in my possession now, dated 1846. It was useful even then to assist us in our consultation. I engaged him and never had reason to regret it.

Occasionally, I would run up to Town for an hour or two on business matters. Sometimes to the Coal Exchange and take note of prices, or to the people at that time (David Hulett & Co.) in High Holborn, who were supplying me with meters, gas tubing and other material we required. Passing along, I would occasionally look in at 76 King William Street in the City where R M Christie then had offices and was agent to several companies. One was the Horsley Iron Co., a well-known company in the gas world, Mr Broad being the Chairman and often to be met there. Years later, I shall have occasion to refer to this gentleman again, a real good man.

Alfred Williams, who was with Christie at Putney College, was there, but I never could understand whether as Clerk or Partner. Poor Alfred, I am afraid he was on the 'down grade'. He had just started to build a large house for himself at Chislehurst, which he never could occupy.

If time, I would also call on H P Stephenson at this offices just by in the Poultry. Here there was also a gentleman in charge, a Mr Robert King (brother of Dr King at Eltham). Robert King became a great friend of our family; there was never a dance at The Grove without King, and I remember my brother John on these occasions used to take great delight in chaffing him, but he was a good-natured fellow and always took it pleasantly.

Christie and Stephenson were intimate friends of Stephy Clarke and were greatly amused when I told them of my first interview with him. My father being always such a good-natured man, much too good-natured in this respect, was induced to act as a steward on many of the charitable societies and would send tickets to several friends and frequently to myself. I never much cared for these Festival dinners, although held at such places as the London, the Freemasons, or the Albion Taverns, but I always liked the singing after, it being the custom to engage some professionals to sing between the speeches.

At one of these charitable dinners I remember a man named Himes, who had a lovely tenor voice, sang a song that quite took possession of me. I got a copy of it the following day and made it my favourite, for the words and the melody seemed to me very pretty. I cannot now, after all these many years remember the name, but curiously I can remember the words of the first verse, which recurred to me a short time ago in a very strange manner and which I will try and relate at the proper time.

"Look always on the sunny side
'Tis wise and better far
And safer through life's cares to glide
Beneath Hope's beaming star."

My father would also, now and then, give a dinner himself to a few business friends, perhaps as many as forty or fifty guests and generally at the Trafalgar or the Ship Tavern at Greenwich. For Auld Lang Syne these little dinners were always enjoyable and real, not by contract as the Festival dinners were necessarily obliged to be.

Then there were Stephy Clarke's sumptuous dinners also to be deal with, once or twice in the year. These dinners were something remarkable and about 150 guests in number, all business friends or clients, generally at one of the large City taverns, or if during the summer months, at the Artichoke Tavern at Blackwall, a favourite place of his. It may be remembered that I met my father, when I was quite a boy, with the clerk John Taylor in a boat, to bring them home to Greenwich from that tavern. How the world has moved on since then.

The menu of these dinners was composed of every luxury that could be conceived. The wines, liquers and spirits always of the finest vintages and cigars the choicest brands.

At first I always declined the invitation, but later on, as we got to know each other better and becoming acquainted with his brothers Harvey, Robert and Henry, all very good fellows and a partner, Tamburini, who had charge then of the foreign department, I started to attend.

Robert Clarke seemed to have the control and arrangement of these gigantic feasts. The cost must have been enormous, but it was evidently amply returned. Although so young a man, he had built up a business that seemed marvellous. I do not think there was a single gas, water or railway company in the kingdom, or even in many of the large foreign cities, who were not his clients or customers. Anyone, then travelling north by our great railways, would see any number of great railway wagons loaded with coal standing in the several sidings, with Stephenson Clarke painted on them, in the largest of letters.

Stephy married about this time and had, I believe, 13 children. It might perhaps be interesting to mention now that when Stephy Clarke died some years ago, he left his widow and large family with a very large fortune in a beautiful estate at Worth in Sussex. The family have all grown up now and only some few months ago I observed in the fashionable part of the newspapers that his second son is engaged to marry the eldest daughter of Lord Henry Grosvenor, brother of the last Duke of Westminster. How the world moves on!! Since Stephy's death his huge business has, I believe, been most successfully carried on by Mr Kent who was, when I first knew Stephy, one of the clerks in his outer office.

Stephy was a great sportsman and his presents of game were numerous. Some few years later Robert Clarke became a director of a gas company on which I also had a seat and we became very friendly. At any of these dinners when I was present, he would always induce me to sit at the table with him, having some magnums in reserve by his side. But such kind attention was quite lost on me as I never cared for wine, particularly champagne.

This seems to have been a long digression but the half year was just near with, to me, its heavy responsibilities. I was to be called upon presently by the Secretary of the company for my cheque for this first half year on the capital expended up to this time. I had in the meantime to prepare my own account against the company for new mains laid for the workhouse and for additional public lamps, new services to new consumers and the number of meters sent out. All this was a rather formidable matter to prepare, requiring much of my time and attention as the Secretary had to measure it all up and check it before being put before the Directors. The quarter's accounts at the same time had to be prepared and sent out to get in as much money as I could, for it was terrible hard work, I remember, to finance that first year safely.

There was also a report to be drawn up for the half yearly meeting of the shareholders (not very many at that time) which was always held every half year at the Court House, where the Assizes were held. All this was a very anxious time for me, but thank goodness I managed it all successfully, and no one knows how glad I felt when it was over and I could get a little time to breathe and rest, before entering into another winter campaign. Speaking of the Assizes, when on, it was very interesting to see the many Counsel in their wigs and gowns running about the Market Place.

It being now summer, and the river looking lovely, although there was plenty of work getting things in order for next winter, I could not help thinking how nice it would be to have a little boating exercise on the river in the evening. The idea grew, and one beautiful evening, after my usual dinner tea, I walked quietly down to Parker's boathouse at Seething Wells, hired a boat and pulled steadily up to Hampton Court and back. It really was wonderful how naturally the art, or trick of rowing came to me, never having been alone in a boat before. There was no clumsy splashing of water and the sculls were used and feathered as if by a waterman.

This unexpected success greatly pleased me and throughout that summer, weather permitting, I never failed in the evening to go out again about twice or three times a week, and there can be no doubt the exercise and the pleasure did me a lot of good. The fishing at Kingston then was not very good, the river being so much disturbed by the operations of the water works dredging for ballast. Still, there was always a punt ready at the Anglers' Inn, for an hour or two on Sunday if so disposed.

Things seemed to move on week after week, month after month, always busy. The chimney shaft had been strengthened. It, of course, looked very unsightly, but would last long enough. The work at the workhouse was completed, and it turned out very satisfactorily all round. The meter for it was so large, I remember, that it had to be specially made.

Although summer, new consumers still came and new mains had to be laid for additional public lamps. These lamps I always supplied and fitted and charged to the corporation. All public lamps then were charged at so much per lamp for gas supplied from sunset to sunrise - to the Kingston Corporation, the Surbiton Commissioners and the Bridge Estates. I think I have already mentioned that the Surbiton Commissioners were independent of the Corporation, and were elected from the principal residents in Surbiton – all gentlemen. I shall never forget their clerk, James Bell, a young man and a solicitor in the City. He was most kind to me throughout my lease in many ways, although we were not intimate friends. His son, young James Bell, later on became a prominent member, with the son of my neighbour by the gas works, Mr Gibbons, of the well-known Kingston Rowing Club. I am almost sure this same son is now, at the present time, in the position of the City Solicitor.

Among the many consumers coming in I can remember Mr Dunnage, then the Chief of Cubitt and Co. the great builders in London, and who had just finished building himself a house here. There was also a large house, Norbiton Hall, not far from the workhouse, then occupied by Admiral Sir George Lambert. What a splendid old man he was and how well known on the Bench, and indeed everywhere. His handsome son, Captain Lambert RN stayed with him when on leave.

I had thought it better not to compete with any of the so called local gas fitters, not wishing to create any ill feeling among the trades people, but simply exercise some sort of supervision over their work. If any new consumer, however, expressed a wish that I should undertake this work, I of course did so.

A gas works is, of all places, the most difficult to keep in decent order; but all through my life it has always been my strong desire to have all things always so. Outsiders may say – Waste of time – but I have never thought so. Surely it is better to work in a system of method and order, than in muddle and dirt. When I first came, I remember giving orders that the road approach from the entrance gates to my office in the yard should always be swept and in order every morning before I came, and the men very soon knew that when I said a thing I meant it.

One morning, sometime in October I think, I was busy with my account books, when two people passed by the window of my room and tapped at the door, seeing 'Office' on it. They came in and to my great astonishment, who on earth should they be but my dear father and our dear old friend, Mr Munro.

It took me some little time to find words to receive them properly, but once started, what happiness to describe and tell them everything I thought might be of interest. I got out all my books and showed them, all pasted up as good luck had it, and even a little profit and loss account for my first half year, or as near as I could get it, and spoke also of the cheerful future I then thought before me. I could see they were both greatly pleased and interested; and like my old friend Mr Mercer, amused at my giving out the coke tickets at the same time.

We then went round the works. My new retort setter was just then finishing the resetting of the new iron retorts and furnaces and I was very pleased, I remember, to hear them say I was right in not going for clay retorts just yet. It then being time for them to have a little bit of lunch before returning to London, I took them round to the old Griffin Hotel in the Market Place, where they had a good dish of hot mutton chops ready for us. I knew to take them round to my poor Quaker lodgings would be a very cold sort of comfort.

During luncheon, I was greatly grieved to hear from my father that he had met with a terrible disappointment as regards my brother, Alex. After all the trouble taken and all the arrangements made for him to undertake and make a great and profitable business in Berlin, instead of pulling himself together and setting to work, he invented some paltry excuse to come home. He then idled about and told my father and mother that he had, some time ago, made acquaintance with some people named Keith. They were tenants of my father, living in Morden Terrace and with whom we were not then acquainted, but that he was about to marry a younger daughter, Janet, at once, crippling himself at his start. This must have been an awful blow to my father. Everything in the past had gone so well and successfully with him, both in business and at home, and now the first serious trouble had arisen. I know it hurt me very much to see how deeply he then felt it. However, after lunch, as they were both anxious to get back, I saw them into the old omnibus for the Surbiton railway station, with a few kind parting words, and returned to my work. Dear old Mr Munro, I never met him again.

I could not, however, shake off the news about my brother. I remember when last in Berlin with my father, I thought he was incurring great risk and that Alex, clever as he was, could never be a practical businessman. It was not in him and so it ultimately proved; but I must not throw stones. He married almost at once and returned to Berlin with his bride. I really do not know whom of the Aird family were present, I only knew I could not possibly be there.

The days were now getting shorter and the evenings too chilly for boating in the river, so I did a lot of work in the evening at my lodgings. I also thought a good deal of the future, for my poor brain never seemed to rest, always rushing along. The idea in my mind just then was that, as it was quite certain to all my friends and certainly to myself that I would never marry, why should I not, as things ahead looked cheerful, take a little cottage somewhere near to the river and make a little bachelor home of it. My lodgings were comfortable and quiet, but very dull. Therefore, when I had an opportunity, I looked about the district.

In describing the district I had to supply with gas, I have mentioned the Portsmouth Road running alongside the river, from the High Street in Kingston to Seething Wells where the two large water works were being constructed, with the pretty promenade between the river and road. This came to a dead stop by a piece of land upon which the two companies, in excavating for the foundations of the engine house buildings and several large filter beds, had piled up a huge mountain of surplus earth. This was an awful eyesore and shut out all view of the river or anything else beyond it from three pairs of semi detached cottages on the road between Westfield (Mr Simpson's residence) and the Brighton Road, causing the rent of them to be merely nominal. The end cottage, by the Brighton Road, happened to be to let and I went over it. It was a very common place, but quite good enough for me, and as I should only have to cross the road to Parker's boat house, and as the rent asked suited my pocket, I ventured to take it on a yearly tenancy from Xmas next.

It contained eight rooms, a tiny bit of garden in front with a narrow little piece at the back, and a little path from the gate to the front door. Inside there was no hall and a little narrow stair. There were two little rooms to the right and left and two behind; one a kitchen and the other a room for the servant. There were also a couple of little places tacked on behind for scullery and WC. On the floor above were four little bedrooms. It was all in very fair order, but I got my handyman James Macrostie to do the little bit of papering and painting required to suit my ideas.

When my dear mother heard of my intention, she begged that she might lend a hand in the equipment of my new little home for me, such as house linen, china and pots and pans, as she always seemed to revel in anything in this way. I gladly accepted her kind help. Now and then, I picked up many little bits of furniture second hand and by Xmas it was got into some sort of shape for entry. However, I little knew or dreamt then of the Xmas I should have and the fight through that year.

It was now November and the Mayor for the Borough was elected. I think it was Mr John Shrubsole for the second time. I, of course, received an invitation to the usual banquet always held at the Griffon Hotel and although old Williams, mine host, did his very best, it never seemed to me a success. The rooms being so small and with so many guests, the atmosphere was rather trying. However, most of the great people in the county had accepted – the Lord Lieutenant, the High Sheriff, Sir G Lambert and other magistrates and many others. I accepted simply for political motives and to keep on friendly terms with my consumers. Mr Shrubsole was kind enough, I remember, to ask me to take a seat at the upper table, but I did not accept, preferring to sit with the other guests.

Winter was now coming on and making me begin to feel very anxious, for in my calculations I had assumed it would in all probability be somewhat like the last. But I made a grave mistake, for it happened to be the most severe winter experienced for many years past. Intense frost, the river covered with ice and with dense fog and darkness day after day. The consumption of gas was extraordinary: as much went out in the day as in the night. The gas holders could not rise in spite of all our efforts and during the two or three weeks at Xmas, it was simply a matter of life or death; break down or get through it. The shops all kept open so late and the Xmas festivities went on at the same time. Fortunately, I had in stock, in case of emergency, a few tons of that useful, but costly, Cannel coal. Without that, we must have collapsed all together. The Directors were all as anxious as I was during such weather, but I could see they one and all felt that I was doing everything a man could possibly do under such circumstances, to avoid breaking down.

I really almost seemed to live, day and night, in the Retort House during that anxious time, to cheer up the stokers. They worked like men possessed and were not forgotten afterwards. When drawing the charges every five or six hours from the retorts, removing the coke outside and then recharging them again as quickly as possible, every moment of time was of consequence. This was then done by an iron scoop the length of the retort, already filled with coal close by. In a moment two stokers would put an iron bar under the end of it facing the retort, the foreman stoker took the handle at the other end and it was lifted and rushed right into the red-hot retort. The foreman twisted round the handle which emptied it, another stoker waited ready with the lid of the mouthpiece already luted and prepared, directly the scoop was drawn out, to clap it on and screw up tight for another charge and thus go through all the retorts in action. During this work all the stokers were nude to the waist, the heat being fearful.

Many times during those anxious days and nights have I, in my excitement, thrown my coat aside, rolled up my sleeves and with a shovel helped to load the scoop, so much depending on speed. Then I would run outside to count the rivets left in the gas holders and see whether we were gaining or losing in height. In addition to this, it was very important as, under the circumstances the gas was necessarily sent out at a high temperature, that the condensers and purifiers should be carefully worked. But how primitive then was this method of making gas to what it is now. Great benches or ovens *dos-a-dos* (in line) and all labour superseded by machinery. What a marvel it was that I did not then break down in health altogether after this, but thank goodness, I was spared that. I was young then and fairly strong, but I remember it took a lot out of me.

Before this pressure came upon us, I had told my landlady of my intention to vacate her rooms at Xmas and she was evidently very sorry; Rachel also. I explained why, and some little time after, Mr Mercer's friend Mr Penfold occupied them. But I was obliged to ask her to let me stay one more week until we had got over the terrible pressure at the works and because my new home at Surbition was half a mile further away.

Having got matters once more into shape, however, and without any break down, there was indeed plenty of work for me to do and think about. It being the end of another half year I had to send a cheque again to the Secretary, prepare my *contra* account for them for meters and services to new consumers, all having to be measured up and taking much time. The Xmas quarter's accounts for gas, coke and other supplies had to be made out, delivered and collected for we could not get along without money and there were many, many other accounts to be paid. But taking it all round, it was most cheery to me to find, when I could get time to make up my accounts for this my first year, that the rental was just about double that of the year before when under Mr Spice's care.

In addition to all this work, I had to prepare a report for the Directors to include in their own report to the shareholders, stating the very terrible trial we had just gone through and what steps I proposed to take to prevent any such anxiety this next year. When I attended the usual little meeting of the Directors held prior to the half yearly meeting of the shareholders they, in a most kind manner, moved a vote of thanks to me and also at the half yearly meeting at the Court House. Afterwards some shareholders got up and moved the same. Of course, I was obliged to get up and say a few words in return.

Most of this work, writing, and preparing the accounts and reports, I was obliged to take to my new home and sit up half the nights until they were finished. When I was at last able to get to my cottage and see what had been done there, I found a very quiet, decent young person in charge sent by my mother to get things in order and take care of me afterwards. She had done her best, but it was some time before I could get quite settled down. Everything in the little place was of the most simple, inexpensive character; one might say it was rather rough, but I had only myself to consider. I have in the room in which I am now writing these lines, a small pair of china vases with a *pot pourri* of some roses inside, smelling almost as sweet now as when sent at that time by my dear mother, just upon 50 years ago.

Having now started on my second year as lessee, I asked Smith, my lighterman, to come round and see me and told him that as the quantity of coal being barged now was so much greater, he must reduce his price. After a good deal of discussion, for he was an obstinate old gentleman, he agreed to it being 4/- in lieu of 4/6 a ton, on the understanding it should be for three years. This I consented to; so far so well.

I then also began to prepare plans and estimates for the very necessary extensions and alterations for this next winter and wrote to a well-known firm of iron people in Staffordshire - Westwood and Wright. Their speciality was gas holders and who, when I was at Greenwich as a boy, did a lot of work there for the old Phoenix Gas Company under my father. Mr Wright in reply, came to see me himself and having explained to him my idea of telescoping the larger of our two gas holders, that is putting in an inner lift and doubling its storage, it was arranged that he would send me a tracing or sketch, showing how he could manage it and an estimate of the cost.

I also decided, as there was sufficient room in the retort house, to build an additional bench of seven retorts (we then had only five in a bench) and also take down an old bench of five and reconstruct it for seven. With this increase we ought to do well. Also I wanted to make a slight temporary addition to the coal shed to prevent any coal lying outside in the wet, which is always bad. Then to make all complete, to have a couple of new purifiers similar to those now in use with all the necessary connections and finally some alteration and enlargement of the trunk mains outside, to further equalise the pressure.

To make all the necessary drawing for this, prepare specifications and the estimates for it all, added to my ordinary work. I need not say this kept me pretty hard at it; but I never tired and always felt a pleasure and love for it, and yet with it all an incident occurred that took me away from everything for one evening. I have already said what great affection I always had for my two sisters and how proud I felt to take them anywhere that would give them a little pleasure. I noticed in the newspapers an announcement of a grand gala performance at the new Opera House,

Covent Garden, on the occasion of the wedding between the Princess Royal and the Crown Prince of Prussia. Although I never cared in the least myself for these grand events, it occurred to me how it would delight them to be present at it. I therefore wrote to Mitchells in Bond Street, asking if they could get me a box opposite the Royal box, and the cost.

They had one and the cost being reasonable, I took it. I need not say how pleased they were when I wrote to them and as they had a young lady friend staying with them, a Miss Jack, whom I did not know, it made our party four. It was indeed a grand and impressive sight: one half of the Grand Tier had been thrown into a royal box, very beautifully decorated and our box was exactly opposite on the tier above, so we therefore had a splendid view. I cannot now remember whether the performance was a musical or a dramatic one, but there was a magnificent orchestra conducted by Sir Michael Costa.

It was a very interesting scene when the Royal party entered their box - Queen Victoria, the Prince Consort, the King and Queen of Prussia and many other great guests – and it was very, very impressive when the National Anthem was played. Then the performance began, whatever it was. Now comes a strange incident, quite under our observation. I do not remember whether it was between the acts, or when it exactly occurred, but all at once the orchestra commenced playing the National Anthem again. Everyone looked surprised: some people rose up and other hesitated, but Queen Victoria suddenly rose from her seat, or rather jumped, went quickly to the front of the Royal box and very energetically shook her fan at Sir Michael Costa and evidently saying something at the time. The poor man looked rather white and the anthem suddenly stopped. For some minutes there seemed to be confusion all round, but at last some explanation was given, I suppose, and the performance was continued, but I never heard the cause of it all.

On this occasion I was a little vexed and put out, I remember, over a silly matter. Muslin dresses at that time were almost universally worn with the hideous crinoline. Miss Jack, our companion (a very ordinary looking girl) wore a white muslin with bright red spots all over it – anything more hideous on such an occasion cannot be imagined. The poor girl seemed as if she had cut or injured herself in some way and her dress had got spotted with blood. My sisters, I need not say, were dressed quietly and with perfect taste. So ends this little event. We returned home and in the early morning I hurried back to my work at Kingston.

It might perhaps be supposed by this recent little bit of amusement or pleasure to my two sisters that I was inclined to be extravagant, but that surmise would wrong me very much. Although I say it, I am sure there was not a more careful saving man than myself throughout my life, even when a boy. With a little allowance from my dear father, I always managed to put something every year into my money box and when the happy day came, when I really earned money and received a salary, I made a practice of putting aside every Xmas a £5 note and how it pleased me to see it grow. If I could spare two notes that Xmas, so much the better. When my day of reckoning came, where should I have been had I not in those younger days been of a steady, careful nature and never ventured into any private expenditure without first feeling satisfied I was justified in incurring it.

One morning in my office, after my return to work a letter came from the engineers' office at the water works signed James Simpson Jr. asking that someone should attend in reference to some little alteration to be made there in the supply of gas. Although so very busy at the time, I thought it would be better perhaps if I ran up myself. Arriving at the works (which were then nearing completion), I went to the Resident Engineer's office and I saw for the first time Mr Simpson's eldest son, known to me ever after as Jimmy Simpson. He was a very good-looking young fellow, perhaps some three or four years older than myself and with bright, dark eyes, but with a curious manner. Having introduced myself I stated that I had thought it better to come up myself to receive any instructions he might desire to give. He did not seem, however, to be taking any notice of my observation but after a few seconds remarked in a quiet, slow, uninterested way that he would show me the alteration required.

Having walked to the spot where the gas supply entered the works, he explained what was needed. After taking two or three measurements and telling him I would send men up at once, I said Good Morning. But as he continued walking with me to the gates saying nothing, for the sake of a little conversation I remarked that I should now be quite handy and near if he should wish to see me at any time. I told him how I had given up my lodgings in the Fairfield and taken and fitted up a little cottage just outside that I might be able to get a little boating during the summer evenings.

I also remember giving the two water companies a pleasant little rub for piling up that mountain of surplus earth opposite and spoiling my view of the river. Several years later, long after I left the cottage, it was fated that I should be the very individual to cause the removal of this hideous mound of earth. Then hurrying back, I sent some men after their dinner to do what was required.

I think two or three days after this I had just finished my dinner at the cottage between 7.00 and 8.00 and had crossed over the passage from one front room to the other that I had made my parlour or study. I wanted to smoke a cigar and look at the newspaper before going to work again with my writing. The door bell rang and my housekeeper announced Mr James Simpson. In he came with almost a little smile on his face and remarking that if I felt as dull as he did, we might together make it a little more cheerful. I most cordially assented to that proposition and, although he had the same seemingly quiet slow manner as before, we managed to get through a couple of hours very pleasantly chatting and he left quite a different man.

I had gathered that his family at Westfield lived a very quiet, retired sort of life and consisted of his father, mother and himself. Also his mother's sister, a maiden lady - Miss Nancolas, and a brother who had rooms in one of the roads just outside, but almost lived at Westfield.

This gentleman was then Chief Cashier at Williams Deacon and Co Bank in Birchin Lane. James Simpson had two brothers, John who was not very strong, had been staying in Cape Town for some two or three years and was expected back that year. The younger son, Arthur Telford, was then a boy at Dr Arnold's school at Kingston. There were no daughters which I thought was unfortunate for himself and his brothers, as they were just the very young fellows who would have appreciated a sister's companionship, myself as an example.

After this first evening, during the whole of that year he never failed to drop in nearly every evening in the same quiet way. Before leaving he had mentioned that at Westfield they always dined at 3 o'clock on a

Sunday and would be glad if I would come in and dine with them. I, of course, thanked him very much, but was obliged to say I was just then so awfully busy, getting out plans for new work, but when that was done and off my mind, nothing would give me greater pleasure.

We were not quite out of the winter yet, only the end of February, and there was a lot of gas to be made before any alterations could be considered. All this additional work, preparing the drawings, specifications and estimates for all, when added to my usual work, was rather heavy on me. Somehow, I seemed to glory in it and take the greatest pains with it all, for I did not know what steps the Directors might desire to take to carry out the work.

My kind old friend, Mr Mercer now and then dropped in, to see how I was getting on with the proposed new work and taking the greatest interest in it. Well, by the end of March I had got everything into shape. The two new purifiers with the their connections required, I remember, a lot of anxious thought, for very many serious explosions occurred now and then in what was called the Purifying Shed. We were then working the dry lime process for purifiers, not the wet lime my father did when at Greenwich.

I remember our lime was supplied to us from time to time by contract from a well-known man in that part of the world called California Joe. He had been out in the gold diggings there, made some money, came home, bought a piece of land at Sutton where there was a good lime pit and used to go round himself with a horse and cart and supply customers. Although he could neither read nor write, he made many friends and was always very straight and businesslike and I often used to like having a chat with him.

Well, the evening came for the meeting I had called for the Directors to receive my work of the last two months – the several drawings, specifications and estimates. How well I remember my nervousness and anxiety on that occasion. Every member of the Board was present. Mr Fricker the Secretary, read my short report and placed the drawings and particulars before them. I remember someone kindly remarked how very neat and clear everything seemed to be and my old friend Mr Mercer also said Yes and without any help whatsoever. (He had never been able to understand when he came round why I should do so much myself) but that was the charm of my work.

Everything was carefully looked into by them, however, and many questions asked of me. Mr Simpson said nothing, but I noticed he was carefully considering the estimate for it all, which I had divided into separate items to enable them to form a better opinion. To Mr Simpson, all such matters were at his fingertips, and were a mere flea bite to the vast amounts he no doubt had to deal with day after day.

When the several Directors seemed to understand what had been put before them, I took an opportunity of saying a few words, not exactly making a speech, but putting our present position strongly before them. I remarked on the great danger we had gone through this last winter and that without these additions and alterations I now suggested, it would indeed be desperate and not a moment should be lost in commencing them. I also expressed my regret that the estimate should have arrived to so large an amount, but assured them every single item had been cut down to the very lowest figure.

After a little pause, Mr Simpson spoke, remarking something thus. "Well, gentlemen, you have now seen the several plans and estimates for the immediate future Mr Aird has put before us. I have also looked through and examined them and cannot put my finger on any one item to alter or object to. Feeling therefore, that you have the same confidence in him that I have myself, I will move that the plans and estimates put before us be approved and accepted and everything put in hand at once and completed for the coming winter." Carried unanimously. I may as well here say, that with this result, I as nearly as possible broke down altogether, such unexpected kindness and confidence being too much for me.

The following morning the first thing to do was to write to Westwood and Wright and Charles Horsley, that I accepted their offers for gas holder purifiers and connections, iron retorts and fittings and also iron pipes; the work to be completed by the agreed date. I also ordered the necessary quantity of fire goods for additional retort benches. My estimate for all this was, I think, between £4,000 and £5,000.

Having now got all this off my mind, I set to work with my books, having got a little in arrears, Strachan being out reading the meters for the Lady Day quarter. There was then a knock at my office door and a respectable looking man came in with a boy of about 13 years. He was a quiet looking young fellow, but did not seem to have much to say for himself. The father, in introducing himself said his name was Henry Packham, he was a baker and had a small shop on the Portsmouth Road by West Green. He had married Jane Freakes, daughter of my mother's old friend and nurse; the boy was their oldest child and had just left school. Having heard of my position at the gas works, he thought he would venture to ask whether there was any chance of my giving him a start in my office. I, of course, had heard of the man and passed his little shop many times, but had never met him. However, having asked the boy some few questions, I felt obliged to tell his father I really had not time to teach any youngster and it would be better for me perhaps to engage someone who had some little experience, but I would think it over and write. After a couple of week's thinking, I wrote that I would give his boy a trial. When he came, I put him charge of the coke book and the issue of the coke tickets and after a time was glad to find the little chap was becoming very useful.

How strangely things do happen sometimes in this strange world. That I should myself be at work and so much interested and involved in what went on in a spot where, when an infant only six years old, my brother John and I were put into some little place called an infant school, at Esher. We were under the care of this boy's grandmother, Mrs Freakes, and now it should be in my power to give her little grandson a start in life.

It is anticipating, but should I not live a little longer to complete these memoirs, I may as well state that Packham remained with me during the whole of the lease – over ten years and making himself of great use. I may say it always was a pleasure to me to help him in return and teach him everything I could to give him knowledge of his work. The result was that at the expiration of the lease, the company then appointed him their engineer and manager. He soon after married and I heard that he held the position for over 20 years. When he died at the early age of 47 he had educated his son who was appointed to the same position in his place and is there now, the year 1909. The company has paid a 10 per cent dividend every year since I left and the works having increased in magnitude more than ten times. I would submit that this is a little something to be proud of.

When the 1851 Exhibition had been removed to Sydenham and re-erected there on a far larger scale, with its lovely grounds and fountains, the whole of the country round about for miles developed into building land. Houses and towns sprang up like mushrooms – Upper and Lower Sydenham, Forest Hill, Catford, Beckenham, Penge, Anerley, Nunhead, Upper and Lower Norwood, until next door to Croydon and London.

A solicitor named Downs, who had been looking ahead, purchased some land at Bell Green, Lower Sydenham (a very suitable site), formed a company and erected a kind of gas works. But after a year or two the Directors found themselves in difficulties. My friend H. P. Stephenson hearing of this, soon dropped down upon them, studied the whole matter and made them an offer to clear out. He went for an Act of Parliament, obtained it on very favourable terms and formed a new company altogether. The then well-known physician, Erasmus Wilson, was the Chairman (and a splendid Chairman he made), H. P. Stephenson was the Deputy. Dr Hetley, a local medical man and also surgeon to the Crystal Palace Company itself; G. P. Pocock, a solicitor; Peter de Vere Foster, Secretary to the Society of Arts; and Fred Linging, Secretary of the British Gas Company. were some of the other Directors. The Secretary and Manager of this newly formed company, known for many, many years as the Crystal Palace Gas Company was Magnus Ohren. He was a well-known figure in the gas world and a prominent member in the Freemasonry world also.

One day when in London chatting with Stephenson and Christie and speaking of the very extraordinary growth of this company, they suggested that as one of the Directors was about to retire, a Lieutenant Willis RN, it might suit me to take his place. They would be glad to have a little more practical knowledge on the Board and they had heard I was doing so well at Kingston. If so, Stephenson would see Mr Erasmus Wilson and ascertain his views. It would mean £1,000 to qualify me, but looking into the matter cautiously and carefully, although it seemed at first a large order, I thought it could be managed. Stephenson therefore saw Erasmus Wilson; I also had an interview with him and a very pleasant one it was, I remember. Some little time after I was elected. The meetings were held every Friday fortnight in the works at Bell Green. The Directors' fee then was £100 each and an annual season ticket to the Palace itself, which it was thought desirable for the gas company to take.

The works and the annual revenue at that time were rather more than double that of my own works at Kingston, but the future growth of this company was really something phenomenal. It seemed almost impossible to keep up with the demand. I thought we were doing pretty well in that respect at Kingston, but nothing approaching this. Mains were laid and in 12 months time had to be taken up and replaced by others double the size.

It seemed quite impossible to estimate sufficiently in advance. We were already paying 10 percent and having spent all the capital, or nearly so granted by our first Act, had to go for another. Robert Christie was Consulting Engineer to the company, but his whole time seemed to be occupied in preparing drawings and specifications for enlarging the works. New offices were also included in a large house for Magnus Ohren the Secretary. John Aird and Son were the contractors and did all the work. I can remember Downs, the solicitor from whom the works were purchased, wanted to sell his remaining shares. and the Chairman and I buying them between us and making some considerable profit.

I now used to meet my friend Stephy Clarke very often, as in addition to purchasing coal for my own works, I had, with H. P. Stephenson to act as a committee to purchase coal for the Crystal Palace Gas Co. We knew each other well now and some few years later his brother Robert joined our Board.

We had got our new Act of Parliament through without any important restriction, plus a new railway siding right into the works, the coal being brought alongside Beadle's wharf at Erith by ship and then run straight into the works. All this wanted a lot of consideration and management and in fact it had come to this – I was now really giving my heart and my interest to two gas companies instead of one. This perhaps was rather piling it up, but I never could do any work whatever it might be, superficially, but put heart and soul into it.

This gas works, the Crystal Palace Gas Co. was rapidly becoming as large as any of the London works and so the Directors' remuneration became £200. I remember one of the first things we decided to carry out, as the works were now large enough, was to manufacture sulphate of ammonia from the ammoniacal liquor which had hitherto been almost given away. It did not require any expensive plant and for this sulphate for some time we got £20 a ton, it being greatly in demand for agricultural manure. My own works at Kingston was not then large enough to manufacture it.

My friend, Jimmy Simpson, still never failed to drop in two or three times a week. I think it was the Sunday after I had taken on young Packham and it being a pleasant, spring like day, we walked together to Hampton Court Chapel, going by way of Thames Ditton and crossing the river by the ferry there and then along the riverside to the Palace and Chapel.

What a pretty little village Thames Ditton was then, with its picturesque little inn by the river. I believe my mother's mother was buried in the little churchyard there, or so it used to be said when I was a lad. After the service and strolling about the beautiful gardens a little while, we walked back to Westfield. This was my first visit there and I was introduced to all the Simpson family. It was the most peaceful and happiest home I ever entered.

On that occasion there were only the two elder members, Mr and Mrs Simpson with Miss Nancolas, the maiden lady, sister to Mrs Simpson and her brother, Mr Nancolas, a bachelor, but what dear, delightful people. I felt at home at once, everything was so thoroughly enjoyable. The dinner was at 3 o'clock, plain but so plentiful and good. One thing in particular I noticed – the potatoes were always served up in their jackets which I have always thought the proper way. As it was a little too early in the season that day to sit in the garden, we chatted pleasantly until tea time when I thought it time to leave. But no, I was induced to remain until a bit of supper at 9 o'clock. How well I can see and remember their ever faithful servant, Henry Lott, going round and asking in a stern sort of voice – wine or Bass?

I remember taking this opportunity of expressing to Mr Simpson my very sincere thanks for his great kindness at the last meeting and assured him he would never have occasion to regret it. We should have gone on chatting for hours had the ladies not expressed a wish to talk with me a little. That was my first Sunday in that home, never to be forgotten, and how many, many such happy Sundays did I spend there afterwards, becoming almost one of the family. When John Simpson returned from the Cape and the youngest son Arthur left Dr Arnold's school, what fun there used to be in the evenings in the billiard room, sometimes the elders in their quiet way enjoying it quite as much as the younger ones.

As spring was now really making a start I began to think again of my boating in the evenings. Having been sometimes annoyed last year by not being able to hire a boat to suit me, and sometimes not being able to get one at all, it caused me to wonder whether I could meet with a second hand one at a moderate price. Therefore when in London one day on business, I went round to Searle's well-known boat building yard in Lambeth and had a little talk with him. As it happened when going through the sheds I noticed a very handsome Random Gig, a boat for three rowers if required – one rowing a pair of sculls, the other two an oar each. The boat had been built for a gentleman only last year who had since changed his mind and it was now on Searle's hands. As the price he asked for it was not unreasonable and he agreed to remove the name put on the bow each side in gilt letters, which I did not like, and replace this with the name *Naiad.* I agreed to purchase it and he was to send it up to Parker's boat house at Kingston.

One beautiful evening following, I asked James Simpson to go out with me in my new boat, and this he did frequently. But I was also out by myself whenever I had the chance and felt a little bit proud of my sculling all that summer.

So things were going on. At the works I was doing everything I could to push on with the many new additions and alterations to get them completed and working before next winter. At the same time, I was endeavouring to increase the consumption of gas by sending out some more of the little books.

While thus busy, to my great surprise a letter was forwarded to me from The Grove, to which place it had been addressed, from Vitruvious Wyatt, with whom I was at work at Toronto on the Grand Trunk Railway five years ago. He had returned to England, having had some dispute with his Chief, Mr G. Tate, or more particularly with the Chiefs in Great George Street, Sir Morton Peto and Betts. As I understood later, I think this was about salary. However, he had written saying he would like to meet me again. I at once replied, explaining my present occupation and that I would be very pleased if he would come and stay a day or two with me. This he did and when James Simpson came round in the evening, our conversation became very interesting. He had not altered, but was the same ill tempered selfish man who used to impress me so with his great ability.

How many, many times have I asked myself, looking back 30 or 40 years, what on earth it was in him to induce me to have such regard for and take such interest in Vitruvious Wyatt. He was I believe of very good family, a cousin of the then great architect Digby Wyatt, but the miserable story will unfold itself as we proceed.

I was just then beginning to be very busy with the new work and taking him over the works, looking about him, he seemed much interested and asked many questions. As an old friend, I was quite open in my replies as he had not the least knowledge of the manufacture of gas.

Some days after this I was in London and happened to call up H. P. Stephenson at his office. He seemed to be a little upset and explained that the agent for the Nicaraguan government, South America. with whom he was acquainted, had invited him to send in a design for a large pier, to be erected at Guatemala. Unfortunately he knew no one at the time he could rely upon to send over. It all at once crossed my mind it might suit Wyatt to go, as he was looking out for something and he would be the very man.

Having mentioned him to Stephenson and given some particulars of him, an interview was arranged. Both seemed satisfied with each other and Wyatt left with full instructions. He was away I think some two months, taking all necessary borings, measurements and particulars. On his return, Stephenson and Wyatt together prepared a design with an estimate from the Horsley Iron Company for its construction. These were sent over to South America by the agent and in due course of time accepted, and I had every reason to believe all parties were satisfied with the result.

I very well remember one Sunday when at Westfield mentioning this matter to Mr Simpson, and at his wish when Wyatt came back, taking him round and introducing him.

When the Crystal Palace was being re-erected at Sydenham, the Resident Engineer under Sir Charles Fox and Co. was Mr Frank Wentworth Shields. I think he was an Irishman, a very clever young fellow, but with singular quiet speech and manner. Somehow the Aird family seemed to have become acquainted with him, and recently with many other well known people in the engineering and contracting world about Victoria Street and Great George Street. Among them was, I remember, a Mr and Mrs John Watson.

One day, I received from Shields an invitation to a dance at his rooms, as he was a bachelor at the time. I thought it rather strange, but hearing from home that my two sisters had also been invited and that I was expected to escort them to the occasion, I went.

Mrs John Watson was really the chaperon for him, but there were several other young married ladies and girls. My sisters, of course as always looked very nice and I felt very proud of them.

It was altogether a very pleasant little party, but during the evening I could not help noticing that one of the guests, a very gentlemanly looking young fellow, was paying much attention and dancing very often with my younger sister Jessie. I asked Shields who he was and was told that he was the senior partner of a firm of solicitors in Victoria Street – Hargrove, Fowler and Blunt. Robert Fowler was a young brother of John Fowler the well know engineer, and Blunt the eldest son of Mr Blunt, solicitor to most of the London water companies. After this little dance, some few months later they became engaged and married the following year.

A happier partnership through life never was formed. Shields himself also was married some little time after this and went to live near Lewisham and became great friends with my mother. At the time I am writing this, poor Shields has been dead several years, as everyone else of that time seems to be, but his eldest son of the same name has succeeded him in most of his appointments.

My father through life was always so fond of Scott's, Burns's and Tom Moore's poems, particularly of Scott's *Lady of the Lake*. I remember he made us all read and learn it. About this time Glasgow was going to Loch Katrine for a new supply of water, the work being principally a great aqueduct of masonry between the two points. Mr Bateman of Great George Street was the Engineer. Mr father, hearing that this contract would be in this pretty part of Scotland, thought he would send in a tender and asked myself and young Mr Joseph Quick to go with him to prepare one, my brother John being just then too busy to go. As it was only a matter of three or four days, I thought I could manage it. We went over the line of work between Glasgow and Loch Katrine and stayed at the Trossachs Hotel to prepare an estimate. What a beautiful part of the world; I had never been in Scotland before. Our tender, however, was not successful, some canny Scot living on the line of work got the contract, much to my father's disappointment.

About the same time, John Aird and Son had just finished a contract for a new water works in South Staffordshire and the opening ceremony was held at Lichfield with a banquet at the Town Hall. I was invited to be present and was much interested. Lord Alfred Paget, the great yachting commodore was President on the occasion and proposed the health of Mr McClean, the Engineer. Poor man, I was seated near him to him and he could not say a word in return. He got up to do so and his lips seemed to move, but no words were heard.

It was Mr McClean who had noticed a young lad, George Wallis, on that work, engaged by my father as timekeeper (a son of the clerk Wallis at Belvedere Road). He asked if the firm could spare him for a time to go to Eastbourne and take charge of a small water works he was then about to make for the then Duke of Devonshire. That was the start in life for George Wallis and to his credit, he took advantage of it.

Returning again to my own work at Kingston, I was rejoiced to find that the firms to whom I had sublet the ironwork were keeping their word and sending in the iron for the additions quite up to time. A few days later, men arrived to put it together. I have often thought on engineering work, what fascination it has always been to me to watch riveters at work. The boy with his portable fire stove and bellows alongside supplying red-hot rivets to the riveters up aloft - one with a great heavy hammer as an anvil, the other man with a very light hammer, striking the hot rivet into a tight bolt in a very few moments. It was interesting to notice also the way they signalled to each other with their hammers, when throwing up the rivets. Of course, on such a little job as this, it is mere child's play, but on large works such as boilers, gigantic gas holders, bridges and great steam ships, where hundreds of riveters are at work at the same time, it is very exciting. In future years, what a lot of such work have I seen. Many years later, it was my fate to become a Director of one of the largest iron works in the north – the Parkgate Iron Works at Rotherham near Sheffield. That company was then making the armour plates for the first iron clad naval ship for the government.

I was also well forward with my own work in the retort house, enlarging the benches and laying the larger mains outside. Packham was also becoming very useful and trustworthy in his small way. So far so well, but there is no rose without its thorn in this world.

There cannot be, I think, in this kingdom any borough or council without some of them being a kind of *bête noire*. This was certainly the case at Kingston during the whole of my time there. There was a little, insignificant member of the council who seemed to live only for the pleasure of attacking and abusing the gas company. Being the lessee I, of course, had to take all his abuse on my own shoulders, but I never had any difficulty in putting before the Directors figures to show how very one-sided this man's statements were. He was a baker in the London Road with a wretched little shop, one window only with nothing in it. How the man existed, I never could tell, for he seemed to have very little trade. The man unfortunately had what is known as the gift of the gab, as many of that class have, and endeavoured to make out that the gas supplied was very bad and the price much too high, particularly for the public lamps. Seeing all this rant and outrageous charges printed in the Surrey Comet newspaper, after the Council's monthly meetings, grieved and incensed me greatly. I knew I was working very hard and doing my very best to please all consumers and being of a very sensitive nature, worse luck, it worried me. I could not, of course, be present at the Council's meeting to reply to these attacks, but thank goodness there were many fair-minded men on the Council.

During the friendship that existed between James Simpson and myself, we now and then went up to London to one of the theatres, having a little bit of dinner perhaps at Simpson's in the Strand first. Our favourite theatre was, at that time, the Princess in Oxford Street where Charles Keen was performing in some of Shakespeare's plays with great success. Ellen Terry was acting as Puck – what a success that was, so young a girl with her sweet, melodious voice. We both fell in love with her and went night after night to hear her, but always returning home by the last train at 12 o'clock; we neither of us cared to remain the night in London. There were no palatial music halls and restaurants as now, and we were, I think, of the quiet sort.

John Simpson had now come back from the Cape. Arthur had left school and, having some other friends to visit them from other parts that year, made Westfield to me very bright and cheerful. There were the Cuthberts and the Nelsons, Mr Nelson being a great railway contractor in the north and residing at Carlisle, with a large, and very nice family of both girls and boys. The younger people were always about on the river as old Mr Simpson had just then had a little steam yacht built to make excursions up the river. It had a screw propeller, the screw having superseded all the old paddle wheelers in shipping. It was called the *Kingston* and many pleasant little voyages did I have in it.

By the by, I must not forget to mention my dear old dog Ben, a present from my kind friend Mr Innes – a splendid retriever and water dog. The young Cuthberts were very fond of him and, when going to bathe in the early morning in the river, would come to my cottage and take him with them, much to his delight.

One morning, to my great surprise, Mr Simpson drove into the works with old Mr Cuthbert, to show him round, and just before leaving, Mr Cuthbert asked me whether I could take his eldest son with me as a partner. There could be but one answer to that.

We were now entering the month of June and were soon to take the meter readings for the quarter, make out the accounts and prepare the usual half yearly report for the shareholders. The quarters come, the quarters go and will go on for ever, as the old song *The Brook* says. How dearly I wished it might be so. I should have been quite content, but I knew the company would never renew the lease, they were becoming much too wide-awake to do so.

One evening, chatting with Jimmy Simpson, I told him of a little matter I had on my mind, to give my two sisters a day's pleasure, by taking them to Ascot races on the Cup day, and asked if he cared to be of the party and help me take charge of them. I could see a twinkle in his eyes that assured me he was all there and that it would suit him very well. I therefore called on Mr Charles Bailey, the then host of the Southampton Arms Hotel at the Surbiton railway station, and talked the matter over with him and his manager, Oxlade. Finally we arrange for the hire on the Cup day of a barouche and four horses with two men, our party being four ladies and two gentlemen on the box seat. My instructions were that it should be a very quiet and unostentatious affair, well horsed and with a couple of very stately men, not forgetting a generous well-packed hamper and a big lump of ice.

When I wrote to my sisters about it they were delighted at the idea and with my suggestions that they should bring two lady friends with them to complete our party. The evening before they arrived with two young ladies – Miss Mynott and Lizzie Copland. How they packed themselves in that little, tiny cottage I do not know, but from the shrieks of laughter I heard now and then, they were evidently enjoying it. James came round that evening to be introduced and arrange matters for a start in the morning.

The morning came, promising a beautifully fine and very hot day. Directly after breakfast the barouche arrived at the little front gate as we had arranged to start early and get a good place on the race course. It had not been there for many minutes before it was surrounded by nearly everyone in the immediate neighbourhood. The landlord of the little public house at the corner of the road was out of his house like a shot. He walked round the team with astonishment, admiring the horses and no doubt putting me down as a regular sporting man from whom he might get a straight tip perhaps now and then. Poor man, what a delusion!

There could be no man on earth with less knowledge of horses and horse racing than myself. I never made a bet in my life and the only time I had even gone to a race meeting prior to this was when my father took my brother John and I, when we were boys at Greenwich, to 'Appy 'Ampton races at Moulsey Hurst in what was then called a one horse shay. However, in this case, Charles Bailey had certainly done his best to meet my wishes in every way. The equipage was altogether simply perfect and quiet with not a sign of ostentation or vulgarity about it.

The two men were quietly dressed in dark coloured jackets of some sort of silk and tasselled caps, white leather breeches and smart riding boots. There were four powerful horses, particularly the two wheelers I remember. There was no shabbiness; the hamper and the ice were carefully packed; we were ready. When James appeared on the scene with his white hat and light dust coat, every workman who was employed on the two great water works there, must have turned out to have a look at him. He was a great favourite with them, as indeed all the Simpson family were, and who could wonder at it. The ladies took their seats after the usual friendly squabble as to who should sit on the front seat. They were prettily and quietly dressed with their dust cloaks and light grey veils over the little bonnets (hats were unknown then). The two esquires of the party, James and I, had mounted on the box seat with our white hats and dust coats and our veils in our pockets, not wishing to be too conspicuous. With the cry of "All aboard" as they cry in America, we started. The postillions cracked their little whips, and off we went with, I am sure, the good wishes of those looking on.

The same little thrill of pleasure passes through me now when writing these lines as it did at that time so many years ago. As the weather was so beautifully fine and we had plenty of time before us, we went along quietly through Esher and over Walton Bridge. I pointed out to my sisters, en route, old Mrs Freakes' house and the infant school at Esher to which I was sent, then into the main road to Ascot and the race course. There we found such a crowd of carriages of every sort and kind and so much dust in that sandy road, that we could hardly see one another. At that time nearly everyone went by carriage or coach.

Arriving at the race course, and having paid the entrance fee, we pitched our tent opposite the Grandstand. We then walked across to its private enclosure and strolled about to see some of the great people; the ladies I remember were very anxious about this. After the first two or three races, and in the interval before the Cup race, we had the hamper opened and never was a luncheon more enjoyed.

When the race for the Cup was over (we had instituted a small sweepstake between ourselves), and when the excitement and the awful noise afterwards had subsided, we decided to have the horses brought up and get away quietly before others. It was our intention to return home on the Middlesex side of the river and call at Virginia Water to have a cup of tea at the Wheatsheaf Inn, which we did and thoroughly enjoyed. When we left the inn, nearly everyone seemed to be on the way home. Our 'boys' seemed all right when we started and on their horses, but I could not help noticing just beforehand that they were a little bit unsteady on their legs. However, we seemed to pass everything as we went along and it was most delicious going along so fast in the cool evening air, after the hot day. Arriving at Kingston Bridge (and anyone acquainted with it at that time will remember the very sharp and sudden angle, turning so suddenly round into Thames Street from the foot of the bridge) our men, being sleepy or stupid, did not slacken their speed. As a consequence it was simply by a miracle that we were not all thrown out and seriously injured, if not worse. However, all's well that ends well and in a few more minutes we were once more at my little gate. Having got rid of the carriage and the men, we were all glad to get indoors and rest a little while after this exciting, noisy and happy day. James returned home to Westfield and the young ladies, after a cup of tea, retired to bed. I had previously arranged they should have a good rest and a late breakfast. I would run up to the works early and see if all was well, come back and join them and afterwards escort them to the railway station to make their way home.

When in the cottage again, I noticed among the wraps and odds and ends removed from the barouche, a race card. We each had one and this evidently was the one belonging to Jimmy Simpson. These race cards are always of a large size and it amused and interested me very much to find three verses written in pencil on the back of it by himself, *a propos* the occasion and including us all. Round the verses were some attempts at portraiture, altogether very clever. This I ventured to put away in my safe with some other relics that I treasured.

The more I saw of John Simpson the more I liked him. He looked very slight and delicate poor fellow, and quite unfit to meet the rubs of a busy world, but he was always good company and he had a great gift to fall back upon – he was passionately fond of music. I do not think he had much knowledge of music itself, but he would sit at a piano for hours playing by ear, any tune you might name. I remember taking him and James to The Grove and introducing them to my father and mother, which was a great success.

I often used to think how much dear old Mr Simpson must be handicapped by his great size and weight in his work and profession as an engineer. It was very difficult for him to walk any little distance; as for running or attempting to, it was simply an impossibility. Nevertheless, he managed to get along somehow and was always bright and cheerful, and now and then very amusing. I have reason to remember the big strong cigars he used to smoke. I never was a great smoker myself, but could always enjoy a good, moderate cigar after a meal. Pipes I never smoked except when off work, fishing or on holiday. When at Westfield on a Sunday after dinner, the dear old gentleman would say "Well, come along, Aird, and have a cigar with me". If fine, he would go out into the garden to the summer house; if wet, into the billiard room, pull out an enormous cigar case, offer it to me and take one himself. Oh dear, what an ordeal to go through, but I did my best and never required a spittoon, thank goodness, as my poor dear father always did. After a time I would manage to get into the garden and join the young people. Some 12 or 14 years later, when Mr Simpson died, I was told his weight was about 20 stone.

One other Sunday when I happened to be dining there, I met Mr Simpson's brother, Mr William Simpson (the head of the engineering firm in Grosvenor Road) and his son Charles. Mr William was a very stylish looking man, something like a Frenchman, both in build and manner, quite different from his elder brother Mr James.

I do not remember what gave rise to that particular conversation, but he had been speaking of Venice and its charms. I ventured to mention that, although I had travelled about in a small way, I had never been there. He then remarked that I ought to take the earliest opportunity of seeing it as it would very soon lose all its interesting antiquity.

Shortly after this, another letter for me addressed to The Grove was forwarded and to my great surprise, it was from the Mr Boyd at Toronto to whom I had a letter of introduction and who was so very kind to me, when I was out there.

The letter was to say that he was staying at the Golden Cross Hotel, Charing Cross; that he had come to England for a little holiday and would like to meet me again. The next time I had occasion to go up to London I called upon him and found him still the same kind-hearted fellow. He came to my little cottage to stay a few days and, taking advantage of this, I thought it would be nice to try and make some return for his kind hospitality to myself by giving him dinner at the Star and Garter hotel at Richmond.

This hotel was then in its palmy days and as a rule frequented by the very elite. Although out of my line, I thought also it would be a nice little opportunity of paying a compliment to some two or three other friends at the same time. I therefore made a little party of six – Boyd, Wyatt, George Smith (Boyd's cousin) who had given me the letter of introduction to him, E. O. Coe, the Secretary of the Grand Junction Water Co., John Orwell Phillips and myself. I had previously engaged a private room with french windows, opening into the well kept and pretty sloping gardens, and looking over the lovely river view. It was fortunately an ideal day, intensely hot, a kind of *'dolce far niente'* kind of day. How well I can even now remember the exquisite beauty of the table as prepared with its fruit and flowers, the glass and the real silver and the delicious coolness and shade when we were shown into the room. Certainly Ellis the proprietor then, and also of the Bedford Hotel at Brighton, deserved his success, but alas, how things have changed since then.

The weather being so very hot, and as Boyd and also Wyatt were staying with me at the cottage for the occasion (with the other three guests coming by train from London) I arranged that the boatman, Phil Parker, should row us down to Richmond instead of driving, and very enjoyable it was. The dinner in every way was simply perfect, the three London men left by the last train, and we three drove home under the hill and through Petersham. What an evening that was with the nightingales.

My friend Boyd left a day or two after this and I have never seen or heard of him since. Wyatt, from some cause or other, seemed unable to meet with any constant employment and I only now and then heard of him.

About this time, the young person whom my kind mother had sent to look after me, gave notice to leave, I think finding it too dull for her, and not to be wondered at, poor creature. I had no other course but to appeal to my dear mother once more.

She had great difficulty I remember this time in getting what she called a suitable person. However, it ended in her sending me someone who had been cook with her a few years ago who was a singularly good looking woman like a gypsy and a splendid cook – far too good for me. She had been with me about a month or two when one evening, she came into my room after dinner to ask whether I would have any objection to her young daughter coming to stay and keep her company. She could help in many little ways and should not be an increased expense. At first I did not know what to say about this, but after thinking it over, I thought it would perhaps be better to assent to the woman's wish.

About a week later, one evening when I was sitting at the table waiting for my dinner, there was a nervous little tap at the door and in came a young girl of about 17, the very image of her mother and even more like a gypsy. The poor girl was shaking so much from fear and nervousness that I quite expected to see my dishes and dinner dropped on the floor; had I spoken, they certainly would have been. But I continued my reading and pretended not to notice. It was very awkward for the girl was very pretty, however. I let things take their chance and to my great surprise and wonder they adjusted themselves.

10
A visit to Venice

The extensions and alterations at the gas works were now completed and I felt quite ready to face another winter. About the middle of September I think, or perhaps a little earlier in that year, (1859) I began to feel very strange and seedy. The more I tried to shake it off the more it worried me, for my health for some time past had been so very good. At last I made up my mind to call on Dr Cox, one of the Kingston Gas Directors. The biggest medical man in Kingston was Mr Sudlow Roots, but Dr Cox would do for me. I therefore called upon him one morning and stated my case. He listened attentively and asked a few questions, then said it was simply a temporary breakdown from overwork and the only remedy was rest. If I could not take a long rest and he feared I could not, knowing my occupation, it must be a short one and frequently repeated, otherwise if neglected, the symptoms might prove very serious. There was no organic disease whatever, but the brain wanted rest. He also remarked, I remember, that at the several meetings between the Directors of the gas company and myself, he had noticed the exhausting nervous anxiety that seemed to possess me at the time and he had pitied me.

While thus talking, the remark of the Mr William Simpson about Venice suddenly crossed my mind and I mentioned it to the doctor. "The very thing, try and arrange it" he said. Well, awkward as it was for my work I arranged for being away for about ten days.

The next Sunday when at Westfield I mentioned to old Mr Simpson what Dr Cox had advised and a day or two afterwards I started, calling at The Grove on my way to see my father and mother.

I should mention that during the whole of this year of 1859 there had been a fearful war on the European continent between France (Louis Napoleon) and Austria. They were quarrelling over the pieces of poor Italy with its several kings, the Pope and Garibaldi. The great and bloody battles of Magenta and Solforina had been fought and the Austrians had got the worst of it, but peace had just been proclaimed which to me looked very much like the old fable – the man (Louis Napoleon) and the oyster.

Travelling then was not as it is now. There were no great tunnels through the Alps; they had to be crossed by Diligence, which was always disagreeable and always subject to the autocrat at the bureau as to your place in it, or whether you got any place at all. Passports also were an awful nuisance then.

Arriving in Paris I went to the Hotel du Louvre, still in its best days as of old, but soon to be superseded. After the *table d'hôte* lunch, sitting in the courtyard with its picturesque Eastern character and enjoying a cigar and a cup of coffee (in itself worth crossing the channel for), I got into conversation with a very pleasant sort of fellow who was lounging about with some three or four others. He was a courier of course, and had been travelling for several months with an American family and had just started them back home again. He himself was now about to return to his home in Vienna to rest a little while.

I told him that I was a little out of health from overwork and intended going straight to Venice for a few days and could not be absent more than ten days or a fortnight at the very outside. It occurred to me that, under the circumstances, it would be good policy to engage this man for that short time and relieve myself of a lot of worry and trouble (the regular charge of these men is always £12 a month). I found he would very much like to accompany me so far and was quite willing to receive whatever little present I might be willing to give him. Well, I never had occasion to regret that arrangement. Far from it and it simply made my little trip a great success. I remember his name was Louis Engel, a German Swiss.

He got tickets at once for the express right through, having satisfied himself that my passport was in order. Arriving at Modane, we went at once to the Bureau d'Administration of the Diligence, but this matter I left entirely to Engel. I found I had been allotted a berth inside, Engel preferred one outside. In the middle of the night at about the summit of Mount Cenis, I was much startled and woke up to find myself being forcibly lugged out of the vehicle, with Engel and other passengers, and made to enter a wooden sort of hut or structure. The door of it was then closed and locked and the interior was perfectly dark. There was a large stove of some sort which seemed to emit a most disgusting and overpowering stench, causing all those imprisoned to feel in danger of being suffocated. However, after some fifteen minutes the door opened and we were all allowed to resume our places in the Diligence.

The explanation I obtained afterwards was that we had been fumigated. There had been an alarm of cholera and all travellers at this point, which was the boundary between France and Austria, had to be fumigated. All the northern portion of poor Italy was then in the hands of the Austrians. Once we arrived at Turin we resumed our journey by railway, travelling again to Milan and then on to Venice.

This was a lovely journey passing along the beautiful Lake of Garda, the old and romantic cities of Padua, Verona and others, arriving at the Terminus at Venice in the early evening. The scene to myself was extraordinary, with no cabs or omnibuses, nothing but gondolas. Putting our luggage in one of these which looked very much like a funereal hearse, with all its necessary trappings, we told the gondolier to go to Danielli's Hotel, but we found it and every other hotel simply packed. We wandered up and down the Grand Canal and these many water streets seeking for shelter, but without success. It looked very serious, but at the last moment Engel remembered an old chum of his, also a courier, who had retired and was living somewhere here with his family.

After some trouble he was found and our unfortunate condition explained. This friend at once most kindly took us in and indeed I heard afterwards gave up his own bedroom to me. However, I felt too knocked up by our wandering about seeking some sort of accommodation, that after a little meal specially provided for me, I went up to the bedroom with the intention of going to bed.

I had only been sitting quietly resting a little while, when there was a knock at the door and Engel came in begging that I would not think of retiring yet as I should miss a most beautiful scene. There was a fête this evening, being the most beautiful of all fêtes excepting the real Carnival. His friend was going out at 10 o'clock with his family and a few friends in a couple of the larger gondolas and it would give them all much pleasure if I would join them. I therefore pulled myself together, make myself a little more presentable and did my best to join in the festivities.

Certainly I never could have imagined such an extraordinary scene.

Every palace, every church, every public building and every gondola floating about were all lit up with coloured lanterns, many thousands of them. To attempt to describe or picture the wonderful scene was impossible. Music was about everywhere, both vocal and instrumental and every gondola had one or the other, some both, and the Grand Canal was the centre of it all. The lovely voice of some great tenor would suddenly strike out, simply entrancing one, and indeed, in our own gondola there were one or two voices that enraptured me. Altogether it was really almost too exciting.

After gliding about in this way for perhaps a couple of hours, we landed on some picturesque island upon which there were several large restaurants, much used by the people, I believe, on a Sunday evening. On this particular occasion everything was splendidly illuminated. Having landed, we all had a most pleasant and enjoyable supper, with music still going on here and everywhere. I had by this time come to the conclusion that my kind host was a man of means and accidentally heard that only those who, on this particular occasion, could afford to pay the very large entrance fee to land on this island, might do so. I have some sort of vision or remembrance that, strolling about after the supper with some of our party, we came upon a large circular platform with a floor beautifully prepared for dancing. This was all brilliantly illuminated and a splendid large Austrian military band in a light blue and silver uniform, just about to play.

I cannot really say how it came about, but whenever a waltz is played I am generally done for, particularly when it is one of Strauss's best and played by such a band. As the introduction was being played in that soft expressive way, that only these German musicians can express, I happened to catch a glance from a pair of pretty black eyes belonging to one of our party and with whom I had previously spoken a few words during the evening. Her little glance or appeal was easily interpreted, and I suppose the answer in my own eyes were plain enough also, for without any words, I simply lifted my hat a little, made a slight polite bow, and in a moment we were floating and gliding along with many other couples to the delicious strains of the music, much to the amusement evidently of the rest of the party. The waltz at last finished, never to be forgotten by myself and I really do think that my pretty partner also enjoyed it, at all events she looked as if she did.

Having returned again to the gondolas, we had one turn up the Grand Canal and then to our quarters. It was a red letter day for me. When in bed, and thinking it all over, it reminded me so much of those few delightful waltzes I so enjoyed with some of those southern belles at the Clifton House, Niagara Falls in Canada.

In the morning Engel had been able to get a room for me in one of the hotels as I did not like to go on trespassing on his friend's hospitality, although I knew it to be sincere. When I spoke of payment, it was very distinctly declined but I was made to promise, by all the family, that whenever I came to Venice again I would come and see them. I need not say that I took the earliest opportunity, with Engel's assistance, to purchase some few little presents to send to my kind friends.

The next day I went to the beautiful service in St Mark's Cathedral for this particular Mass. I never heard such singing before; it was simply heavenly. Then there was the Cathedral itself to appreciate with all its lovely mosaics, the Palace of the Doges, the Piazzetta and its pigeons and the gondoliers lounging about at this part waiting for hire. In fact I wandered about in a gondola all that day revelling in all the beautiful scenes, by daylight now, and not forgetting the Bridge of Sighs. I spent the evening sitting on the Piazzetta again listening to the lovely military band.

The next day I did the same, never seeming to tire of it all. However, after the third day as time was passing so quickly, I made up my mind to go on to Vienna with Engel for a day or two, finding it would not make more than one day's difference to the time I had allotted myself. I would then go straight home by way of the Rhine. We therefore took berths on the Austrian Lloyd steamer for Trieste, a short distance across the Adriatic. We then crossed the Semmering Mountains by Diligence and reached Vienna.

At first I thought it a very dull old city, not to be compared to Berlin, but its palaces, museums and other public buildings were full of the most interesting relics of past ages.

The Prata, or their Hyde Park was also very bright and animated, the music of their military bands excelling all others. There were some public gardens of great extent called Sperls, much frequented by the military, always in uniform. They were very much like Krolls gardens in Berlin with everyone drinking beer. But why are all Germans as a rule, so gross in their manners and habits, in spite of their great politeness, bowing and scraping? These gardens seemed immensely popular with their concerts and sometimes dancing. There were many splendid horses and equipages about and some of the shops were very good. It was here that I purchased the few little presents for Engel's kind friends.

During my short stay here, I went to the Archduke Charles Hotel which was then the principal hotel. I remember noticing the waiters, how clever they were and how very short and little, but the soldier officers, who seemed to be always swaggering about in uniform with their swords, in an insolent manner, were every one over six foot.

The time now came to say good bye to my friend Louis Engel. What a difference it made to the success of this little trip, that he should have been with me and we parted with a real true and sincere feeling of friendship. Some years later on when it became necessary for me to have a courier, I tried to find him again. I tried the couriers' clubs and other places such as Lee's Agency shop at Charing Cross and Bradshaw's office, but could never heard anything of him. I could therefore only assume the poor fellow had gone the road we all have to go one day.

Leaving Vienna after a couple of days, I went on to Munich and was much impressed by its public buildings and splendid museums full of the most choice works of art. Also the gigantic Bust of Liberty, its beautiful clean wide streets and certainly not forgetting its splendid Beirish beer. But unfortunately I only had one day to spare for it all. I then went on to Mayence on the Rhine and down that river by steamer to Cologne with its lovely romantic scenery and again its meals on deck with a bottle of Rudesheimer. This in itself would set a man up again.

Home at last, a new man and one day within the time I allotted myself.

I was very glad to find all well in the gas works and the mother and daughter at the cottage still alive; and how glad I was to get to my little office again and settle down to real work. I was glad to shake hands with my friend Jimmy Simpson again and tell him of all my adventures in

Venice. I also called on Mrs Mercer and with very sincere respect gave her an exquisite little card case of the beautiful silver Genoese work, that I had purchased for her en route.

Another little matter now also came to a head. Mr Angus Cross, the well-known gas engineer, had invited me to become a Director of a company he had just formed – the Gas Products Company, which was doing very well. The offices were in 10 Coleman Street and the works on Bow Common; this meant that another £100 a year was added to my income. My friend H. P. Stephenson was also on the Direction and his name would certainly not have appeared unless he had seen his way clear. As a rule I generally took my bearings from him and never once had occasion to regret doing so. I always went to him when I wanted advice on finance or investment and believed in him thoroughly.

Mr Croll had also organised a very large business in a gas meter making company, of which I was a good customer, but I always kept myself a free man in these matters. Just as in my purchase of coal, I always went to the cheapest market, but also to the best. Sometimes to Milne of Edinburgh, and sometimes to Glover or Cross's Gas Meter Co. [Glover was the forerunner of the appliance group Glover and Main, which Charles's son Kenneth joined in later years]. So also for my purchases of wrought iron tubing for services, which were now becoming very large. It should be understood that apart from the profit I was making on the sale of gas made, I also achieved a considerable profit every half year on all these other matters referred to.

About this time I asked H. P. Stephenson if he would mind giving me the name of a good broker to act for me now and then. He very kindly gave me the name of his own brokers, Messrs Foote and Adams and they acted for me in some few transactions for investments. Also in some few others which proposed by carefully watching certain stock, to make some profit. I was really very fortunate in all these little speculations.

I can remember very well at that time buying some Egyptian and Turkish stock at a price which gave me about 13 per cent and with great care in watching the markets, I did very well. But my real work was now becoming so large and important and the consumption of gas increasing by such leaps and bounds, that I felt it would be better to give up these little gambles on the Stock Exchange, although so far they had proved very satisfactory.

How little was I then dreaming of the extraordinary change in my future life and career that would occur before another three months had passed over my head.

About the last week in September, the weather was so beautiful and still so like summer, that I was tempted to steal a day off at the end of the week and row up to Oxford 'random' in my boat. This would be if I could find a couple of fellows to join and make a crew for a last row of the season, little thinking it would be my last in that boat.

I asked Jimmy Simpson whether he would make a crew member, but at the last moment I got a letter from him regretting he could not join me and remarking that events were occurring which he would explain later on. As it all seemed a mystery to me, I put it aside till that time came. I therefore had to fall back on Wyatt for one member and two young fellows I knew could row.

Wyatt could not row, or said he could not, so I got over that difficulty by putting on board a good long length of line for a tow rope. When we got to places where the river became narrow and weedy, Wyatt could get on shore and do his share of the work by towing. How I revelled in noting how silently savage and angry he always was when called upon to do this. How he would suddenly sometimes tug at the rope savagely, inwardly cursing when some little shrub or impediment got in the way, almost pulling him flat on his back. Wyatt certainly was the most ill-tempered, selfish man I ever knew, and yet I seemed blind to it and stuck to him and helped him when no one else seemingly cared to do so. It has ever been a mystery and inexplicable to me why I should have continued to be so blind and foolish.

Rowing up river, we of course had the stream against us, but rowing 'randem' we got along very fast and had no luggage with us, except brush and comb or any little thing we could cram in out pockets. I was supposed to be captain of the crew and stroke oar; Horsey had the sculls and Noakes was bow oar. Wyatt was to steer.

I recollect on this same journey when we had made a fresh start early one morning, Wyatt was steering and he ran us right into a spot where some swans were nesting, causing the old birds to make a fierce attack on us. He, in his fright, got up so suddenly it was only by a miracle that the boat was not upset and all of us overboard. In the excitement of the moment I very well remember snatching up the boat hook and giving him a particularly nasty crack on the head and further telling him if he was going to play the fool like that, he had better get on shore and find his way back. I was awfully angry I know. However, things quietened down after a little while.

Although so late in the year, how lovely the river then was, I think almost more so than in mid summer. All was so quiet we seemed to have it all to ourselves with no house boats or noisy steam launches about. We slept the nights at Shepperton, at Marlow and at Shillingford, enjoying the roughness of the accommodation.

I can remember one day we stopped at a very tempting little inn on the river side at Pangbourne to lunch. A splendid sirloin of cold roast beef was put before us, together with a great large loaf with delicious fresh butter, and an unlimited supply of first class beer. In the meantime, while the meal was being prepared, we had explored the little garden and pulled up some young crisp lettuces, thoroughly washing them in the river from the boat; stealing also a few ripe tomatoes, we soon made a salad. How that lunch was enjoyed, the joint of beef when we left the table was simply a bone. We had well earned it, for we had been pulling hard and steadily since 8 o'clock. What a thing it is to be young, strong and happy.

Just before arriving at Marlow a thunderstorm came on half-drowning us. We therefore had to strip, leave all our clothing for the lock keeper to get dried for us and dress ourselves again in whatever clothing he could scrape together of his own and from one or two friends. We had to walk some little distance up to the town to get lodging for the night and something to eat, but had some difficulty about it, looking such queer objects. However, when we explained, it was soon all right. Early the next morning we got our own clothes again in good order and amply rewarded the lock keeper for his kindness.

At Shillingford the next evening, we were late. The man in the wretched little inn we stopped at had nothing whatever to eat in the house, and had to jump into his cart and drive two or three miles to get something. Our sleeping accommodation was also a dead failure.

Having arrived at last at Oxford, we had a good dinner at the well-known hotel, The Mitre. Our boat, which had come along splendidly, was left with one of the numerous boat builders, with instructions to send it back to Kingston. Then we all returned to London by train, my three companions returning to their homes and myself to Kingston.

The next morning I went to the works very early, for the gas accounts for the Michaelmas quarter had to be prepared and sent out. As they were now so much larger and so many, they occupied nearly all my time and most of my evenings. Indeed, I still helped in the collection now and then, although I could see the youth Packham was taking much interest in it all and becoming very useful.

During that week, Jimmy Simpson, whom I had not seen for some time, came in one evening looking, as I thought, rather solemn and at last I got the promised explanation of his letter, which was news indeed. He was engaged to be married to Miss Nelson, the eldest daughter, and was to be married soon and thus my friend Jimmy Simpson's bachelor days ended.

I could not help thinking he was a lucky fellow. She was a very pretty girl, very clever and intelligent and no doubt had plenty of money. I had always admired her, and years later on Jimmy and his wife were my first friends. Her family consisted of three or four sisters and three brothers. Tom, the eldest, then working with his father; James and Jack were then boys at school somewhere and great friends with Mr Simpson's youngest son Arthur Telford, who had just then left his school.

I remember one day when I had been up to London and was returning home in the evening, seeing two boys on the railway platform at Waterloo in great distress, having lost their luggage. A porter had taken it, but they had in some way lost sight of him and the luggage also. However, after a little hunt about it was found and as it was some time before the train would start, the man had put it aside and was busy in another place. This matter being adjusted, they said they were going to Surbiton to stay with some friends at Westfield, we therefore travelled there together.

At this time there seemed to be great trouble and agitation among gas stokers, particularly in London. At last all the men of the Imperial Gas Co., one of the largest with stations at Fulham, St Pancras, Haggerstone and Bow, struck work and came out. This was a very serious matter and also alarming as the greater part of the north side of London would be in darkness (there was no electric light then). As John Aird & Son were the contractors to that company, knowing all the officers and the work so well, the Directors call on the Firm to render what assistance they could in the emergency. Within 24 hours, having called for volunteers from the numerous large works upon which they were then engaged, within a reasonable distance of London, some 2000 navvies responded and within a week the strike was at an end. I remember being asked to lend a hand in coaching these new men and when it was all over, many of them preferred to remain. I need not say how very kindly the Directors expressed themselves afterwards.

11
Betrothal and Marriage (1860) and a business trip to Palermo

Without any premeditation, desire or thought of such a thing an event was to occur which was altogether fated to change my future life and career. This showed how true is the old adage *"l'homme propose mais Dieu dispose"*.

During the two or three years I had been living at Kingston doing my best at the gas works, I had been so interested in my work, and it was so pleasant and enjoyable to me to spend my Sunday with the Simpson family at Westfield, that I had somewhat lost sight of the many friends at home and elsewhere. However, some little time after the Michaelmas quarter, as Jimmy Simpson was still away in the north with the Nelsons, I thought I would try and make this up and go back home now and then on a Sunday, sometimes taking John Simpson with me for he was always a great favourite. Just at this time, however, my old friends the Quicks gave one of their frequent little dances. Their young family was growing up and there was always something of the kind going on. I received an invitation and of course went.

Looking round on my arrival, I could not but notice how Polly, Mrs Quick's eldest girl, had come on. She had now quite become the young lady, still very petite as she always was, but always a charming woman. While remarking to Mrs Quick in conversation upon these changes, I remember noticing the face of a young girl, who at that very time was speaking with Polly, and asked her mother who it was. The reply being that it was little Ellen Noakes, a great friend and school fellow of Polly's. She must have been one of the many little children I had often helped her to amuse many years past, but she had now grown up as Polly had done. The face I was then noticing so particularly was a very striking one, to myself at all events. It was like an exquisite cameo, cut from a piece of the purest white marble, with an expression serene and quiet, which of course made the beauty of the face more attractive.

So the matter passed for the moment, but although there were many of my pretty friends of old present, I took the opportunity of amusing and paying her some little attention - no doubt to the poor girl's surprise and wonder, that a man so much older than herself should trouble to do so.

Well, the evening came to an end and the following morning early I returned to Kingston and to my work. Fortunately nothing could supersede my love for my work, and all the pretty faces in the world could not do that, but what on earth was the matter with me? Instead of at once putting aside this pretty face and also the many other charming, nice girls of my own age, any one of whom would have made a suitable wife in every way had I been so inclined, but hitherto I had been able to avoid, I now found I could think of nothing but this one girl and wondered how I could manage to see her again. The more I tried to pull myself together and realise the folly of it all, the greater the desire seemed to be, it simply haunted me.

When sitting alone at the cottage in the evening after work, I would ask myself what it all meant. Did I intend to persist in this folly and wickedness by marrying so young a girl, who could not yet know her own mind? She was much too young to be accomplished and not even in the least musical, which would always have been a great pleasure to me. What would she think of me in ten years time? Never was man more worried at the position in which he found himself.

I knew nothing of the family, had never been in their house, but had heard from Mr Quick that the father was rather a strange old gentleman and difficult to get on with sometimes. He was a hop factor, then in partnership with his brother Mr John Noakes, living at Brockley. They had recently quarrelled and he had opened a place for himself in Southwark Street. There was no mother and the family then consisted of the two daughters - the elder one, Lizzie, was something of an invalid - and the son Clarence, a restless and excitable sort of youth. I may have met the elder sister and the son at Mr Quick's, but the father I had never seen. My father and mother did not then know the family and I do not think my sisters did at that time.

I determined I would spend the Sunday after that memorable dance at Westfield, although James was still away, but I am afraid they must have thought me very strange.

I have often thought it very curious that this strange dilemma should have occurred to me so soon after James's engagement to Miss Nelson, but I am sure, indeed positive, that I was not in any way biased by it – indeed more the other way.

The following week I still stuck to my work. Indeed it was a matter of necessity whatever my feelings and wishes might suggest. But on the Saturday I had to give it up and went to The Grove for the Sunday. Instead of going with my father and sisters to St Margaret's Church as usual, however, I found myself in the Quick's pew at Sherman's Baptist Chapel by the Blackheath railway station. I had heard the Noakes family also had a pew there and found myself directly facing them. What a wretched, poor sort of a fool I then felt myself to be. I really almost thought myself as foolish as my unfortunate brother Alex, but not quite as bad as that, thank goodness. This extraordinary attack of nerves, or whatever it may be called, did not happen to me until I had firmly grasped my work at Kingston and elsewhere as a director of substantial companies, already making a sufficient income to marry upon and live quietly. There were 11 years more of the lease yet to come and if I had good health, what might I not achieve? However, to cut this long story short, it had to be settled sooner or later.

For several Sundays I attended the service at the Baptist chapel and ultimately got on friendly terms with the Noakes family, my mother very kindly, at my wish, asking the girls round now and then. Mr Noakes was very busy, as it was hop picking time, and he was looking up clients among the farmers in Kent. The elder daughter had a little pony carriage and used to drive about that part of the country, while he was at his work, and stay sometimes at the Crown hotel at Sevenoaks, then a pretty place with lovely views. Sometimes I would be invited to accompany them, which I would try and do just for one day now and then. Mr Noakes and his son Clarence seemed to be very fond of riding and I remember they had an old hunter called Susan for the purpose.

As I have said before, in spite of all this, my work was still everything to me. Thus the matter seemed to drift along for at least a couple of months. I could plainly see that my father and mother were disappointed, but they never said an unkind word or attempted in any way to bias my mind.

At Xmas, the Noakes family themselves gave a dance, it seemed for the purpose of introducing me to some of their friends. I cannot say I very much enjoyed it, for really there was not then any sort of engagement between the youngest girl and myself, simply a kind of friendly camaraderie which made the time pleasant for both of us when we did meet. As for love or love letters, I never knew anything of that sort of pleasure and do not think I could write a love letter if I tried to do so.

This state of things could not go on, however. One day after the Xmas quarter, toward the end of January in the new year and after the usual half yearly meeting of the shareholders had been held (and so very satisfactorily), I had run up to London on some business purpose. I think it was the Coal Exchange and I was about to cross London Bridge from New Southwark Street, when a hansom cab coming from the London Bridge railway station passed me containing two young ladies dressed exactly alike. I can remember as if it were only yesterday that both were wearing very pretty little bonnets trimmed with blue myosotis flower. They did not see me but I determined there and then to put an end to this uncertainty before 24 hours were over. Therefore, having sent a telegram to my housekeeper, I went home to The Grove and in the evening went to Mr Noakes's residence and asked to see him. My request was granted and in a few words I expressed my desire to make his younger daughter my wife. I remember him expressing great surprise that it should be the younger daughter and not the elder, Lizzie, but that could not be altered; Fate had arranged it.

Having also explained my position and means, I remarked that I had not yet spoken or said anything definite to Ellen, but if he would kindly give his consent, I would much like to do so at once, that it might be off my mind. He left the room and after a little while on his return said that if I would go into the drawing room, she would be there. I did so and it was not very long before we came to an understanding. It was arranged shortly afterwards that we should be married in March two months later in 1860.

In the meantime I continued my work at Kingston and elsewhere, which just then required much thought and attention. Occasionally during the week I now and then spent an evening with the Noakes family and on Sundays I would usually make my appearance in their pew instead of Mr Quick's. They were a very quiet family who seldom or never entertained, with no late dinner, but supper of some sort at about 9 o'clock. The old gentleman was a great smoker and after his tea would sit through the evening enjoying his pipe and a glass of grog and the society of any one of his few chums – Oastler, Palmer, Courtney and Groves - who might drop in.

With this matter on my mind and the arrangements that would have to be made about the wedding, I had to give anxious thought to the gas works. Although I had so far got through the winter very satisfactorily, and far beyond my expectations thanks to the additions and improvements I had carried out in the past year, it was perfectly clear, that to meet the enormous increase in the consumption of gas, much larger extensions must be carried out without delay. Looking boldly to the future and not living from hand to mouth, the time had come to reconstruct the works altogether and abandon the old fashioned cast irons retorts and use clay instead. This meant a very large expenditure of capital, but it had got to be done, or lose the great opportunity that offered itself.

The gas works at Sydenham in which I was now so closely interested were curiously in the same condition. It was so difficult to keep the works expanding sufficiently in order to supply the demand for gas. In the Directors' last half yearly report to the shareholders of the Kingston Gas Co. this was mentioned as most gratifying, as indeed it was, particularly to myself. I was requested to prepare plans and estimates for such further extensions as I thought necessary for the consideration of the Directors, as soon as possible. However, I could do nothing about this until the wedding had passed and gone.

What at lot there was now to think of and arrange, but I remember I was very calm and quiet over it all. First of all I made up my mind to insure my life for £2000. This, I was told, was rather a large sum for a young fellow like myself, but I thought I could manage it and have continued to pay the premium ever since without difficulty (now 50 years). The London, Liverpool and Globe Insurance Co. was the one recommended by my old friend Mr Quick. Having sent in my application, I was desired to attend at the office for examination by the company's medical officers. This I did and after being subjected to what I thought was a very severe examination, I was coolly asked to attend again in a fortnight as he would like his colleague, who was at present away, also to see me. Fancy my feelings being left in this state of suspense under the circumstances! However, the time passed and the second time I was accepted, but what a farce this examination is.

I have previously mentioned in these pages that 14th February (Valentine's Day) was always a favourite day of mine, believing it to be always the herald of spring as it was then. Now you never know where you are, or in what season, everything in that respect seems so mixed up. However, on this particular occasion I sent the young lady a little valentine somewhat different from those generally sent. I had arranged for a pretty little locket to be made, about the size of half a crown, the back of it being of blue enamel and in the front there was a large stone in the shape of a heart. This was pierced with an arrow formed from small diamonds and inside on a slip of paper two or three little verses I had managed, although quite out of my line, to put together. I don't think they would have had much difficulty in guessing where it came from.

It was also, I remember, thought necessary that we should both of us have our portraits taken in the form of a miniature. This I had also to go through.

The next matter to arrange was to find a home for my future wife. I had decided that the cottage I then occupied would not be a fit place for her. There was nothing pretty or pleasing about it, with a public house as its next door neighbour, making it on a Saturday night, rather noisy. It also had that hideous mound of earth shutting out all view of the river. The only advantage being that it was only a few yards from Parker's boat house.

Noticing an advertisement in the local paper that a small farmhouse was to let on the other side of the river between Hampton Wick and Teddington, with about 6 acres of meadow and at a rent of £60, I went to look at it. It was not more than 10 to 15 minutes walk from Kingston.

After crossing Kingston Bridge, two roads went to the right through

Hampton Wick to Teddington parallel with the river. Along the lower road there were some half a dozen small villas here and there facing the road, while on the upper road there were only meadows. A little less than midway along each of these roads could be seen a swing gate with a privet hedge and a farm road leading from one of these gates to the other, thus giving a double approach to the little cottage or farmhouse as advertised.

This stood about half way between each road and really was the funniest little place imaginable – only a dolls' house. As the ground sloped very much from the higher road to the lower one, it was perched up on a little level mound, necessitating some five or six wooden steps to get into it.

There were four tiny little rooms on the ground floor and four above, with a very small staircase leading thereto (difficult for any very stout person to negotiate). The place was in a rough but pretty fair condition. There was no garden, but a little bit of shrubbery in front of the wooden steps. The cottage faced south and to the river which, although so near, could not be seen; the only way of getting a boat was by walking to Teddington Lock or to Kingston. It would therefore be of no use to me in that respect, but the place would be near my work and that was everything.

In the little room to the right on entering, called the dining room, there was a side window looking directly into a small farmyard, quite perfect in its arrangements. There was a barn, some cowsheds and cows, piggeries and pigs and the usual dirty, smelly little pool in the middle of the yard for the ducks. There were chickens and also pigeons. The old lady who had the farm lived in another farmhouse half a mile away in Teddington, but two strapping, good-looking girls (her daughters) looked after this yard. They were always hard at work at something, milking the cows, feeding the pigs and poultry and taking the milk over to the other farm.

After some consideration, I decided to take the place on a yearly tenancy. Although so ridiculously small, I reasoned to myself that for so young a wife, it would be much better to come to a little home that would perhaps interest her and give her pleasure, than to go into some place, perhaps one of a row of houses at a big rent with the responsibility and worry of servants. Having thus decided, I sent for James Macrostie, the builder who did many little odd jobs for me and who had very good ideas and taste. I gave him the order to paint, paper and decorate the little place throughout. All the woodwork was to be prettily grained and varnished in maple and the walls were to be papered with the prettiest of the French papers. There were to be new little fireplaces and also a tiny little conservatory was to be built outside the room that would be termed a drawing room, although I very much doubted whether a cabinet piano could be got into it. I also gave him the order for some few little outbuildings at the back: coal cellar, knife house and so on.

The farm road leading from one gate to the other was to be levelled and kept tidy, as the approach to the cottage. Also at the back a piece of the meadow was to be levelled and turfed for croquet (which had just come into fashion) and a few shrubs and standard roses were planted. The cost of all this, when totalled up, really amounted to very little, for it was all on such a small scale. I should mention that Eliza Cook, the well-known poetess of that time, had lived here some time ago and in the shrubbery there was a little grave for her favourite dog.

There was hardly any room for furniture, but I took some few of the smallest things from my bachelor's cottage at Surbiton, which I still held. My mother kindly selected and purchased for me such remaining things she thought it necessary to get, to have the house quite in order.

My bachelor cottage at Surbiton I had let for a month or two to friends of the Simpsons, the young Cuthberts. My boat, which would be of no use to me now, I gave to Arthur, Mr Simpson's youngest son, who had just left school. Unfortunately, when out with it one day, he got in front of their little steam launch, the Kingston, and it was cut down and wrecked.

It was not an unusual procedure, at the offices of the large contractors, for some foreigner to call with a concession for some public work granted to him by his own government - for example, a railway, dock or more frequently gas or water works. This gentleman would endeavour to induce the contractor to take it off his hands and form an English company if it was found to be *bona fide* and of a satisfactory nature. Of course this would be a speculation to some extent, but some very good things of the kind have proved successful.

About four or five weeks before the wedding day, which had been fixed for 29th March, to my great surprise, I was asked by my father, whether I could manage to go to Palermo in Sicily. This would be to examine and report on the particulars of a concession for a gas works in that city which some titled Italian gentleman had brought to them. As a rule I knew the firm did not then much care about foreign contracts, but one gets a big fish now and then. This gentleman and this particular concession, however, seemed to have impressed them. They had also received some favourable information regarding it from a gentleman in the City, a Mr Wheeler, who was doing a large amount of business in that part of the world. The Wheeler family in fact became great friends with us a little later on.

The firm thought it desirable to look further into the matter, particularly as this gentleman offered to pay all the expenses of the journey. As neither my father, nor my brother John could go away, (they were then so busy) they thought of me. Seriously inconvenient as it was, I consented, feeling bound to help my father at any time, if called upon, in return for his great kindness and generosity. I only agreed to leave on the understanding that Mr Wyatt should accompany me as my assistant, but really only in case there might be any delay in negotiations which could prevent my returning on time for the arranged wedding day, the 29th March; in which case Wyatt could remain and, if possible, complete the matter. Fortunately, this was so arranged and as the gentleman holding the concession had agreed to accompany us and pay all expenses, it made matters a little easier. I cannot remember the name of this gentleman, but he proved to be a most agreeable companion.

Having said Good Bye to all and promising faithfully to get back in time for the wedding, the three of us left for Marseilles, taking the night steamer for Naples. Immediately on our arrival we would then take another steamer to cross over to Palermo. Speeding and getting our journey over quickly was everything to us, and yet how cruelly Fate seemed to treat us. It was blowing great guns when we left Marseilles and a storm in the Mediterranean is never a pleasant time.

I have already mentioned that at this time the European Continent was very much disturbed by war between France, Austria and Italy. Poor Italy! The French then had possession of Rome, therefore irrespective of the elements, travelling in that part was somewhat uncertain. I remember also when starting on this journey there had been a very serious earthquake some two or three years before at Palermo.

Looking back over these pages, I seem somehow to have omitted to describe a very interesting little holiday I much enjoyed in this part of the world many years ago, I think before I went to America. I visited Genoa, Pisa with its Leaning Tower and Baptistery with its beautiful echo, then on to Leghorn, Florence and Naples: a most enjoyable time, but what changes since then. After leaving the harbour at Marseilles I tumbled into bed and in spite of the bad weather, was soon asleep. When I opened my eyes again, it was broad daylight and although I could hear the wind howling and roaring as in a tremendous gale, the steamer itself was quite stationary as if on shore. Dressing myself as quickly as possible, I went on deck and to my great astonishment found that the steamer, for safety, had run into a little, tiny, natural harbour in Bastia on the island of Corsica. It was so small that, by leaning over, you could almost touch the wet and slimy perpendicular rocks surrounding us. Looking upwards on the rocks I saw the town of Bastia and here we remained for two days before the Captain would start again; I need not say to our intense disgust. These French boats were then so ridiculously small and badly equipped that they seemed afraid of a little wind, in spite of the Captain wearing a kind of Field Marshal's uniform with gilt epaulets and white gloves.

The great steamers of the P&O and the Orient Companies were not then running on this route as they do now. There was no upper deck and cabins, but a flush deck with a saloon with berths on each side and I can remember well a hard, uncomfortable horsehair couch going round the stern end of the saloon. On the third day the gale abated and the Captain plucked up courage to come out and go on to Naples, but as ill luck would have it, we just missed the steamer crossing to Palermo and had to wait two further days. This made me seriously anxious and I could see our companion was in much the same state of mind. The weather, however, was now fine again and although bitterly disappointed at this continued delay, I need not say how much impressed I was again with the beauty of Naples and the entrance to the beautiful bay. Vesuvius was then in eruption and at night when dark, it was an extraordinary sight to see even from the windows of the hotel the red-hot lava so solemnly and continuously rolling down its sides; it was really very impressive.

Our companion, in spite of his vexation and annoyance was very kind in showing us many interesting places and things, taking us to Pompeii one day. This was awfully interesting and what impressed me more than anything there were the deep ruts made by the chariots running along the narrow streets, paved with blocks of lava and with only the one fixed track. Then the museum again, the opera with a dinner before it at one of the best restaurants and being, I remember, much impressed by a cockle soup.

Going into our hotel one morning and passing by the side entrance, I was reminded of my favourite old proverb: "Where ignorance is bliss, 'tis folly to be wise". I saw two old women squatting down by the door and engaged in skinning alive a quantity of frogs they had in large baskets, for our dinner.

Looking at some of the shops which were very pretty and good, but awfully expensive, I tried to find some little thing in pink coral that I could take back with me for a wedding present. On my mentioning this to our companion, he remarked that he knew a man from whom I might be able to buy something first hand. We went, therefore, into one of the old streets at the back part of the City, so narrow that one could almost jump from an upper window into another on the opposite side. In nearly every one of these old narrow streets, many makers of macaroni lived and it was interesting to see in nearly all of them strings or lines across from one side to the other with long lengths of macaroni drying on them. However, entering one house and ascending I do not know how many floors, all in a very dirty and smelly state, we entered a room where there was a man, evidently a working man, in the trade. Our companion explained the object of our visit to the man, who said something in Italian, went to a drawer and placed some few pieces of coral on the table. Seeing that they did not particularly strike me, he went to another drawer and brought out a very large brooch of the purest and most lovely pink coral I have ever seen. It looked like a bunch of grapes formed together on gold wires and it was almost too beautiful for me and my pocket.

The man of course asked a price that was out of the question, but at last I obtained it for £13. Several years after this the brooch got a little bit damaged and I took it to Phillips, the well-known jeweller and coral merchant in Cockspur Street to be put right. Phillips himself received it from me and was most anxious to know what it cost me. When I mentioned to him how I had obtained it, he said he had nothing in his stock of coral to equal it and when I told him the price, he shrugged his shoulders and said that if he had it, he should want 40 or 50 guineas for it. If I had then been a little richer and could have spared the money when I was in Naples, I have no doubt now, looking back, that I could have picked up a few good pearls in the same way. I have often since regretted that I did not try.

In the evening of our second day we embarked on the steamer for Palermo. Never shall I forget the exquisite view before me when we arrived at about 7 o'clock the following morning. It had been beautiful enough when we entered the Bay of Naples, but this, although on a much smaller scale, beat all creation (as the Yankees say). It was like one of those fairy scenes at a pantomime, not real. As we entered the lovely bay or harbour, the mountains seemed to form themselves into an immense cockleshell at the back, and as the sun rose, it tinted the edge of them with a lovely golden rose tint. Palermo has always been called, I believe, the City of the Golden Shell, but what with this beautiful effect, the picturesque view of the city itself and the lovely blue sea, it really was almost too much to absorb.

On landing we took up our quarters at the Trinac Mia hotel, a very comfortable place in the leading street. The first thing to be done was to call upon the Viceroy, Prince Pallavachini, at the palace to pay our respects, leaving cards and writing our names in the callers' book. Having done this, in company with our companion who introduced us, we were ushered into a very palatial looking reception room and the object of our visit to Palermo was explained to him. As this was only a preliminary and formal visit, we did not remain long but returned to our hotel and a very short time afterwards the Prince himself called and left his cards. These foreign noblemen are, I believe, very particular about such matters; his cards were so magnificent that I kept them and on my return home put them away with many other little relics. After this I called upon the English firms to whom I had letters of introduction, Messrs Morrison, Rose and Co. being, I think the principal. They were all very kind and, in our view, did their best to assist us.

As time was indeed pressing, I got to work at once collecting all the information I could to enable an estimate of the cost to be made, seeking also the most suitable site for a gas works and interviewing the owners of the land. As to the cost of it: price of materials, price of coal, probable revenue from sale of gas and the price that could or should be charged per 1000 cubic feet, and having many interviews and discussions with the local council on the subject.

I should mention perhaps that at this time I could manage a little of the French language, enough just to get on with. In my own way I could understand and read a newspaper or book and, if spoken to slowly, I could understand and even speak a few words to be understood. How often have I regretted not having followed this up.

After a little while it seemed to me that there was a strong adverse feeling about the project in the Council and also among many of the citizens. The whole country was so disturbed just then that it was an unfortunate time to attempt to bring out such a matter. Dining with the members of the English firms on several occasions I found that this was quite their view. They would one and all be glad to have gas introduced but they feared that the prospects of an English company formed at this time would not be a success. I also fancied our companion was beginning to have the same opinion, but he was careful not to show it in any way. However, it would be for the Firm in London to decide, after they had seen the information we had collected to put before them and to enable them to prepare a prospectus, if so decided.

We had two or three further interviews with the Viceroy, but they were really only matters of form. On one of these occasions I remember a rather amusing incident occurring. After the usual bowing and compliments and so on, two gorgeous footmen entered with large old silver salvers, upon which were some little cups of the most exquisite dainty china. They were filled with something very dark, thick and stiff which one could not drink or sip; and with no spoons, so it was a little awkward for Wyatt and myself for the moment. But what a great sigh of relief when another footman entered with a silver basket containing some delicious long finger cakes in lieu of spoons, and thus put us out of our misery.

While at work collecting all the information we possibly could and attending the several meetings and discussions with the local authorities, (sometimes very warm between them and our companion), my eyes were always wide open to take in anything and everything that would give me a lasting remembrance of this beautiful and interesting city. On every morsel of land round about oranges, lemons and delicious melons were cultivated and the air seemed to be simply saturated with their perfume. As for the beauty and brilliancy of other vegetation, flowers particularly, it would be impossible to attempt a description of it all.

On the first Sunday, Wyatt and I drove to Monreale, a suburb on very high ground at the back of the city, and what a magnificent view there was!! Never to be forgotten. Here also was their beautiful and interesting monastery and the wonderful, but awful, burial place, the catacombs, connected with it. In these catacombs underground were a series of corridors of many miles in length, we were told, in which were many, many thousands of bodies. These had simply been embalmed and then, in their clothes, fastened standing up in some way on each side of the walls of these vast passages or corridors, side by side with a large label fastened in front giving name, age and cause of death. It really was a most extraordinary and repulsive sight.

I do not remember how many miles of these corridors then existed, but to walk down one was enough for Wyatt and myself, for a deadly faintness came over both of us and I believe we were carried outside again by the monks. It really was too ghastly to see these dreadful objects standing in rows along the walls, leering at you, some with only half a face.

In the hall, or entrance chamber to these vaults, I remember the walls were pasted all over with innumerable rough paintings, depicting in the most realistic manner, any death caused by accident, quite sickening enough without going below.

It was also very interesting to observe the strange effects of the earthquake that had occurred here some two or three years previously. We could see the enormous rents or cracks made in many of the buildings, in the roadways and particularly in the large heavy granite stone seats fixed round about the city. These seats consisted of a slab of granite six or seven feet long and six or seven inches thick placed on two little square blocks at each end. A number of these seats could be seen broken or split in half.

When dining with the English merchants, I was much interested in observing that it was always the custom before soup that macaroni was served round. It was most beautifully cooked, each person taking, I thought, enough for a meal, but it was only an introduction. In the centre of the table there was always a large centre dish filled with oranges and it astonished me to see how many were eaten by the guests. I also noticed that the peel of them was carefully put aside and afterwards heard that many liked it put into their morning bath; not a bad idea!

By this time Wyatt and I had collected sufficient information and matter for our purpose and we seemed now to be waiting only for our companion and the local authorities to agree to arrange matters, if they could.

Apart from my anxiety to get back in time for the wedding day, I was most anxious to get back to my work, for although I had taken the greatest care to leave everything safe for the time I had arranged to be absent; in the supply of coal and other matters, my work was still my love. I therefore decided to return alone at once, leaving Wyatt to await events one more week and giving him written directions. Our companion was quite satisfied with this arrangement, but evidently grievously disappointed at his failure so far. I said Good bye to my kind English friends and left.

The next morning I once more entered the Bay of Naples and on landing, secured a berth in the Messagerie Imperiale steamer, leaving that evening for Marseilles. A boat precisely similar to the wretched one I came by, as I was sorry to find. The evening came and with it the wind which looked very like another storm. This is the worst of the Mediterranean; you never know where you are with the weather. We left at the proper time, however; there were very few passengers and I think only one Englishman beside myself. I soon found out he was a military man, a Captain in the

Royal Horse Guards and a Queen's Messenger. Whoever he was, he was a really good fellow and the largest man in size for a soldier I had ever met, but I cannot for the life of me remember his name. It is possible that in the position we found ourselves in later on, I never knew it.

Soon after leaving Naples, it went round that the steamer was to call at Chivita Vecchia and embark 400 troops, returning from Rome. This was dreadful news as it meant further delay and the weather was getting worse and worse. We arrived at Chivita Vecchia the following day and these soldiers embarked. They looked a very poor and sad lot, very young and mostly conscripts I should think. Every man when he came on board wore a rough, light blue flannel overcoat down to his ankles, the usual little peaked chapeau and all his accoutrements, knapsack, rifle, pots and pans and so on, all of them looking very miserable and sickly. "But where on earth are they going to put them?" my new military friend cried out, who no doubt had seen a lot of military movements. Their officers had managed to get some kind of protection for themselves from the awful weather by arrangement with the ship's officers, but for the rank and file, there was absolutely nothing but the open deck and there they remained huddled together like sheep for two days and nights.

The English captain and I got very friendly, being alone, as it were, and although the awful weather made it difficult to get on deck, the vessel rolling so, and the rain coming down in torrents, we managed to do so and see how those poor soldiers were getting on. We were, however, so shocked at their miserable condition that we determined they should have something to help to cheer them up and, if possible, get a little warmth. We went round among the passengers and asked a contribution for that purpose: we all subscribed and I hope with some benefit to the poor fellows. I remember when this meal was distributed, how grateful they all looked.

Some really good, strong soup had been made in spite of the weather, and served in some large iron tanks, and the wine we had selected for them was of far better quality than the poor, sour thin stuff they usually had. Taking it all together, this little gift to those poor creatures was a great happiness, both to donors and recipients. The weather was so bad that our own meals were prepared under great difficulties and at irregular hours;

I do not think more than eight or ten passengers appeared at table. The English army captain was evidently a bad sailor. He was at the table for every meal and enjoyed it, then a short time after, having relieved himself by sickness, he would be quite ready for another. He was always cheerful and pleasant in spite of the threatening, dangerous weather.

Having been out now, I think for three days, it was clear to everyone that we were too deeply laden. The vessel could not rise with the heavy waves, but could only drag along slowly like a waterlogged ship, and no wonder as we were shipping a quantity of seawater in addition to the awful rain. When night came on we were shipping some tremendous waves. A great wave would make a rush at the vessel, lift itself very high and then, everyone looking very anxious, would plunge down on the deck, taking all the breath out of the poor fellows huddled together upon it.

Another good meal was provided and given to these troops under great difficulties, for it was very difficult now to move about at all. The night of the third day came on and the gale having increased to an alarming extent, the troops were put on to the pumps. The vessel was now taking in so much water it was considered necessary to batten down the two hatchways and the passengers were therefore prisoners.

Whatever others on this wretched ship may have thought, I myself sincerely believed we should never see land or home again. Neither did the Englishman who had much more experience than I had. Neither the Englishman nor I had taken off any clothing, but lay stretched on the hard horsehair bench at the end of the cabin, waiting for the last final plunge. Even now I can feel that awful sensation of down, down; neither of us cared to speak. But as the Scriptures say, "Joy cometh in the morning", for when the next day came, the hatches were opened and although there was still a fearful sea on, we were told that if things were no worse, we should get into Marseilles harbour during the day. I need not say what my own feelings were, coming to life again after the certainty of death.

After a most welcome cup of coffee, I went on deck. What a sight of destruction and misery! The poor soldiers were still lying huddled together on the naked deck. Whether any of them had been washed overboard during the night I did not learn, but they all looked, I remember, more than half dead possibly exhausted by the pumping. As for myself

I seemed almost paralysed and thought it was all a dream. All the boats and most of the officers' deck cabins and nearly all the bulwarks had been washed away, and yet after all we had been spared. When the English captain and I found ourselves on shore once more, the sudden reaction and joy of coming to life again was almost too much. Indeed it was a great wonder that I did not break down altogether with it all. However, after a comfortable quiet meal at one of the hotels and a good bye to my friend the captain, I left for home by the express, not stopping a moment in Paris or anywhere if it could be avoided.

At last, home again and what a time it does seem since I left, but at all events I was in time and only just a few days before the wedding day. Having given my father and brother John all the information I had collected relative to the Palermo expedition and presented my little wedding gift of the coral brooch, which seemed to give a good deal of pleasure, I rushed off to Kingston, very, very anxious to see how matters stood after this long absence.

As regards my future little home to be called "Broomhill", it was very satisfactory to find that a very decent-looking woman, selected to be our one servant, was in charge. But in a couple of hours I soon had a few things altered to my way of thinking and arranged for certain tradespeople to call and so on. To give an idea of the size of the rooms, if four persons ventured to sit at the table in the dining room, they would have some difficulty in moving their arms and our poor maid could not possibly go round the back of the chairs, but had to stand by the door and pass the dishes across. It certainly was a very pretty little dolls' house or, as some of our friends thought proper to call it, "Love in a Cottage". But that was a mistake, there was no time to do anything then, with nothing but work, work for me.

I was certainly not pleased at conditions in the gas works and it became quite clear to me that what was to be called our honeymoon must be as short as possible. When I had once more got settled down, there would be no more going away until things were put right. The prospect at the works at that time of carrying out the plans for the great alterations and additions that must be made as soon as possible, was alone enough to make any man serious and thoughtful, quite apart from the existing unsatisfactory state of things.

In the face of all this trouble and work before me, I could but again realise how cruelly I must be sacrificing all the poor girl's future happiness and I determined to do whatever was in my power to make it otherwise.

If Fate had arranged that we were not to come together, Ellen would in all probability have remained disengaged and enjoyed her young life for another five or ten years, and then married some clergyman. She would then, no doubt, have been a very happy woman instead of being tied to a man, much older than herself, and who for the latter part of his life, because of a serious breakdown in health, never knew what happiness meant. But it is no use moralising now. I have simply to write down matters from time to time as they actually occurred.

It was a great mistake my having accepted the invitation to go to Palermo for the firm, but it was done for the best. About ten days after I had returned, Wyatt came back alone, having had beautiful weather, (such is luck) but without any further progress having been made as regards the concession. My father and brother therefore decided to withdraw altogether from the negotiation, and it was well they did so. Some four or five months later, Garibaldi, with about 1000 of his Red Shirts, (volunteers or rather adventurers and many of them English) advanced from the back part of the country and made a desperate attack on the City. After several days fighting, he drove all the Neapolitan troops out of Palermo. As we know now from history, a short time afterwards he followed up his success by also taking possession of Naples and finally making Victor Emmanuel, then King of Piedmont, King of Italy.

I think my wedding to Ellen Noakes must have been the most matter of fact one that ever took place; at least I seemed to look at it in that light. Leamington had been suggested as a quiet place to go to, but I did not myself care where. Rooms were therefore taken at the old Regent Hotel. On the wedding day itself I remember feeling as unconcerned as possible with not a particle of excitement. After breakfast, having quietly read the newspapers, I went up to my room again to put on another necktie; my sisters thinking the one I had on was not suitable. While engaged on this, I remember hearing a great noise downstairs and presently my bedroom door was banged open and Clarence, the brother, who was to act as best man, rushed in. He was in a great state of excitement, shouting we should not be in time and that he had left his sisters at home almost in hysterics at the parting. Having tried to put him in a better state of mind, I went downstairs and drove with him to Lewisham Old Church. There were but few guests and the church was very quiet. After the service, which seemed very soon over, we returned to Mr Noakes's residence for the usual breakfast and view of presents.

Wedding presents at that time were not much thought of, I fancy. Except for a cabinet piano from Mrs Noakes, a drawing room clock from my father, some house linen from my mother and a silver claret jug from Mr Quick, there was really nothing to speak of. After the usual breakfast and absurd speeches (in this case, very much so) I need not say how glad I was to get into a carriage waiting to drive us both to Euston station. We were met by one of my father's clerks, old Mr Wallis (the father of George and William Wallis) who had already taken tickets for us and reserved a compartment. We thanked him for his kindness and kind wishes, which I felt were sincere. I remember well how quietly and pleasantly we travelled along, arriving at the hotel in time for dinner, as previously arranged. It was a comfortable, old-fashioned place and judging by appearances, was well-known to young married people. It had been a very tiring day and although I had taken it very coolly and quietly myself, I could see after dinner that my wife was pretty well done up and suggested her early retirement. An hour afterwards I found I had tumbled off to sleep in my chair, for which I was very sorry and vexed.

The following morning we strolled about the place and had some pleasant little drives round about to Coventry, Kenilworth, Warwick Castle, Stratford on Avon and so on. As it was only a few hours to get to the lovely scenery of North Wales, we decided to move on to the Penrhyn Arms at Bangor for a few days. We visited the Menai Straits, Carnarvon Castle, Beddgelert, Capel Curig, Llanberis, Llanwest and other pretty places with their interesting little inns and trout streams. They reminded me of days gone by when I did all this alone with a knapsack on my back.

We then returned straight home to our dolls' house of a cottage. Although it was so ridiculously small, I remember it pleased me very much to see how amused and delighted my wife was with it all, and her one servant maid.

FAMILY TREE OF CHARLES AIRD - born 1831

Charles (1831-1910)
married 1860
Ellen Louisa Noakes 1843-1945

Ellen (1861-1947)

Edith (1863-1954)

Charles (1863-1927)

Lily (1864-1943)

Margaret (1866-1938)
married 1896
Alfred Dent Kt. K.C.M.G.
(1845-1927)
- Leslie (1897-1977)

Leslie (1868-1876)

Mabel (1969-1956)
married 1905
Westgarth Duval (Rev.)
(he died 1946)
- David (1914-2001)

Sybil (1870-1916)

Kenneth (1871-1941)
married 1913
Dorothy Helen Miller
of Belvedere, Glasgow
(1890-1966)
- Peter Stirling-Aird (1915-2004)
  married
  I. 1942 Penelope Stirling
  (1918-1971) : 4 children
  2. 1973 Lady Margaret Boyle
  (1920-)
- Penelope (1918-1994)

Winifred (1873-1931)

Agnes Mary (1877-1950)
married 1909
Ernest Jorgensen
(1875-1947)
- Mary (1910-2001)

Our kind relatives had got everything in order and it did not take us long to settle down; but how glad I was to get back to the gas works again and see how matters really stood. The first thing I ascertained was that Strachan, my working manager, taking advantage of my unfortunate absence, could not properly account for the money he had collected. This was serious, and I had no alternative but to give him notice to leave my service and put the matter in the hands of a solicitor. I therefore asked Mr Hargrove (the gentleman who was engaged to my sister Jessie) to act for me. A few letters with threats of prosecution caused him to put money matters right again.

This trouble was bad enough, but it upset me much more to find so many complaints about the supply of gas, showing that proper attention had not been given to the works generally: so much for trusting to others. However, I determined I would not be disheartened, and went round among the principal consumers expressing my regret. I also took the opportunity of mentioning the great enlargement of the gas works the company was about to undertake and that in a few weeks I hoped to have a more suitable man residing in the works as Manager. I had already inserted an advertisement for someone suitable in the well-known journal of Gas Lighting; all this took time, but it was worth it.

It was now, I think, about the middle of April when I began to prepare the many drawings for the proposed large extensions requiring much care and consideration.

There was a little room attached to the manager's house which he had used as an office. This I had cleared out, and put Packham in it to take charge of coke sales or payments for gas, so that I should not be interrupted in my work. Packham was still very young, but very trustworthy. How I did work then from early morning to late, never going away to London, except to my Directors' meetings and perhaps the Coal Exchange. There was no time for boating in the evenings, or even proper meals, more frequently a sandwich in my pocket for lunch. Sunday was the only day to get a little rest.

I do not know what would have become of my wife just then had it not been for the many kind friends about, who were anxious to see and know her. Dear old Mrs Simpson, Miss Nancolas her sister, young James

Simpson and his bride, and Mrs Mercer who, when she called, brought with her the little silver Genoese card case that I had given her on my return from Venice and presented it to my wife as a wedding present. Some two or three of the young married ladies, who preferred to call at my office and pay their gas accounts, found their way to the Dolls' House and made friends with her; one particularly. Looking through some old albums recently, I came across some interesting old photographs, among them one of this lady, which I have placed with some other few odds and ends connected with these memoirs.

How well I remember the first Sunday walking with my wife along the riverside from Kingston Bridge to the Chapel at Hampton Court Palace, this took something under the hour, on a lovely spring morning in the middle of April. On arrival at the Chapel, the Verger, seeing my very lady-like companion was, I noticed a little more attentive than usual in securing seats for us, of course in the arena. I could not help fancying that some of the very aristocratic-looking ladies in the raised side benches or pews, were taking stock of my wife. However, after the service and a stroll round the beautiful gardens, we walked home by the Home Park road to our first Sunday dinner. Whenever the weather was favourable, this was always our occupation on the Sunday, perhaps now and then dining at Westfield.

I might mention that when I did have occasion to go up to London after my marriage, I got considerably chaffed by my business friends, Stephenson Clarke, H. P. Stephenson, Robert Christie and fellow Directors, but it was all taken in good part.

12
Enlarging the Kingston Gas Works

The plans I was now preparing for this great enlargement of the works were to lengthen the retort house and build a new chimney shaft. To erect inside three new retort benches for seven retorts in each, instead of five as before, and these were to be of fire clay instead of cast iron, with a larger hydraulic main, ascension pipes, mouthpieces etc. There was to be a new coal store to hold at least 400 tons and a much larger condenser.

The small, old gasholder and the iron tank were to be removed to the level of the yard. The seven feet remaining below the yard to be covered by iron girders or brick arches to be made into a tank for stowing tar and giving larger space in the yard for storing coke. The purifying shed was to be enlarged and two larger purifiers were to be fitted. Near this a small building was to be erected by the engine and boiler house to work an exhauster which would take off the back pressure on the clay retorts. There was to be a larger station meter governor and bye pass with all necessary underground connections. Finally a gasholder tank and gasholder were to be built of such dimensions that would store sufficient gas for 24 hours use, and meet our demand for several years to come.

I felt convinced that it is but poor economy to tinker on and work from hand to mouth as we seem to have been doing. I did not intend that any quantity of water in the soil should this time prevent a tank being made to a suitable and proper depth. I cannot quite recollect the exact dimensions of the one I was then preparing for, but I think it was about 120 feet in diameter by 28 feet deep – no doubt a large tank for what one would then term a country gas works, but justified by the rapidly increased demand for gas.

In answer to my advertisement for a manager, there were nearly a hundred applicants. Having weeded them out to about six, I finally decided upon a man called James Watson, who was then manager of the gas works at Hexham in Northumberland. A rough diamond perhaps, but a thorough, practical all-round man, about 35 years of age with a pleasant quiet manner. He was married, with 3 little children; I think we both liked the look of one another. I never had occasion to regret his appointment and this was a great weight off my mind.

He had to give a month's notice, but thought he would be able to make a start at Kingston in about a fortnight. He had a brother also engaged with him at the same works who would be glad to come south, if he could be sure of getting some employment. As I then wanted another main and service layer, I told him he could bring his brother also and found him almost as useful. Some time after this the brother got an appointment as manager to the gas works at Buenos Aires.

In the meantime I had the manager's house thoroughly put in order and made as comfortable as possible for them. Some three weeks afterwards, when entering the works, I noticed three little children playing about in the dirty muddy road in front of their new home without any shoes or stockings. This I remembered was the habit of children in the north.

During this busy two months additional work, my old friend Mr Mercer would still trot round once or twice a week to see how I was getting on, taking a great interest in everything as it developed. When I had completed the drawings for the new gasholder and the surrounding tank, he seemed somewhat alarmed at the large size and particularly at the depth. He remarked that with the existing tanks (which were of cast iron) they could not get any lower than seven feet in consequence of the water level and he felt sure I could never accomplish the depth I intended. Having explained to him how different things now were in this respect, he gradually came round to my views and asked me whether I would mind, when all the plans were completed, bringing them up one evening to show Mrs Mercer which I was of course very pleased to do.

It was, I think, about the middle of June when I was prepared to submit all my plans, specifications and estimates to the Directors. The Secretary having called a meeting as usual for 8 o'clock, we all met to discuss this, to me, very important and serious matter.

How well I remember that particular evening. It was a pouring wet night when I left my little home and walked to the meeting. Earlier in the day I had sent all the plans and papers to the Secretary, that they might be seen and examined beforehand, if desired. The expenditure of capital the previous year seemed large, but was a mere flea bite to that now proposed, for all the several items, when totalled up, came to a large sum of money and made me very anxious. But there really was no help for it. In spite of the soaking wet night, however, the Chairman, Mr Simpson and every one of the Directors were present and all the plans and papers were on the table for their inspection.

Having been carefully examined and many questions asked and afterwards discussed in a very kind and pleasant way, there really did not seem to be any opposition whatever to any single one of my proposals. The size of the gasholder, particularly the depth of the tank, evidently staggered some of them, but having explained the great advantage and the immense saving with it, and also how I should go to work and take all responsibility, there came a pause for a little while. I remember that made me feel very, very shaky. One of the Directors then said it was a great responsibility also for them, and although they one and all had the greatest confidence in me, the estimate for all the work was very large and they felt that in justice to the shareholders they ought to have the opinion of an expert upon the matter.

I think it was Mr Ranyard, Mr Mercer's son-in-law, who then suggested there could not possibly be a greater authority or expert than their own Chairman, Mr Simpson, who was then President of the Institute of Civil Engineers; Mr Simpson, in reply to this, in his usual quiet, slow manner, remarked: "Well, if the Secretary will send all the plans and papers to my offices at 29 Great George Street, I will have them carefully examined and checked and if he will be good enough to call another meeting in about a fortnight's time, I will give you my report upon the matter". Of course, I expressed my willingness to this and it was so arranged. It had, I remember, to be postponed one more week.

During that interval, it would be impossible to express my anxiety as to the result. Not in any way as to my plans and figures being disputed, but as to what course the Directors might feel bound to take under the circumstances, the work being so very large. I need not say I never went near Westfield or Mr Simpson in the meantime and I had plenty of work to occupy my mind. The time soon passed, however, and the Directors once more met to come to a decision.

Mr Simpson handed in a report something to this effect; that he quite approved of the alterations and extensions proposed by Mr Aird. If they erred at all, it would be on the safe side, that is to say looking well forward. The figures regarding the quantities and the estimates had been all carefully gone into and carefully checked and found to be correct and reasonable.

The gasholder tank, in consequence of the quantity of water likely to be met with at such depth, must necessarily be an anxious work, but he saw no reason why it should not be carried out successfully. There was but one little alteration he would suggest to the plans, that the new chimney shaft should be built somewhat larger and look more important than the one proposed. He always thought any tall chimney shaft on a public works might be made ornamental as well as useful. It might cost a little more, but it was always worth it. He had therefore directed his people to send a sketch to Mr Aird of a chimney in the form of a tower, similar to the two he had recently erected at the water works.

As far as I can remember, this was the pith of Mr Simpson's report. A short discussion then took place about the tank and I explained that from some borings I had taken, I hoped to build the wall round it on the blue clay. There seemed nothing more to say after this and it was moved and seconded that Mr Aird's plans and estimates be approved and accepted. This was done and when I had said Good Evening together with a few other sincere words, and left that little room for my home, there was no prouder or happier man than myself. The first thing I did was to write a few lines to my father, telling him the result of the meeting, knowing how pleased he would be.

The next morning I got to the gas works, I really think almost before my breakfast, so anxious was I to make a start with it all. The first thing to be done was to peg out all the work, then write to Westwood and Wright, Charles Horsley and Mr Simpson and Co. accepting their several tenders and begging that no time should be lost in the delivery of the iron work.

My first order for the clay retorts and all the necessary fire clay tiles and bricks etc. I sent to Joseph Cowen of Newcastle, (the then well known radical Member of Parliament). A few days after, to my surprise, he himself called at my little office to thank me and came again several times after during my lease. Although my political view were quite different, I certainly liked the man, but I gave orders to other manufacturers as well.

It was always the custom for contractors to sublet a good deal of their labour as piece work, such as brick work, excavation etc. Therefore, I sublet the labour in erecting the new retort benches and setting the clay retorts and furnaces to my retort setter. When writing to my father telling him of the kindness of the Directors, I asked him if he could recommend a master bricklayer and also lend me a ganger and a couple of timbermen for the gasholder tank. The timber for the trench, such as struts, runners and wallings I should of course have to purchase; as the trench, in consequence of the loose gravel soil and the quantity of water expected, would have to be securely and strongly timbered (what is technically called "close-timbered"). It meant a very large quantity, but a good deal of it could be sold afterwards.

The timbering of large works on deep excavations for docks, dams or trenches is really quite an art and it is most amusing and interesting to watch one of these man at this work. Each man has his assistant or labourer up above the trench, waiting for the order for some special piece to be cut and lowered down, perhaps a great heavy strut eight or nine feet long. It is fine then, to see him take up a great, heavy wooden beetal or hammer and drive it with enormous strength right home. How fascinating this sort of work then used to be to me when young; years afterwards one ceased to take notice, there was so much.

Although so many years ago, I can see as if only yesterday these three men coming on the work. The ganger named Gregory I can remember with the slow, heavy manner these men always have, and his two large timbermen. Gregory was a little fellow, but quite all there with his work. At this time I knew many of the men working for the firm of John Aird and Son and was known to the men also, having accompanied my father or brother sometimes on their visits to any contracts that would be interesting.

How enormously their business seemed to have increased lately, both at home and abroad. They were just then completing a large contract in Calcutta for a new water works in partnership with Mr Thomas Brassey and Mr George Wythes. My younger brother Joseph, I remember, was out there on the work for some two or three years, having previously worked sometime in Messrs Lucas Brothers' shops at Lowestoft.

Returning to the work at Kingston: when writing these lines, I ask myself why was I then so interested and anxious about this new work? The answer of course was, because it was my first effort in construction. I was then young and anxious for success and gave heart and soul to it.

I little dreamt or imagined at that time that some few years later I should be responsible for the successful completion of a number of gasholder tanks, all being built at the same time in various parts of the kingdom of far larger size and far greater difficulties than the one I was now about to construct at Kingston.

The first thing to be done, having set out the trench wide enough for the wall, the puddle behind and the short struts in front to keep the wall in position, was to sink a shaft or sump alongside. This had to be at least a couple of feet deeper than the foundations so as to have perfect command of all water that might filter in and to keep the trench as dry as possible while the wall was being built.

I therefore went up to London to a firm of engineers near Blackfriars Bridge, Henry Sykes and Co. who let out on hire powerful pumps and portable steam engines of all sizes. I selected a large 20 horse engine, a large double headed 18 inch pump and two large hand pumps, to be sent to Kingston as soon as possible with the engine driver.

The ganger and his two timbermen now made a start with the sump, about ten feet square, closely timbered. Very quietly I started feeling my way down about six feet and when they reached the gravel, there was no doubt about water; but the men put on their great water boots over their thighs and stuck at it. Two large double headed hand pumps were also hard at it, keeping the water just a little under control, which enabled us to get down to nine feet. Pumping by hand after a certain time, however, is very exhausting and very expensive, as it has to be done in shifts and relays; in spite of the frequent shouting of the men below to pump up you -- blank -- , they were really very good natured and plucky over it all. The water was increasing so fast now, over their boots, that they were really half drowned, so after thinking it over a little, I decided to have them out of it and try another way. I knew it was no use putting in the steam pumps yet, until the sump was much deeper, so I decided to send a messenger up to Belvedere Road and ask them to lend me one of their divers for a few days, plus his assistant, dress and so on. I had a good deal of trouble with water while getting the sump down; but there was no occasion to say more, they would know only too well on their own works, what the difficulty meant.

In the meantime, the men could make a start on the trench itself and get down the top frame of timber, independent of any water. Several other timbermen had found their way here having followed the ganger, Gregory. It is curious that on works when a ganger is shifted from one work to another, how men will follow him.

The next morning early, as I arrived at the works, I found the diver already there waiting for me (this looked like business) and two or three hours afterwards a light cart drove into the works straight from Belvedere Road with the diver's assistant, dress and the necessary apparatus. Directly after the dinner hour, he dressed and descended into the sump which was then about nine feet deep. The water since the pumping had ceased, had risen to within a couple of feet of the surface, showing what a tremendous head of water had to be beaten back in the area round about. The diver when at the bottom, then commenced the further excavation and timbering; it was of course slow work, as he had to come up now and then to rest a bit.

The townspeople had understood the gas company were going to enlarge the works considerably and construct a much larger gasholder and tank, some 28 feet deep. As the two last tanks were only about seven feet deep in consequence of the water level, cast iron tanks were constructed about ground. It was therefore a general subject of conversation, how as it going to be done? Was Mr Aird going to pump the Thames dry and so on? But when it was known that a diver had arrived and was at work endeavouring to sink a sump down to the required depth, the excitement and curiosity was indeed great. I was obliged to have the entrance gates shut and to give orders that no one was to be allowed to come inside, except a director or shareholder of the company.

After three or four days' work, the diver had managed to get down another eight feet; this I thought would now be deep enough to fix and work the large pumps. As a diver's work is always very slow and perhaps more expensive than any other method, I decided to send him back, giving him and his man a little *douceur* and a letter to the firm. We then had a strong timber staging put together for the double headed 18 inch pumps, got these down the sump, fixed them with the steam engine and started pumping.

Although throwing a splendid stream of water, which was conveyed to the Thames by a large pipe temporarily laid across the bit of meadow intervening, there was no perceptible diminution in the level of the water. I felt certain, however, from the quantity of water that was now pumped that the head of it was being reduced and I must therefore have patience. After a few more days, just to make sure, I ordered another set of pumps and engine from Sykes's firm. I might mention that, when preparing the estimates for this work, I had put in a liberal sum for the pumping, anticipating some trouble, therefore so far, all was right.

Having got the additional pump and engine to work and steadily pumping, after some days, perhaps a week, to my great joy I fancied the water in the sump was getting lower. For several following days came letters from people all over the district complaining that their wells had become quite dry, which they assumed was caused by the pumping operations going on the gas works.

One letter was from my neighbour Mr Gibbons, or rather from his solicitor threatening to take out an injunction at once, but I was not much frightened, having had several letters from him complaining of this, that and the other. However, I sent this letter from Mr Hargrove in Victoria Street (who was then engaged to my sister Jessie) to obtain Counsel's opinion, which he did, sending the case to Mr J. H. Lloyd, the well known lawyer. The reply he gave was that the Courts had already decided on several occasions that water in the earth was no man's property or something to that effect. A copy of this was sent to Mr Gibbons and nothing more was heard of it.

All this clearly showed I had conquered, and presently the water in the sump was so reduced as to enable the timbermen to go in again with their water boots and get it down to the depth required, which should be two or three feet deeper than the wall foundations. I intended that this sump should be made in what was then called the dry well, about five feet in diameter in brick work outside the tank. The inlet and outlet pipes were to go under the foundations into the gasholder itself, so in case of trouble, they could be got at readily. Finally, to my great joy, it was found that at the proper depth of 28 feet, firm, solid blue clay was there to build the wall upon.

Feeling satisfied that I had now really got control of the water, I wrote a line to Mr Simpson, mentioning this and also my kind friend Mr Mercer. He trotted into the works the next day, looking as pleased as Punch and was kind enough to say a few complementary words on my perseverance and success so far.

As many men as possible, both timbermen and excavators, were now put at once into the trench to get that down to the blue clay and also to enable us to get out and make use of the splendid sand and ballast excavated for the brickwork and concrete generally.

As it was not proposed to go deeper into the clay, it would be necessary for me to look about at once for some good stuff suitable to puddle behind the walls of the tank as they progressed.

I ascertained there were some larger works going on by the riverside lower down the river, where I could have as many barge loads of clay as I wanted: this was very satisfactory. When it arrived on the ground, any foreign matter, stones, stray pieces of wood or rubbish of any kind was carefully removed and the clay worked and trodden into a soft plastic stuff ready for use. This was of the very greatest importance.

As the excavation of the trench proceeded, there was of course some water, but not enough to interfere with the men working, so long as the engines and pumps kept at it. Now and then, one or other would have to stop a short time to clean or repair, and the water would gain on us.

As things went on a little, it seemed clear to everyone that all the loose talk of my attempting to pump the River Thames dry was all nonsense and the river had nothing whatever to do with the flooding.

While the excavation for the gasholder tank was going on, the foundations for all the additional buildings had been got in and brickwork started, including the new chimney shaft. I had given James Macrostie the job to make and fix the timber roof for the new coal store in accordance with the contract drawing and therefore so far things looked fairly cheerful.

Well, after a certain time the gratifying news came that the blue clay in the trench had at last been reached and found right round it at an even depth, firm and good enough to commence building the wall. It was therefore arranged, I remember, that I should go down very early the following morning, properly equipped in water boots and so on to satisfy myself that it really was so and that there were no pot holes or weak places to trouble us afterwards.

I went down with Gregory the ganger and the master bricklayer, to whom I had sublet the brickwork (labour only) providing him with all necessary material – bricks, lime, cement etc. I got one half of the bricks for all the work (many thousands I remember) from the Kentish brickfields. The other half came from the brickfields at Southall, the finest and strongest bricks made.

A nice, wet sloppy job it was, almost in a forest of timber. However, having gone right round crawling over and under the many wet, dirty struts, I found the clay in good, sound condition; the brickwork was then commenced, the clay being first worked up a little on the surface so that the first course of bricks might subside into it.

The specification I had prepared for this work was as follows. The foundations were to be in Portland cement 5 to 1; the wall itself to be in blue lias lime (hydraulic), but two rings of it at equal distance between bottom and top to be in cement four courses thick and two rows of hoop iron bond to be built in with it.

It is curious how, while writing these lines, so many little incidents of that time come back to my memory and so vividly. The day we started the brickwork of the gasholder tank I can remember several of the Directors were on the ground seeing for themselves that, after all the evil prophecy about it, the trench had been brought down to the required depth and onto the blue clay in spite of all the water. Standing alongside the work, talking with one of the Directors, I overheard a remark made by someone behind me. "He certainly is doing his best to make a satisfactory job of it, they are putting in the foundations with neat cement instead of a 5 to 1 mix".

Yes, that was so. I had given those instructions when the brickwork of the foundations commenced, being very anxious no risk should be run as it was very wet and sloppy. Turning round I could see the remark had passed between two of the Directors, Samuel Gray and Michael Walls, and those two men during the many years afterwards when working out the lease, never once raised any question or differed with me. Indeed, I need not have mentioned this trifling matter, for throughout the work it was always my first desire to make it a success beyond any thought of profit.

It is only a matter of detail, but as it will give me a little pleasure to mention and explain it, I will do so. When I pegged the work out, the centre peg of the gasholder tank was of the greatest importance during construction so that the tank when completed should be handed over to the gasholder men, with a perfectly true dimension.

It was therefore not the simple little wooden peg of ordinary use, but a large piece of timber seven or eight feet long and nine or ten inches square – in fact one of the struts. This was sunk into the ground and embedded in concrete to stiffen and make it immovable, leaving about 18 inches sticking up. Into the end of this was inserted an iron pin and this was the centre peg. A long wooden stiff rod was then made with the precise radius of the tank marked on it and with an iron ring at the other end to slip on to the pin of the centre peg. This was the trammel or traveller, in constant use to move round the trench and check with a string and heavy leaden plummet the radius of the wall as it was built up.

In addition to this, the leading bricklayers who always worked on the inner face of the tank would always have at hand a light wooden template made strictly to the radius of the wall and checked it frequently as they worked. This was also of the greatest importance as the men putting the puddle in behind the wall would sometimes push the latter a little out of the perpendicular and there would be serious trouble with the gasholder men if they found the tank not perfectly true.

I think there were six columns to be erected for the support of the gasholder, perhaps eight, with heavy girders securing them at the top. These columns would stand on brick piers about 3ft 6 inches in diameter, built up with the wall, which would be 18 inches thick. Built in the foundations of each pier there would be a York stone landing for the gasholder to rest upon when empty. At the top of each pier there would be a block of granite as a base for the columns with four holes drilled in them for the iron rods built into the brickwork to secure them.

When the tanks were finished, guide rods would be fixed to a small block of stone built in the wall on the face of each pier on the inside of them, to enable the holder to work up and down securely. As the work progressed, I hardly ever left the work, watching the puddlers and the bricklayers. I was anxious that the wall should be made perfectly solid and true, frequently testing it myself with the trammel or traveller. Of course, as the wall grew higher and higher, the timbers were drawn out.

At last it was finished. The granite stones were fixed ready for the columns and between each pier at the top of the wall, I can very well remember having a very large blue moulded brick put on for a coping and very neat it looked.

We had now to excavate and take out all the loose earth, gravel and sand in the middle right down to the blue clay, that the whole area might be carefully examined. All round the base the foundations were made secure and finished ready for the gasholder man, a good deal of whose material was already on the ground.

Although perhaps there was now no necessity to continue the pumping, yet as I was very anxious not run any risk, I carried on with one of the pumps so that there should not be any undue pressure behind the wall while the work was green. There would then be room to build the dry wall and put in the inlet and outlet pipes. When that had been done, I remember going down to the bottom one day with the Foreman of the gasholder men as they were about to start work. Looking around it there was not a sign of water anywhere and as the man said after testing the dimensions, as true as possible. I do think then I was entitled to feel a little bit proud.

Except to attend my sister Jessie's wedding that was in August, and also the Board meetings of the Crystal Palace Gas and Gas Products Company, I never left the works.

This happy wedding was a great pleasure to us all. I remember so well, after the return from the church when busy appreciating the wedding breakfast, the guests were all startled with the horrible noise from three bagpipes suddenly playing up and down the garden. My father had the idea of engaging three of the Queen's pipers to give a little music on that occasion, but I am afraid it was not a success. I think there was a dance afterwards in the evening, but my dancing days were now all over. The happy pair, after their honeymoon, resided for a time in Bayswater in one of the squares or terraces in the Pembridge or Westbourne districts that I knew so well from my Grand Junction Water Company days. How often since that time have I had occasion to think what a fortunate girl my sister Jessie was to have met such a man as Hargrove for her partner through life. A gentleman in every sense of the word, clever with a cheery, amiable manner and whose friendship everyone appreciated.

I could not help thinking at the time that this separation of the two sisters would be a great trial for my sister Agnes, for they had been everything to one another in the past and now she would be the only one at home with my mother. The two youngest boys, Henry and Walter were away at school and all the others were away at work. Joseph was in Calcutta and William had just taken possession of one of the prettiest farms in Kent, the Cage farm near Tunbridge, of 400 or 500 acres, well known for its fine growth of hops. My father always had a weakness for country life and thinking, and I know anxiously trusting, it would give William a chance to work and make something out of it. Perhaps at the same time it would give my father himself a little pleasure by running down there occasionally and see how he was getting on. Unhappily, later on it became a disastrous failure.

The reason why I feared my sister Agnes would feel the separation from her sister Jessie was that my mother's health was causing us all some anxiety. The old pains in her head reappeared, causing her at times to be very irritable and somewhat difficult to get on with. Had it not been for the loving kindness of our dear old friend Mrs Keeble, I really do not know how things could have gone on. If ever there was a good angel in the world, I am sure she was one, and I can see her now in her prim Quaker-like dress and cap and hear her speaking to me about my mother in her quiet, sedate way. If anyone cared to open Charles Dickens's book *David Copperfield* and see among the illustrations one showing David's aunt, they will see a portrait of dear Mrs Keeble. Curiously, later on in years I found that I was the only one in our family who could control and soothe my mother when she had these sad attacks. My father, in his frequent visits to Berlin (as a great deal of his attention was required there) would sometimes take my sister Agnes with him. However, I must return to my work at Kingston and continue to relate my movements there.

All the new work was proceeding satisfactorily and every month I would receive a payment on account of it from the Secretary. From the commencement I had been very careful to keep the accounts of this contract work quite distinct from the general book keeping; it gave me a lot more writing, but I did most of it at home in the evening.

The new shares issued for all this work had been readily taken up by the present shareholders, showing their confidence. I thought I would also show mine, by asking for a few which I received and hold now, giving 10 per cent. Charles Horsley was well forward with his work erecting the new purifiers, the condensers and the fittings to the new retort benches. The new chimney shaft was nearly finished and I can remember, even now, how easily I used to run up and down the ladders (one not being long enough). I was anxious to see that the fire clay lining was not neglected and I held on with only one hand. How different things are now in that respect.

I had been wondering now and then what had become of Wyatt these several months past, but one day I got a letter from him saying he had got some temporary work for a time somewhere in the North. Curiously about this time and to my great surprise, I also received a letter from my old colleague and friend when in America, Thomas Mahoney Quill. I had not kept up any correspondence with him since I left there several years ago. He said he was staying with little Horace Coombe, the draughtsman who came back with him, at the Old Hummen's in Covent Garden. What quaint, dirty old-fashioned hotels there were then about that part. The New Hummen's, the Tavistock at Evans's, what queer stories they could tell if they could speak. I remember having stayed the night once or twice in my bachelor days when going up to London to the theatre with Jimmie Simpson and losing the last train. How well I remember Paddy Green, Herr von Joel and the glee singers at Evans's and the supper of kidney, potatoes and stout going on at the same time. The next time I was in London, I called upon him and very pleasant it was to meet and chat over old times.

Soon after Quill's return to England I was glad to hear he had got an appointment on the East Indian Railway. As for Horace Coombe, I had the good luck to get him an engagement as a draughtsman in my old friend Mr Quick's office, who seemed to be very busy.

I may as well now mention a most sad matter that occurred to poor Quill some two or three years later. He had a younger brother William and an opportunity occurred on the railway where he was engaged, for this brother to get an appointment also. He wrote to his brother to that effect and arranged for him to leave home and come straight to Calcutta, he would come down from his work in the upper country and meet him at a certain hotel there.

The particular day arrived; going to the hotel to meet his brother he was horrified to find the place had been burnt down that very night and only the charred remains of his brother William were left. It appeared that he had arrived the previous day, secured a room and had gone to bed, but being frequently guilty of the evil practice of reading in bed, it was assumed it had caught fire, with this sad result. The shock, however, was so great to poor Quill that he threw up his appointment and returned to England. I was glad to be able to get him into Mr Quick's office for a time.

The Michaelmas quarter had now arrived, requiring all the consumers' accounts to be made out. This I always continued to do myself, for it was always a pleasure to me to note while doing so, how steadily they were increasing. Of course, it was not all *coulour de rose*, and as things went on, there was worry and anxiety more or less every day. The little baker Wenman on the Council continued his attacks on something or other and the great brewer Hodgson never seemed satisfied, do what I could to please. I can remember that the most difficult member of the Council to deal with on general subjects was Mr Frederick Gould, the chemist, but who always took up the cudgels for myself and the gas company whenever attacked. Why, I knew not, as he was not even a shareholder.

Sometimes one of my ships of coal would from some mistake go astray, necessitating my rushing up to London and seeing my kind old friend Mr Innes at the Phoenix Gas Works, Bankside. I would then borrow one of his ships in the Pool.

It is now becoming clear to me when writing these pages of my past life that I must hasten and somewhat condense the many matters I have yet to refer to, as my health and strength is now becoming very feeble and I may not be able to complete them. I will therefore simply remark that during Xmas of the year 1860, everything went on very satisfactorily at the works and I had now arrived at the age of 29 years. My wife came round to the works on Boxing Day with a large plum pudding she and her servant had manufactured between them, with a bottle or two of gin and beer to distribute to all the men – men and masters were good friends in those days. When the half year's account up to Xmas had been made up, I remember feeling more than satisfied with the result of the year's working, the credit for which I attributed entirely to my new manager Watson, my attention having been so much taken up with the new works.

The half-yearly meeting of the shareholders was also very gratifying, every one who spoke expressing their satisfaction with the new work.

Before the next half year at Midsummer had come round everything in my contract had been completed and was in use. The gasholder, when erected and the columns and girders painted was something then for this young fellow to be proud of. The last portion of the work was the removal of the little old gasholder and the portion of the cast iron tank above ground, then covering the area left below, being seven or eight feet, with iron girders and concrete forming a very useful tar tank and space above on the new surface for the coke. Perhaps the most difficult task in all the new work and the most complicated was the forming and making of the different connections underground from one point to another and making them work smoothly, but with a little patience even this was accomplished.

I should mention that Messrs Simpson & Co. who supplied the engine and the boiler to work the exhauster, sent a beautiful specimen of engine work. The young fellow who came to take charge of the filling of it was one of their senior pupils, working in their shops. I cannot remember his name, but evidently he was a very clever young fellow and I remember how pleasant it was to watch and chat with him when he was at work.

I took the opportunity of getting a thorough knowledge of the engine's mechanism and use.

For a time I had some difficulty in drilling the men in charge of it, but after a time got things right and understood. The difficulty was that if driven too fast, too much air was pumped into the gas, causing it to deteriorate and of course causing complaint, but in the course of time all these difficulties were got over.

I may perhaps again have been too prosy and gone too much into detail, describing all this new work and probably it may be very uninteresting to the reader. It must be remembered, however, that when writing these pages of those old days, everything seems to come back to me and makes me feel I am still there. It should not be forgotten that at that time this was my first attempt or real experience of practical work and therefore to meet with success under such adverse circumstances, perhaps caused me to be a little vain, but I hope I was never one to blow his own trumpet.

Some few months after the completion of this new work at Kingston, the firm of John Aird and Son had a contract with the Crystal Palace District Gas Company to make a large gasholder tank at their Lower Sydenham works. This was somewhat larger than the one I had finished at Kingston, but how different the circumstances. The directors at their Board meetings every Friday fortnight always went round the works first and, being one of them, I had the opportunity of seeing the work in progress.

There was not a spoonful of water in the ground, so the contractors had to get water laid on from the water company for their brickwork. The ground from the surface to the depth of the tank (30 feet) was all clay, not the blue clay of course, but stiff brick-making clay, therefore no puddling or timbering was required. A trench was excavated only just wide enough to allow the wall to be built close to the soil as it had been excavated, or rather cut, like a cheese. I did not then know what my experience would be regarding gasholder tanks later on.

As the Kingston company would soon be wanting more capital, the Directors decided to go to Parliament for a new Act and some further powers. This they obtained, there being no opposition to it.

The more I knew of my new manager James Watson, the more I appreciated him. Packham also was now becoming very useful. In fact as everything at the gas works was completed and up to date, things seemed to pass along like clockwork.

It also seemed to me at that time that Kingston upon Thames was really becoming a large and important town. The South Western Railway had commenced making a new line from their Twickenham Station on the Windsor line to Kingston. This went through Teddington by Bushey Park gates, then passed along the north side of our little home to Hampton Wick requiring a bridge to cross the river into Kingston and the terminus close to the gas works.

This, when finished, was a great convenience all round and obviated taking the dirty little omnibus to the Surbiton station on the main line. It also seemed certain that this development must increase the consumption of gas.

The engineer in charge of this work for the South Western was a Mr Tolmie whom I had met some little time before in Westminster. Meeting him accidentally one day on the ground, I asked him to come home to lunch with me, apologising at the time for the small accommodation I had to offer. Being always a most good-natured man, he kindly agreed to do so. One would almost think Tolmie was an American the way he spoke. He was also very tall, over six feet and very thin. When he attempted to sit down to the meal in our miniature dining room, I never laughed so much in my life and he also, for the poor fellow's legs were so long, he did not know where to put them. Finally finding there was no room for them under the table he had to sit sideways. How many, many times we laughed and joked over that little incident afterwards. Poor old Tolmie, dead many, many years ago.

I must now refer to a little domestic matter that occurred at this time on 19th March, 1861. This obliged me hurriedly to jump out of bed in the early hours of the morning and run as fast as I could into Kingston for Dr Cox and bring him back with me. He had arrived but a short time, I think, when I was told that a little daughter had made her appearance into this troublesome world. Not then being accustomed to this kind of thing, I did not quite know that day whether I was on my head or my heels, but I remember the first thing I did was to send a messenger to The Grove to my mother and also to my wife's sister. Later in the day both ladies arrived, glad to find all was going on well. The house being so small, it was rather a tight fit for an occasion of this kind, but somehow we managed to get through with it. The young lady was afterwards christened at the old church in Kingston and called Nellie. I much regret, however, to say that if it were really possible for parents to hate a child, I think we must have had those feelings at times when there was no possibility of any sleep or rest. Night or day, there was no pacifying the poor little thing. During these extraordinary fits of crying and passion everyone seemed to be worn out, and how often did my thoughts then go back to my bachelor days.

In the summer I took a little holiday for a fortnight, the first since my marriage, and went to Teignmouth in Devonshire with my wife, child and nurse. We occupied a couple of rooms in a lodging house facing the sun and endeavoured to make the best of the cooking. I can very well remember, when there, having a fondness for cherry turnovers. Teignmouth was then a small place, but very interesting. In the one street there was a fruiterer's shop and immediately opposite was a small confectioner's. Nearly every morning we would purchase some cherries, cross the road and leave them to be made into turnovers – I really felt quite like a boy then.

Finding that my wife had not read very many books, I tried to see if I could not raise in her some little desire for reading and nothing gave me greater delight than to read aloud some book that I had delighted in reading to myself. At Teignmouth there was a long length of sea wall, built to protect the railway from the sea, with a rough sort of pathway along it. We would therefore go out there nearly every day and sit for hours reading. This I did, and also delighted in doing so at home whenever I had the leisure and opportunity.

I never could take the same interest in or appreciate the companionship or conversation of the other sex. I always found it so difficult to sustain the conversation any length of time, but with a man, assuming him to be educated and somewhat of a kindred spirit, it was quite different. I could never tire, chatting, arguing and discussing any and every subject that might arise; that was the one reason why I continued to be so friendly with Wyatt and liked and appreciated his society, in spite of his disagreeable manner and ill temper. He was a clever, well-educated man, far more so than myself and I was never ashamed to pick up any little information I could, particularly on mathematics, a subject often referred to by him. Wyatt was never at a loss for a subject either to interest or amuse, in fact a born raconteur.

Feeling that I would like to have a good long walk and see something of the country, I wrote to Wyatt and asked him to join us for a few days, which he did. I remember one day we all (including my wife, child and nurse) made a little excursion to Antis Cove, a pretty, well-known place between Teignmouth and Torquay. It was an attractive drive and we took a picnic sort of lunch with us, not forgetting the cherry turnovers.

We had not long started, when Miss Nellie began her pranks and continued, in spite of all our efforts to pacify her, until we arrived at our destination. I remember Wyatt rather took to the child and did his best to amuse her. Leaving the carriage, we had to descend the high cliff to the white, pebbly beach. This was then a very difficult and awkward matter for females to attempt, there being no direct path and the cliff very precipitous. At times, it was rather exciting with the nurse and child, but with great care it was accomplished. After this we were all quite ready for the lunch and cherry turnovers, which Wyatt thought a great institution. Each one was very large and required both hands to hold it and he much preferred to eat his in that way. It certainly was very vulgar, but the sea air I think must have increased our appetites and I am afraid I followed his example.

It was a lovely day and there was not a soul to be seen. The sky and sea were as blue as in Italy. After lunch we concocted a game to pass the time away, pitching a stone at some small object a reasonable distance away; whoever knocked it over won sixpence. The result was that I was the richer by several shillings, to Wyatt's disgust and my amusement. I had told the coachman to meet us at a certain time the other side of the cove, being told the ascent there was much easier. Thus ended, taking it altogether, a very pleasant day.

Another pleasant excursion I made during Wyatt's visit was to Dartmouth, but as that involved a walk of about five and twenty miles, my wife and child were left at home. Having previously made a few enquiries, we left by a very early train in the morning for Torquay and then started on our walk. We passed round the beautiful bay by Paignton to Brixham and then somewhat inland with pretty scenery. It was a long walk, but walking on chatting and smoking, we thoroughly enjoyed it and felt no fatigue. Presently, however, we began to wonder whether we were really on the right road, and also we began to feel a little bit peckish, when suddenly the road turned almost at a right angle to the left and before us was a scene never to be forgotten, it was so beautiful and unexpected. We were very high up and looked down on this strange scene. To the right was the lovely little river the Dart, meandering along in the distance, to the left nothing could be seen, but the projecting cliff with the ruins of Dartmouth Castle on it. On the opposite side of the river was the town and there was not a sound or soul to be seen. It really looked like a dead town, uninhabited and forgotten. However, the road continued downwards in a zigzag sort of way to the river and there we found a ferry to take us across. We could then see the little land-locked harbour, with the training ship *Britannia*, and above on the cliff we had just left was a good view of the castle.

The town itself was most quaint and interesting and so foreign-looking. The little shops were mostly ship's chandlers shops where they sold everything. Our first desire was to get something to eat after our long walk so, finding an inn or tavern, we soon did justice to the lunch they put before us and then considered how to get back. When speaking with the landlord about it, he remarked that an excursion steamer from Torquay had come into the harbour that morning and was to return at 3 o'clock. Nothing could have been more opportune, so we walked down to the harbour, embarked, paid our fare and after a most pleasant sea trip, arrived at Torquay just in time to catch the train for Teignmouth. Wyatt left the next day and our holiday being over two or three days afterwards, we also returned home.

It was about this time that the news came of the commencement of the Civil War in America between North and South, on the slavery question. I was remembering so well the ground and the places where many of the great battles occurred. I became greatly interested in it all and I do not quite know why, but I sympathised with the North throughout.

I do not remember whether it was this coming winter or the following year, for my memory is not very good now as to exact dates, but I must not forget to refer to it as perhaps it was the most anxious period of all during my lease. This particular winter (1861) was a most severe one. An intense frost lasting some six or seven weeks caused the River Thames to be frozen over at Kingston, so that people could walk across the river, and some six or seven of my coal barges were imprisoned on their way here. This was dreadful, as I had only about four weeks stock in hand; however, thank goodness the barges were released with the help of a tug, which forced a way through the ice before the coal was all used up. In the meantime I had been all over London buying whatever sort of gas coal I could pick up and having it carted to Kingston, of course at great expense. It was an awful, anxious time while it lasted and I can remember that Mr Innes lent me a few tons and had them sent by road.

Of course there is trouble and anxiety in all work and business, but upon my life I thought then that the management of a gas works had a good share of it, for some little time after this, a man came from the works one evening to tell me that Mr Watson, the manager, had got his hand crushed while doing something to the engine and had been taken to St Thomas's Hospital in London. This was terrible news; of course I immediately went back to the works to take charge during the night. In the morning when Packham came, I went home for a few hours to get a little rest. I had earlier sent Watson's brother, who was then our service layer, up to London to see how his brother was getting on and to express my great sorrow to him.

When the brother returned, I was told that two of the fingers of his right hand had been amputated and that he was doing well. We managed between us to keep things going and in about a week Watson was able to resume his work. Of course, for some time longer his right hand was off duty, but in the meantime he could do very well with his left, and later on with his right also, mutilated as it was. I had a great regard for that man.

I can now come to a matter a little more cheerful. I have already mentioned that my father and brother John had made a beginning with railway work, in addition to their general work, embracing water and gas companies, and so large that I think they seldom entered into anything where there was competition. My surprise and astonishment were indeed great, saying nothing of the immense pleasure it gave me, to hear that the firm had just entered into a contract with the South Western Railway to continue the line the company was just completing into Kingston from the other side of the river. This would then run through Kingston, Norbiton, New Maldon and connect with the main line at Wimbledon. The line would also connect with the London and Brighton to Croydon, so that when all was finished, I had only to step out of my office, walk a few yards to the railway station and go anywhere. Mr Stannard was to have charge and my younger brother Joseph, who had just returned from Calcutta, was to assist him.

When all this became generally known in the town, I seemed to become rather an important man; many erroneously supposing I was part of the said firm. When the work was about commencing, my old friend Mr Mercer came to my office one morning to congratulate me. He was in some little state of excitement about some land and property he owned, through which this railway passed, and he was very anxious about some matter regarding it which I was fortunate enough to get arranged for him. This new line must also lead to a lot of building and of course consumers of gas.

It was already wonderful how much the consumption had increased, and the great results from having clay retorts instead of iron, so much so as to induce me to lower the price from five and six to five shillings, to the great satisfaction of the consumers I could see and hear.

Occasionally we still dined with our kind friends the Simpsons at Westfield on a Sunday. I can remember we were much upset one morning reading in the Sunday newspapers of the death of the Prince Consort and the upset following. At about this time my wife and I were much grieved to hear that Mr William Simpson, Mr James Simpson's brother and chief of the engineering works at Grosvenor Road had died very suddenly. Mr Simpson had then to represent the family and put his own son James in charge, I think to manage the works. I can remember my wife and myself dining with James and his wife some little time after, when they were living in some square off the King's Road, Chelsea and meeting a great friend of their's, a Mr Best in charge of the Geological Museum in Jermyn Street: Little Best, Mrs James used to call him.

13

The family move to Strawberry Hill

We had now been living in our little home at Teddington for about two and a half years. One morning to my great surprise and disgust I received a letter from a Mr John Guy, a solicitor, giving me six months notice to vacate it, as he had become the owner of the farm and also intended to build a larger house. This was a serious and very annoying matter to me, not at that time knowing of any suitable place into which to move.

I had heard of this gentleman before from Mr Mercer, and that he was a very awkward customer to deal with, and so it turned out. I replied to his letter quoting my surprise and regret receiving such a notice, and hoped that at least some portion of the money I had expended on the place might be returned, especially for the little conservatory. To this I got a reply that there were no fixtures of any kind that could legally be claimed. As I did not care to go to law with this individual, I endeavoured to make the best of it and looked about for some other home.

Not a great way from my bachelor home at Surbiton and the other side of Westfield in the Portsmouth Road, there was a house then to let that had been occupied by some friends of the Simpsons, named Stroud. The father died and the family left it. On going over the place it seemed a very convenient and comfortable house, very different and larger than the little place we had to leave. It was the corner house in the Uxbridge Road, the entrance being in that road, but the side of the house with its windows faced the Portsmouth Road. As my old friend, that hideous mound of earth, was some little distance to the left, there was an uninterrupted view of the river and the Home Park opposite. This pleased me more than anything; one only had to walk across the road to be in Parker's boat house. I think the rent was £85.

The agent with whom I negotiated was a local builder named Boxall and an odd circumstance appeared in this matter. Between the side of the house and the Portsmouth Road itself, there was a tiny piece of seemingly waste ground; it occurred to me it would look more ship-shape if it was enclosed by a little dwarf wall; but the area was too small and too public to be used as a garden.

The agent raised no objection whatever to my suggestion and therefore, after thinking it all over, I agreed to take the house, agreed with Boxall to build the wall, put the house in thorough order and repaint. I then felt quite happy. I remember also my kind friends the Simpsons and Mr and Mrs Mercer were very pleased at the prospect of my making my home near them again. Everything looked pleasant and nice and the work commenced. One morning I got a letter from the agent to the effect that he had received a notice from a gentleman occupying the house next to this in the Portsmouth Road, a Mr Knyvett who was Chaplain or Secretary to the Archbishop of Canterbury. He said that if this dwarf wall was built, he would at once apply for an injunction to have it removed as we had not legal right to erect it. This was an awful pull up and disappointment after all the trouble. I had never spoken to this individual, but had seen him about; a fussy, pompous little gentleman and knew him also as one of my consumers. There was a lot of correspondence between them I believe, but as the agent or owner of the house evidently did not see his way to make my tenancy secure, with great annoyance and disgust I withdrew from the matter altogether and began again.

When having my usual walk on a Sunday afternoon, I had noticed that three pairs of semi-detached houses were being built in a meadow alongside the river between Teddington and Twickenham. They were also opposite Strawberry Hill, once well known as the residence of Horace Walpole, then occupied by the Countess of Waldegrave and her fourth husband, Chichester, Lord Carlingford. The meadow indeed formed part of the grounds. The ground, being very low, was subject to floods at times and the builder, whoever he was, had cocked these houses to avoid the water.

To get into them, an occupier would have to climb up some seven or eight stone steps and to descend at the back the same number of slippery iron steps from a veranda or balcony running along each pair, with only an iron rod between them. There could not be the least privacy – in fact the two families might almost be living in one house. The narrow slip of ground at the back to the river was too small to attempt a garden, so all that could be done with it was to put down a little bit of turf and a gravel path all round. The dividing wall between each house was ridiculous and was so slight it was hardly needed. Summing it all up, they were cheaply and badly built with no comfort and no care as to sanitary precautions, as the future would show. Yet I was, however, foolish enough to agree to take one of the two middle houses, and for months afterwards it was a case of civil war between the owner (a man living at Richmond who had recently returned from California) and myself. But beggars can't be choosers and I was sick and tired of hunting about, after the great disappointment of the house at Surbiton. There must have been other foolish men as well as myself, for all the six houses were taken and occupied, almost before completion. My neighbour was a Mr Booker, a house agent and auctioneer in London. Even now, I can almost laugh myself into a fit seeing this individual endeavouring to moor or fix a fishing punt in mid stream.

It is curious also to note that although my wife and I resided here some three years and lived so uncommonly close to this gentleman and his family, we never exchanged a word. I always had such a dislike of knowing strangers.

The two houses near Teddington were occupied one by a Mr Burrowes, a member of Lloyds, the other by a well-known Counsel, Mr Harry Poland, QC. The other two houses nearer Twickenham were also occupied, one by a Mr Frisby, a well-known man, I believe, in Society and a judge of horse flesh. But he and his wife only came for a few weeks in the summer to potter on the river with some friends living near.

Before moving from what I call the Dolls House into our new home at Strawberry Vale, I remember we went to Tunbridge for a week to stay with my brother William at the Cage Farm which he was then occupying. He had also asked Mr and Mrs James Simpson (my old bachelor friend) which made it very pleasant. What a beautiful farm that was and what a lot of money my poor foolish father must have spent, putting it in order, trusting that my brother would steadily work and do his best to make it a success. I think it was about 400 acres in extent and noted for the hops grown. The farmyard was ideal in its arrangements and the farmhouse, a plain, comfortable old place. What pleased me more than anything was the bright, clean and cool dairy attached at the back with its row of large, white shallow china pans on a bench with the milk. There were also several large orchards and a beautiful extent of wood round about. How I should have worked and revelled in such a home had it been my fate, so loving the country. Unhappily, my unfortunate brother was too fond of pleasure and associating with the several aristocratic county families round about and joining in their various amusement and pleasures, leaving the farm in great measure to his Bailiff. One family he seemed to have become very closely acquainted with was the Hardinge family, residing at Bounds Park, a beautiful place between Tunbridge and Tunbridge Wells. The head was the Reverend Sir Charles Hardinge, Bart, his brother being the Viscount. There were two sons, Charles and Edmund and I think one or two sisters. The son Edmund seemed to be my brother's great friend, or may I venture to say, enemy.

I happened at that time to be rather busy at Kingston and obliged to go back to my work for one or two days, but the visit altogether was very pleasant and nice. I remember one lovely day we drove over to Tunbridge Wells and I can recollect stopping at the then celebrated Tunbridge Ware shop and presenting the two ladies with a little cabinet.

Another day my brother gave a kind of garden party to some of these gay people, having cleared out a large barn for a dance. Well, it was all very nice and enjoyable, but I could not help thinking my brother was not doing his best to make that beautiful farm a success.

When we got back to the Doll's House, or Broomhill as I think we called it, we had to arrange to pack up and remove to our new home at Strawberry Vale. We also said goodbye to the two girls of the farm who had been so very kind and so very useful to us during our tenancy, which we all did very sadly, I remember.

I certainly did not spend more money than I could help about the new house, when we took possession, but our troubles seemed never to come to an end and it was really almost like civil war with the owner. Rainwater appeared in many of the ceilings and the sanitary arrangements were anything but perfect, but the deed was done and we had to try and make the best of it.

Since I had managed to get the Kingston works into such good shape and every year turning out to be better and better, I had been thinking seriously of trying to find some additional work of a similar kind. If not that, then some other work or business, so that by and by when the Kingston lease expired I should not be left adrift and have to begin again. I knew very well that there would be no chance of renewal, gas companies were getting too knowing.

Curiously, however, one day at a meeting of the Gas Products Company, Mr Angus Croll, the Chairman, asked me whether I would care to take over the lease of another gas works, to which I replied yes, certainly, if it is a workable concern and we can agree upon terms. He was kind enough to further remark that he had heard how well I had done at Kingston and as he found he had got too much to do, he could not really give sufficient attention to this other company. He therefore thought that perhaps a younger man than himself, who could always be about the works or nearly so, might be able to get a little life into the town he was referring to, namely Winchester.

There were yet ten years of the lease unexpired and after some further conversation, it was arranged that he would direct his secretary and managing man, Mr George Rait (Mr Croll's offices were then at 10 Coleman Street, City) to get out such particulars as would enable me to form some idea of the concern. For example, the number of consumers, the present yearly rental, coals consumed and the present price of gas; then afterwards I should go over the works and the district with him.

Having received such information, a day was named for our meeting at Winchester, and as it was an important matter, certainly for me, I asked my father if he could kindly spare time to run down there with me and give me the benefit of his advice. Always to be relied on, he assented at once. Mr Croll was then one of the Sheriffs of London, living in a nice place outside the City.

We met at the gas works and went over them. A more dead or alive place than Winchester I had never seen, quaint and interesting no doubt, but there did not seem any life or growth in the place whatever and with a gas works to match. Poor old antiquated shops burning very little gas and always shutting up very early. There were no surrounding suburbs or residents as at Kingston, growing up like mushrooms in the night.

There were some few large residences scattered about, but without gas as it would never pay to lay a long length of main for them only. There was not the least sign of any building going on and in fact everyone seemed asleep. Before returning home Mr Croll was kind enough to ask us to dine with him and talk the matter over. It was finally arranged that the subject of a transfer of the lease should not at present be brought before the gas company, but that I should take the works in hand and work them for six or twelve months, representing Mr Croll. This I did, going down there two or three days each week, thoroughly looking into everything and doing my very best to increase the consumption, but it was no use; there was really no life in the concern as far I could see and never would be.

In the first place, a serious error had been made in the yearly rental in the statement given to me, the book keeping having been very carelessly kept. During the whole time I had been in charge, there had been no increase in building and no likelihood of any that would be profitable. Someone might build a villa or mansion some long distance apart, but it would never pay to lay a main to supply it with gas. Therefore, after about six months trial and summing it all up, it really was simply flogging a dead horse and I felt obliged to decline going on with the proposal. It was unfortunate and a disappointment on both sides. Mr Croll, however, was a gentleman and a fair dealing man and having arranged matters amicably as to expenses incurred, continued working the lease himself. This probably would not affect him very much for he must then have been coining money with his other speculations, particularly with his Dry Meter Company. At that time I was a very good customer of his company in that respect.

While this Winchester matter had been going on, it had been arranged that my wife, self and our little daughter Nellie (who seemed to be getting over her passionate fits of temper) should spend our Xmas day with the family at The Grove. As Xmas time is always the most anxious time at a gas works, so much gas being consumed by shops kept open late and with Xmas festivities going on, I could not possibly leave until there was a good stock of gas in the gas holders and I was satisfied all would be well. We were therefore rather late for our arrival on Xmas Eve and were surprised and astonished beyond measure to learn that some two or three days prior to this my brother William had brought his friend Mr Hardinge there. Shortly after his arrival Mr Hardinge had been taken very ill and the doctor had stated that morning that he believed it would develop into typhoid fever and that his family should know of it.

Under these circumstances I suggested that we should return to our home again and take with us my two youngest brothers Henry and Walter, just home from school for their holidays. My dear father and mother seemed relieved at my proposal for if any of Hardinge's relatives came, there would not be much accommodation. My mother, in spite of all this trouble so suddenly brought upon her, would insist on packing up some few things for the two boys to take away with them and enjoy on the morrow. I can recollect also, how very angry my sister Agnes, also at home, was at our leaving, but goodness knows I did it for the best. I most certainly did not consider my own pleasure or convenience, particularly as it turned out, my brother John was I think, then living at Moray Villa, only a few yards from The Grove in Belvedere Road.

It was getting late when we turned out and I think it will be admitted the prospect before me was not a pleasant one; travelling back to Twickenham at that time of night on Xmas Eve, with my wife, the nurse and our little daughter (not yet two years old), together with the two young and unruly boys. However, I did my best, arriving at London Bridge station I engaged a couple of cabs to take us to Waterloo. As we had made no arrangements for a Xmas dinner for ourselves when leaving home, on arriving at London Bridge I stopped the cabs at the large fishmonger's and poulterer's shop at the foot of the bridge in Tooley Street, Smithers I think was the name. I can remember quite well that they were then shutting the shop up, but I went in and bought, I think, the only turkey they had left and some sausages, jumped back into the cab and just caught the last train from Waterloo.

One would naturally think my worry and misery was at last over, but no, taking cabs again from the station to the house, we find the two servants had of course gone to bed (at the top of the house) and no ringing or shouting could waken them. My wife, the nurse, the infant and the two boys were standing about in the cold all the time, but fortunately it was not snowing or raining. Just as it was getting serious, I went round to the back of the house and threw some heavy stones at their bedroom window on the top floor and succeeded in smashing two or three panes of glass. This brought them down, but almost frightened out of their lives. We were, however, only too glad to get inside, chilly and cold as the house then was. After having arranged some accommodation for the two boys and had a mouthful of food, we went to bed and were glad enough to do so.

The next morning, the cook was found to be too ill to leave her bed from the shock and excitement of our sudden return; we had therefore to do the best we could and cook the Xmas dinner ourselves very much to the amusement, I remember, of the two boys. But I did not see much amusement in it myself, neither, I think, did my wife. However, after a few days, Cook got all right again and things seemed to settle down.

I had resumed my work at Kingston and Winchester, but I felt a little nervous leaving those two boys behind, fearing they might be up to some mischief. The youngest, Walter, I had heard was a terrible young pickle. He had been sent to a school in Cornwall to be well looked after, but the first thing he did was to go to the only butcher (it was a small town) and give an order that when the butcher killed, all sheep's kidneys were to be sent to him at the school. He also gave a sovereign to one of the local boatmen for a little toy boat, obliging my poor father to send down one of the clerks, Mr Wallis, to put the matter right. Henry was at Tonbridge School and a steadier boy. The news we had from The Grove was that Mr Hardinge continued very seriously ill, that several of his family had been to see him and it had been arranged that his sisters should remain and nurse him. The result was that within a fortnight Mr Hardinge was well enough to be removed to his own home and then the boys went back.

There was one nuisance in this change of residence I did not much like which was that I had double the distance to walk to my work at Kingston. When fine I did not mind one bit, but when raining hard and no cab within reach, it certainly tried one's temper. There was no help for it then, but the new line to Kingston would soon be open, thank goodness.

There is no rest for the wicked it is said and therefore one thing I now found it was necessary to arrange was to find some medical man to look after my wife's interests, as we had got beyond Dr Cox's district.

I therefore wrote to a Dr Clark in Twickenham, being the nearest and seemingly the best known, to ask him to call and see her. This he did, but she was evidently not much impressed, and a little later I was told she could not possibly have him to attend her when the time came. I had therefore to look about again and how often my thoughts went back to my old bachelor days.

The next Sunday, when taking a little walk through Twickenham, an unmistakable doctor's pair horse brougham stopped at a house only a few doors from where Dr Clark resided. A gentleman got out and while waiting for the bell to be answered I had the opportunity when passing to have a look at him. When he had gone inside, I turned back and asked the coachman the gentleman's name and was told that it was Dr. Julius of Richmond; a lucky find as it turned out. Being impressed by his appearance I wrote to him and when he called, my wife also took a great liking, and for many years afterwards there was a strong and sincere friendship between the two families. Dr Julius was then about 50 and I should think the very ideal of a doctor in appearance, always with a cheerful and pleasant manner and undoubtedly clever. There are probably very few now living, if any, who would remember the great Palmer poisoning case, in which he was so prominently engaged just before this. He was very fond of a quiet game of billiards and in the summer would take a little holiday. He would hire a suitable boat and with his son Churchill and his two pretty lady-like daughters would wander up some of the smaller rivers and canals leading into the Thames. One of these young ladies was married to Mr Bateman who became his partner shortly afterwards.

Although matters at Kingston were going on so very satisfactorily, there was one thing at this time that troubled me. The Town Council had requested the overseers to make a reassessment throughout the borough and to engage an expert surveyor on such cases as railways, water works, gas works and any other large trades to assist them. They therefore engaged a Mr Castle, a man then well known in such cases and a very hard customer to deal with. This being my first case and coming so soon after all my large extensions on the works, I knew I should have to gird up my loins for a fight. Before coming in contact with this gentleman, I took the opportunity to get a few hints from some two or three friends who had gone through a similar ordeal – Mr Innes and H. P. Stephenson. Thus fortified, I was ready some two or three weeks later when I received a letter from Mr Castle naming a day for calling and viewing the works and requesting certain information regarding them. To this, my reply was that it would give me much pleasure to meet him on the day he had named, but I could not possibly give him the information he wanted as being only the lessee, the figures he wanted were entirely my own private property and not the gas company's. In that case I remember he remarked he would have to make a guess and state an amount that might be far larger than the actual sum, leaving me with the responsibility of appealing. I could not, however, help that and serious as it might be, I would have to take my chance.

The day for our meeting having arrived, we walked round the works, Mr Castle closely observing everything and taking notes. At the same time he asked many questions to which I replied in a friendly spirit whenever I could do so without trespassing into my own private figures. Before he left, we sat for some time in my little office, discussing several points and as it seemed rather dry work, I can recollect producing and opening a bottle of sherry (that I knew to be first class) and some biscuits. Although so many years ago, nearly fifty, how well I can remember that at that time sherry was nearly always the wine drunk. I myself seldom ever dined without my half pint, but today at the time of writing (1910) it is never seen nor heard of. Whisky and soda is now the universal drink at meals, but mine being simply a glass of lager pilsener beer, which is everything to me.

Well, at last our meeting came to an end, and, asking him when he thought he would be able to send in his report to the Town Council, he replied probably in about three weeks, an anxious three weeks to me. However, work went on as usual in the meantime.

One day I met Mr James Bell, the Clerk to the Surbiton Commissioners and also a solicitor in the City, upon some little matter about the public lamps of that district. He was a very nice fellow and always very kind whenever any little difficulty arose about them, but on this occasion he asked me if I could help him in a matter of which he was acting as the solicitor.

The Great Western Railway Company, having for some reason quarrelled with the gas company in London supplying Paddington, decided a year or two earlier to erect a gas works at Wormwood Scrubs, make their own gas and supply Paddington station, the hotel and all stations, workshops and signals within reach. So far it had not been a success as the lights frequently went out causing much inconvenience and also danger. Mr Bell was therefore authorised to endeavour to arrange with some responsible person to take a lease of the gas works and get things put right. He therefore thought of me and asked me whether I would care to take it in hand. At first I thought I saw my way to do so and make something out of it, so I thanked him and agreed to look into the matter and see him again. Now the London gas company with whom the railway company had quarrelled was the Imperial, the largest of the London companies and John Aird and Son were their contractors. The firm did nearly all their work and there was at the time a very strong and good feeling existing, particularly with the officers. Knowing this, it occurred to me that if I took a lease of the railway company gas works, it might perhaps upset that good feeling when they heard John Aird's son had done so. I therefore thought it better to mention to my father the proposal that had been to me. He suggested that he should see Mr Henry, the Secretary and Mr Hersey, the outdoor engineer of the gas company and get their views before I took any decided step. They both came to the conclusion that it would be very unwise for any member of Mr Aird's family to be connected with any works in opposition to the interests of the Imperial Gas Company as the Directors would probably consider such a step seriously. Rather than cause any trouble, I agreed to see Mr Bell again and explain my difficulty and how matters stood: Mr Bell was sorry but quite understood my position. I therefore thanked him again for his kindness, feeling very sorry also myself as I felt sure the company would have answered my purpose and I could have made it a success. Curiously, the next day when in London I mentioned the subject to my friend H. P. Stephenson and he thought he would like to take it in hand and work it, as their offices were almost next door to one another, so I took him along and introduced him. They soon arranged matters and H. P. Stephenson worked the railway company gas works for some three or four years afterwards, and I think to their mutual satisfaction.

The result of the reassessment at Kingston had been sent in somewhat later than anticipated and the amount at which the gas works was rated was, as I feared, very high. After some consideration, however, I felt that it would not justify an appeal with all its expense and time; I would therefore make the best of it and certainly not be in a hurry to reduce the price of the gas again.

Now I have to mention a domestic matter that, at that time, disturbed and caused me more worry and anxiety than ever I had before experienced in my life. On 5th April 1863 it was necessary that Dr Julius should be summoned to attend my wife as soon as possible. I hurried to the livery stables at the Twickenham railway station, having previously arranged with the stables that a conveyance should be ready at any time in the night, when summoned. The proprietor kept his word, for I had hardly pulled the bell when a man came out leading a horse ready and in a very few moments I started in the cab, it being between two and three miles to Dr Julius's residence at the Old Palace at Richmond. The doctor was very soon ready and we came back together, thinking to myself what a strange state of things all this seems to be. The doctor immediately went upstairs, but shortly returned, remarking that all was satisfactory and we should not have to wait very long. When we had been chatting for a little while, someone tapped at the door, asking him to go up at once.

After some time I began to get very nervous and anxious, the interval seemed to be so very long and there was so much talking and running up and down stairs. But as I could do nothing, I wandered up and down the room until I had worked myself almost into a sort of fever. Then to my great relief there came another tap at the door and a female came in, who I think must have been the cook, saying the doctor had sent her down to tell me all was going on well. One child had arrived, but there was still another one yet to come.

Well, since that particular time I have never been able to remember clearly what happened to me. I must certainly have dropped down onto a near chair or on the floor, one or the other, and it is also certain that the shock and the sudden news had upset my poor brain and senses for a time. It was too awful and seemed so difficult to realise but after struggling a

One of the 'little strangers' (page 358): Edith Aird (1863-1954), second daughter of Charles Aird (1831-1910)

little while to pull myself together, my first thought was for my poor wife, not much more than a child herself, having these two babes to nurse and rear and our first little one still little more than a baby. How on earth was it to be managed? Sufficient baby linen for one child had of course only been provided, but the little nurse girl who had charge of Nellie would be of no use under the circumstances. I was completely distracted.

While in this state of mind Dr Julius came into the room looking as if he had had enough of it, poor man. Having said a few kind words and that the little strangers [Edith and Charles] were quite in form and already very noisy, he then left remarking that when he got home Mrs Julius would send some few additional items of linen that might be useful at that moment. I can remember it took me sometime after he left to realise it all and what I could do to help and get things straight.

I felt quite ashamed to send such unfortunate news home to The Grove, but there was no help for it. I believe on arrival it caused great excitement and much sympathy. When my brother John heard of it, he telegraphed the news to my father who happened to be away on the Continent at Warsaw with Mr Thomas Hawkesley the engineer, about some new water works. I heard afterwards that when the telegram arrived at their hotel, my father happened to be out and Mr Hawkesley, however, thinking it might refer to the business upon which they were engaged opened it and found it thus: "Charles has twins" which my brother in his excitement had worded. Mr Hawkesley was rather bothered as his eldest son was named Charles. However, when my father returned to the hotel the mystery was solved.

As the weeks passed along, we seemed to be getting accustomed to our unfortunate position and difficulties. My dear mother had been down and in very short time had arranged matters as to the nursing and so on very nicely that in my mind I felt much relieved. I had of course to get away to my work during the day for I now felt I must work double time to find more ways and means for the future. When in London and other places on business I remember being very much roasted by my business friends about the twins, particularly by Stephy Clarke. My poor wife was, however, completely deprived of any little bit of pleasure going on when she got strong again and for some little time after. But really I think at that time she was much happier being with her children as they interested her more than they did me. But neither of us knew then what further trouble was coming.

Although sorry and disappointed that my attempts to get additional occupation had failed at Winchester and for the railway company, I still persevered and negotiated with several other gas companies. The first was Croydon whose district was next door to the Crystal Palace District, of which company I was then a director. Had I succeeded here, it would have sufficed as it was as large as Kingston. It was then leased to an elderly Quaker gentleman whose name I cannot remember, but the lease he held had ten years yet to run. We had several interviews, but he wanted too much.

The next try was at Reigate. One of the directors of that company was a Director of the old Phoenix Gas Company in London, where my kind old friend Mr Innes was Engineer. I do not quite remember how the idea originated, but Mr Mason, the Director of the firm of Shand and Mason (then the well-known manufacturers of Fire Engines) in Blackfriars Road was talking to Mr Innes one Board Day at Bankside. He remarked that they were not getting on very well with the gas works at Reigate and he had suggested they should lease the works as some other companies had done. I believe Mr Innes replied that it was the best thing he could do and that he happened to know the very man to undertake it; the son of John Aird, at present working the lease very satisfactorily at the Kingston upon Thames Gas Works. It was then arranged that I should be communicated with and was invited to visit the works and this I did in company with Mr Innes. I remember I was very pleased with them and had I the luck to obtain a lease, felt I could do very well with it. But worse luck, when the subject was brought before their Board, after considerable discussion it was unfortunately decided to carry on by themselves.

It is hardly worth mentioning other companies with whom I negotiated and failed but there was Tunbridge Wells, Southborough, Crawley and lastly Swanage in Dorsetshire. This latter was brought to my notice by my father. At that time there was a firm of paviors – Mowlem, Burt and Freeman – having wharf and offices by the river at Westminster and they knew my father very well, as everyone did. It was my father's business to be always pulling up the roads and laying mains for the gas and water companies, and Mowlem and Burt's business to repave the roads afterwards. Indeed the two firms seemed to have a kind of monopoly at that time, certainly in the City. Mr Mowlem, I remember was a fine-looking old man residing at Swanage in Dorsetshire where the Purbeck quarries are. Burt, I think, was his son-in-law and managed their paving work in London. He was a strong, perhaps rough looking man, but a very pleasant one and evidently a good man of business. Freeman the third partner was a very meek and mild gentleman. Mr Mowlem one day mentioned to my father that they had no gas at Swanage and he wished he could find someone who could come to that place and build a works. My father was kind enough to mention myself and it was arranged that we should go there and see what could be done about it.

My father and I went there on a day appointed and with Mr Mowlem went round the town to find out the future prospects of a gas works. I must say at once I was not impressed myself, the place seemed to me so very chilling and depressing. I remember we all went round the district, endeavouring to find a suitable site for a gas works and eventually came upon a place that would be in every way suitable. In accordance with Mr Mowlem's own wishes to go thoroughly into the matter, we arranged to remain another day and meet some of the people and hear their views. But it was evident that there was a strong opposition to any gas works and particularly to one being erected on the site we had suggested, so we returned home, disappointed. I was now really getting so sick of all these failures with regard to gas, that I told my father I would not try any more and would seek additional occupation in some other channel.

The idea came to me to try and find a business as a wharfringer for the sale of building material to contractors, such as bricks, cement, lime, sand and gravel plus fire clay goods and retorts to gas companies. I thought then I could make a very good connection in such a trade, either as a partner in some going business or start a business myself and if I could only succeed in this I should not require to shift about any more which would be a great blessing.

When I told my father of this idea, he said he would speak to a man he knew very well, a Mr Forest, who had a large wharf at Nine

Elms, Vauxhall and who had a thriving business of the kind. But the old gentleman seemed to prefer to job along alone and I could not blame him if he had sufficient strength.

At this time, there was a man who was very well known to contractors and builders, a Mr George Smeed, an oldish man who would have some difficulty in writing his name. I believe he began life as a kind of scavenger collecting ashes and making bricks in the large Kentish brickfields near Sittingbourne: Kentish stocks were always the best in the trade. Beginning life a very poor man, by the time I became acquainted with Mr Smeed he was considered wealthy, as he had acquired a very large cement factory on the river Medway and was doing an immense trade. My friend, Mr Innes, the Engineer of the old Phoenix Gas Company, was very friendly with this old gentleman, as they had considerable dealings with the breeze, which was the refuse of the coal heaps. The old boy would send up a barge load of bricks to some contractor or builder, call at the gas works at Bankside and load up with a return cargo of breeze. I have often wondered how many thousand barges of bricks and cement were delivered to John Aird and Son by him.

Mr Innes, knowing how anxious I was to find some further occupation to Kingston, arranged a day with Mr Smeed, to take me down to Sittingbourne to have a look round the several works. I remember being immensely interested and thought if I could only get a start I need not worry about the future. Apart from the interest in visiting the several brick making and cement works in this part of our country, it was doubly so being so closely associated with Charles Dickens and his many enjoyable books, particularly *Our Mutual Friend* and *David Copperfield*. However, I found that nearly all the largest works had become limited companies and would be of no use to me. What I sought for was a young business, but on a good foundation that I might, with energy and hard work be able to make and supply material to builders and contractors at a remunerative price. At West Malling, I remember we came across a small concern that took my fancy immensely. It was in the hands of an old gentleman whom I felt would not be indisposed to retire if it was made worth his while. It was a most compact little business and the little residence near to in which the old gentleman then resided was so pretty and so prettily situated that, loving the country as I have always done, it made me fall quite in love with it. When we had finished our day's roaming and parted, Mr Smeed very kindly said that as he was very friendly with this gentleman, he would take an opportunity of sounding him one day about it and let me know.

It must not be supposed that in thus seeking further occupation I should be neglecting my work at Kingston, not a bit of it. Kingston was always my first love, whatever happened. From nothing, I had somehow managed to realise a return far beyond my expectation and as I had yet some seven years to work it, I was not very likely to throw it aside. The new piece of railway from Twickenham on the Middlesex side of the river was now finished and the portion on the Surrey side from Kingston to Malden and Wimbledon, in the hands of John Aird and Son, was so far advanced that new roads were being laid out already for buildings as far as Malden. That place was the end of my district in that direction and indeed, I had been thinking very seriously of inducing the company to allow the mains to be extended to Malden and also to Esher on the other side. For some time it would be a long, naked length of main, but the people of both places had been very anxious to have it.

When writing these lines, I pause for a few minutes to give the old brain a little rest, for I am now 78 years old and the year is 1910. While thus resting, my memory seems to get replenished by events and the condition of things that occurred in the year 1864 when I was about 33 years old. London then was a very different city to what it is now. How well I remember when the power that was to rule the future of London was formed. It was called the Metropolitan Board of Works and I think numbered some 30 members elected by ratepayers in the several parishes or boroughs. This body at their first meeting elected, as their Chairman, a Mr John Thwaites, a linen draper in a small way of business at Deptford. I can see him now when he used to attend the discussion class at the Greenwich Literary Institution, of which I was a member when a young lad. Also the present Lord James, the great lawyer, who was a lad a year or two older than myself. Perhaps both of these young fellows had a wonderful gift of the gab; what a help that is to any young fellow entering into the busy world. However, there can be no doubt that the linen draper was the pioneer or cause to some extent of the great undertakings then commenced for the advance of London. The Engineer appointed by the Board was Mr Joseph Bazalgette, afterwards Sir Joseph, who was instructed to prepare plans for a first section of the now much appreciated Thames Embankment from Westminster Bridge to Waterloo. He was also asked to plan a large, comprehensive scheme for the main drainage of all London, and carrying it away beyond Barking to be discharged into the sea, enormous pumping stations being erected en route. Apart from these two gigantic undertakings, costing several millions, another great work was commenced by a company newly formed, an underground railway called the Metropolitan. The Chairman of this was then, I think, Mr Edward Watkin, the Engineer John Fowler (both of these gentleman were afterwards knighted). The builders were the well-known Lucas Brothers and Sir John Kelk was the contractor. Their agent over the work was Mr Walker who afterwards started as a contractor himself and very successfully carried out many great contracts at home and abroad.

As these underground railways were spreading all over London, they gave an immense lot of work to John Aird and Son in the way of pipe laying as a vast number of mains, large and small, in the way of the railway work, had to be removed and relayed. The gas and water companies required that their own contractors should do all such work.

My reason for introducing such matters here was to show and impress on my mind what an immense amount of material of every sort of kind would be required for these great works and tended to make me more anxious that ever to find some business to enter into and share in the supply that would be required.

Going back to my work at Kingston for the moment, I decided to call a meeting of the directors and endeavour to get their consent to the main being carried on to Esher and Malden. If they assented, this would make me very busy for a time, but after, I would very seriously look round again for some additional work.

14
Ellen, Charles's wife, catches Scarlet Fever; convalescence and afterwards.

One evening, getting home after my usual day's work, I thought my wife looked very unwell. Speaking to her, she remarked she had not been feeling very well for some days past. I at once wrote to Dr Julius and asked him to try and come round in the morning before I left for my work. He very kindly did so and my wife, at my wish, did not get up until he had seen her. The report he gave me, after he had done so, was indeed serious enough and for the moment took all life out of me. She was evidently very ill; he could not then speak positively but he came again in the morning and was much afraid she was going to have an attack of scarlet fever. In the meantime he would advise that the three little babes should be kept away from her. This sudden news was really awful and I need not say that there was very little work done by me that day. The next morning he again came very early and having seen her, said that my poor wife was indeed to go through the fight with that miserable illness. He also, I remember, spoke very seriously of the importance of the three little ones being taken away from the house at once, then giving certain directions as to treatment and so on, he left. Now, what on earth could I do for the best under such sad circumstances, with two little babes only about five months old and Nellie our first born only between two and three years old? To have to send them away to some strange place and with the probability of their having already taken the disease; it was indeed a problem to solve.

I however felt obliged to let my people at home know of this great trouble that had so suddenly come upon us and also to inform my poor wife's relatives. This not only brought immediate relief to my almost distracted mind, but almost unheard of generosity and kindness from both families. My father and mother both insisted that under the circumstances the proper and only place for their grandchildren to be taken to was The Grove where they could be properly looked after. At the end of that day the children and their nurses were comfortably located there. It would be impossible for me to express on paper what a relief this was to my mind. I had hard work to pacify their poor mother at their having been sent away from her. At that time, I hurried back every day from my work to sit with her and try and make her a little more cheerful.

Before my marriage, when visiting my wife's family there was a lady sometimes staying there, a Miss Ellen Chase, who seemed a great friend and companion to them all. When that family received the news of my wife's illness they were of course very much disturbed and anxious. This lady, being with them at that time, most kindly offered to come and take charge and nurse her during her illness. This was really a most kind and charitable suggestion as it would enable me to give more time to my work and help to make Ellen more cheerful and regain her strength.

I can remember well how pleased she was when the arrangement was made and Miss Chase arrived. Of course in an illness of this kind my poor wife was very much worse before she began to get better and what a strange object she looked with all her hair cut off and wearing a wig. I cannot quite remember whether her head was shaved as well, but I do not think her hair was ever so nice again. However, illness of this kind will, as I have said, take its own time and after the oiling and skinning process was over, things began to look a little more cheerful. The time came when she arrived at the convalescent stage and we began to think of where she should be taken to recoup and become herself one more.

I should mention that during all this anxious time, the report we had from The Grove (almost daily) was most cheery and satisfactory. There had been no evidence whatever of the little people having taken the wretched disease and they were a constant source of pleasure and amusement to them all. I need not say how anxiously my wife longed for the day when they might be allowed to return to her. After some pleasant discussion as to the best place to go to for a change, it was decided that Brighton would suit best and it was arranged that I should run down one day and engage a couple of rooms somewhere.

I took an early opportunity to do this and visiting many houses, finally decided on the drawing room suite (first floor) of a little house at the eastern corner of Bedford Square, between the Bedford and Norfolk hotels. The rooms were small and very unpretentious, but bright and cheerful and faced the sea. The dining room suite below was then occupied by a Lady Grantley who was kind enough to make herself very pleasant and agreeable to my wife. The owners of this little house were a Mr and Mrs Gosling, cook and butler as usual, and were most obliging people.

How different the Brighton then was to the Brighton now. The only hotels I can remember at that time were the Bedford and the two in the Steine, the Royal Oak and the Albion. The Norfolk was a very little place, but very select, there was no King's or Queen's Gardens, no Avenues - in fact no Hove district at all. How little I dreamt then of knowing Brighton, as I afterwards had the opportunity of doing, but if spared a little longer in this world, I will come to that by and by.

Having thus arranged this matter, I returned home the same day and on the way I concocted a bit of amusement out of it. Of course on my return the two ladies were most desirous of knowing what success I had met with, but I put on an air of great disappointment and was afraid they were not likely to be pleased. However, having arranged to take my wife down the following Saturday, our very kind friend Miss Chase said goodbye and returned home – what a kind nurse and dear friend she had proved herself to be.

I had also arranged with Macrostie to disinfect the rooms and paint and re-paper several of them, but of all things to carry the pipe for the drainage of the house further into the river. There can be no doubt as to the cause of my wife's unfortunate illness, having three little babes to nurse and look after, she was unable to go away from the house or take any exercise whatever and the summer of that year was a particularly hot one. So much so that there was not a sufficient flow or stream of water in the river sufficiently to flush the drains of the several houses; indeed, in many cases, leaving the outlet of the drain exposed. Our residence was among that number and it was a shameful state of things, but rectified by the authorities some two or three years later by a sewage farm. My poor wife, to get any air at all would sit on the bank of the river most of her time, no doubt inhaling the foul air and bringing on the illness. I also have reason to remember that time as many of my barges of coal were aground, as there was not sufficient water. This caused me much anxiety at the gas works.

Having made all these arrangements, we left for Brighton. On

Agnes Aird (1837-1939), eldest daughter of John Aird (1801-1876), married Henry Gill.

arrival I was amused and glad to see how curious and anxious she was, after what I had said about the apartments and what they would be like. When we arrived at the house and she had been introduced to Mr and Mrs Gosling, and afterwards to their cooking and kind attention at our well-deserved dinner, her surprise and delight amply rewarded me. After a day or two she made friends with the old lady who occupied the ground floor rooms (Lady Grantley) and seemed so happy and comfortable that I thought I might now and then go back to my work. What a bright, cheerful place Brighton was - not half the size it is now and as I said, no King's or Queen's Gardens and no Hove; there was also no Aquarium, no Grand or Metropole hotels and best of all, no trippers, only the very elite of society. How well I can remember seeing the Princess Mary of Cambridge walking about on the front with her somewhat loud but happy voice, chatting with the several ladies and gentlemen in attendance. I little knew then or thought that, in a very few years, I should be associated with many large works carried out in that area for improvements.

I cannot remember how long we stayed at these lodgings, but I can remember we were very sorry to leave them and have always since felt there is no place like Brighton.

The time came for our return home; it had been arranged in the meantime that our little ones should be restored to us on our arrival. I need not say with what delight and happiness this event was looked to, particularly by my wife who, during this long and wretched illness and separation from her three little babes, had borne with it all so patiently.

Having again come together and after a little while quietly settled down to our usual ways, things looked a little more cheerful, but I could not help noticing that the little ones could not quite make out what it all meant and their mother, whose head had been shaved, was still obliged to wear the hideous wig and looked very comical. The difficulty now was how we both could sufficiently express our heartfelt and sincere gratitude to my dear father and mother for their generous and loving kindness in taking charge of our little ones during this long and anxious time. It was a great happiness to know and feel this was quite understood on both sides.

I remember it was also a great happiness to myself, to find that my work at Kingston, in spite of all this domestic trouble, was still going on satisfactorily and prosperously. During our stay at Brighton I had managed to run up two or three days a week and keep things in order. During this time also, the Directors of my gas company, after some little discussion, had assented to the main being extended to Esher and also Malden. This involved a lot of pipelaying, some six or seven miles, and they were afraid it would be a long time before being remunerative, but when the work was finished they found they were mistaken. I had no idea myself that the inhabitants would respond so generally and so heartily to our invitation to supply them with gas. Going along the gang of men laying these mains to Malden, I would frequently come across my younger brother Joseph, who was supposed to be assisting Mr Stannard on the new railway the firm was then making for the South Western. Poor Joe, what an odd and mysterious being he was; as years went on he never would willingly mix or meet with either friends or relatives, but always seemed suspicious of everyone he met of ridiculing or laughing at him, and yet he was as kind hearted as anyone living. He was a very good-looking young fellow and he knew it, certainly far more so than any one of his brothers. Later on he became very proud of an enormous moustache he had cultivated, very similar to one grown by Sir Robert Peel MP, the well-known son of the great statesman. One day, years after, when staying at the Bedford Hotel at Brighton he was taken for Sir Robert who much frequented that well known hostelry at that time; Joe was never tired of mentioning this little incident afterwards.

Some years after an opening occurred at Birmingham to take a lease of a small tube works. My father, hearing of this through the kindness of an old friend J W Gray, then the Engineer, purchased it and put Joe in possession to do the best he could with it. He took to the work at once, the firm later became the Wellington Tube works and was a main supplier of gas and water pipes for John Aird and Co.

I was glad to see how quickly the firm was progressing with the railway, as it would when completed, lead to an increased demand for gas. Even now new roads were being made and even some few houses already built. How different things looked now to when I first started gas making at Kingston. I have already many times mentioned how my own work occupied the whole of my love and attention, but as time went on I was always interested to hear any news of my father and my brother John's progress. About this time I think they competed for one of the large main drainage contracts and were successful in obtaining it. Curiously, this contract was to erect a very large pumping station with immense deep wells or chambers alongside the old Deptford Creek bridge. This was where, when small children, my brother John and I were living at the gas works at Greenwich and had that same bridge before us all day long. It was, I believe, very heavy work, there being so much water to contend with.

The young fellow put in charge, I remember, was a Mr Samuel Gale, a friend of Mr F W Shield's. My kind friend Mr Simpson's youngest son, Arthur Telford Simpson, was there also to get some knowledge of practical work. I believe the number of pumps required to conquer the water was something extraordinary.

It was about this time that another railway work was entered into by the firm. I only refer to it as it brings Wyatt's name in again. Sir W. Clay, whom I have mentioned as being Chairman of the London Water Companies, resided at Fulwell Lodge, Twickenham. As he owned a great deal of the land about it, he conceived the idea of a railway being made to pass through this from the Twickenham station of the South Western to Shepperton, that it might open up this land for building. It was to be called the Thames Valley Railway. This idea he suggested to our old friend, Mr Quick (who seemed to be his factotum in all matters) to work out. The railway company was, I believe, first approached but they would have nothing to do with it, believing it could never pay. Mr Quick then came to my father and after some time, it was arranged that the firm should construct the line and finance it altogether. It was only a short length and with no heavy work on it at all. The stations were to be Fulwell, Hampton, Sunbury, Kempton and Shepperton. Mr Quick and Mr Tolmie were the Engineers but the contractors, I think, prepared the plans. When all was arranged the firm had some difficulty in finding a suitable man to take charge of the work, so hearing of this I suggested Wyatt, being still blind, unfortunately, to the man's evil nature. I knew neither my father nor my brother John had ever much cared for him, their instinct being stronger than mine in that respect and I did not press the matter, but finally he was engaged.

Some little time after this I remember him coming one day and asking that our eldest little one, Nellie, might go on the work and cut the first sod. Of course it was only a little farce for the child's amusement and was, I believe, carried out. The railway was completed and passed by the government inspector and worked thereafter by the South Western, but was by no means a success financially. These sort of contracts never are I think. Unfortunately for me Wyatt was found to be a most expensive man in his management of the work, yet even then my eyes were not opened regarding him. I remember there were a number of iron bridges with road approaches over the line and that this portion of the work was sublet by the firm to James Simpson and Co, and of course this brought Wyatt to their notice.

This year there occurred two weddings in our family. My dear sister Agnes was married to Mr Henry Gill, the Engineer of the English Water Company in Berlin. For some time past she had frequently accompanied my father on his visits there to look into business matters and I suppose during those visits their partnership through life was arranged. Mr Gill was a very nice fellow, every one seemed to respect and like him and I know I did. He was a widower, left with a little daughter, very quiet and gentlemanly and certainly clever, though perhaps a little austere in his manner and very much a German. This wedding was a very quiet one.

The other wedding was my wife's brother Clarence with Mr Quick's daughter Polly. Poor little woman, how little she then knew how sad her future life was to be.

Since our return from Brighton things seemed to drift back again to the old groove before my wife's unfortunate illness. I had been over to Richmond and had a talk with Mr Albert Chancellor, the well-known house agent, a most obliging man, with reference to seeking another home for my wife and family, for I certainly did not feel safe in our present one. He gave me a list of some few, but most of them were too far from my work. I remember going over some very nice newly-built houses on the St Margaret's Estate by Richmond Bridge and one or two at Petersham just outside the Park, quaint, old-fashioned houses and Robert Fowler, Hargrove's partner was then living in one of them. I have often regretted

I did not take one of these for it was a very pretty situation and with any number of nightingales about and really not very far from Kingston.

I also began again to consider about a future occupation after the Kingston lease and continued my enquiries. So far my inclination was towards the cement and brick trade and the little works at Malling in Kent had much impressed me. If nothing better turned up I would try my luck there, if I could come to terms with the owner. Looking back to those days I almost think I should have been a much happier man had I done so.

I remember on one occasion going over a large brickfield at Arlsey in the north of London with Wyatt and George Hall (he was the firm's right hand man then, who lived and died in their service). Knowing of my wishes he thought they were worth visiting and might suit me. Their brickfields were of great extent and held by a Mr Dennis (a real Irishman); the brick they made was a white brick, very heavy and more for ornamental work than general use.

Having wandered about these fields for some time, in a hot sun I remember, we all (at least I know I did) felt a little done up. There was no sign of any place where we could get so much as a biscuit, when Mr Dennis, who had been showing us over the place, produced from some pocket in the tail of his coat some chops carefully wrapped up in a piece of paper. I can very well remember this incident, even after all these years, and his going to one of the men's shanties and asking the old woman in it to cook them for us. With a piece of stale bread, a scrap of salt and our pocket knives, we came to life again. But in the end our day was wasted for the idea never came to anything.

I will mention one more similar expedition that offered itself some time later and if that failed I had decided to come to close quarters with the old gentleman who owned the small cement and brick works at West Malling in Kent.

Before describing this idea, however, with much sorrow I found the time had once more come round for Dr Julius to attend on my wife. After the last unfortunate event I became very, very anxious wondering whether there would be couplets again; in the Scriptures it is said "Blessed is the man who hath his quiver full". Well, that view of it is no doubt correct, but it made me think very, very seriously of the future and if such a calamity should occur that would be an end of all things to me; but thank goodness I was spared that trial and another little daughter made her appearance. I can remember some few days before the child was christened some of our dear Simpson friends came round for tea in their usual friendly way, Mr Nancolas being one of the party - what a man he was for tea. A discussion arose about a name for the little one, Mr Nancolas suggested Lilian and this was unanimously adopted.

Speaking of tea, it occurs to me to mention that, although I never cared for it myself, I was always anxious that my wife and friends should have the best china tea, indeed there was then no other. I used to purchase it myself at Fortnum and Mason, a well-known establishment in Piccadilly, and was obliged to pay 5/6 the lb (in 1865). How things have altered in that respect as now (1910) it costs £5. [Today, 2004, the cost is over £10].

I will now mention the particulars of the idea my kind friend Mr Stannard brought to my notice some time later. The firm, having completed the new line from Kingston to Wimbledon, had now to make another line for the same company near Bishops Waltham and Southampton. Mr Stannard was in charge of it, and, knowing that I was looking out for something in addition to my present occupation, wrote to me that there was a brick business near his work that might be worth while looking into. I therefore went down to see him about it. The brick fields in question were in the hands of a small syndicate of gentlemen living in that particular neighbourhood – Sir Theodore Martin, Sir Arthur Helps and Mr Stone - MP, I think, for Southampton.

They were not doing very well with the business and were desirous of letting it on lease to some practical man who had knowledge of that work and who had some sort of connection with contractors and who could bring in customers. At first I thought something might be done with it, if a special rate could be arranged with the railway company. These gentlemen were exceedingly kind, going over the works with me, dining with me, putting me in possession of all their books and giving me information on every point required and altogether making things very pleasant while trying to arrive at some understanding. But with all our mutual efforts to accomplish an arrangement by meetings with railway people, builders, contractors and others round about and even shipping agents at Southampton, there really seemed no possibility of creating a business that would give a sufficient profit to make it worthwhile. The brick made was the same heavy white brick that I had met with at Arlsey and I had to abandon the project.

To give up this idea was a great disappointment to all of us, certainly to me, as it would have been a great pleasure and very nice to have had the opportunity to working with or for such men. Arthur Helps was then Queen Victoria's private secretary and Theodore Martin was the head of a well-known firm of parliamentary agents in Abingdon Street, Westminster. He afterwards married Helen Faucit, the great actress, and later on Queen Victoria sent for him, knighted him and made him a confidential secretary and writer of her private memoirs.

One day, when house hunting, my father came down to help in the search. While chatting as we wandered about, he mentioned the firm were rather anxious about a law case that was to come on next week at the Croydon Assizes. As the case was rather a singular one, I will try and give a little sketch of it. They had entered into a contract with the Metropolitan Board of Works to carry into the Thames at Horsley a large pipe for some drainage purpose; this work could only be carried out by means of a coffer dam. The piles were carefully driven under the eye and under the direction of the Clerk of Works and this work was being carried out in the busiest part of the Pool. One evening, the day's work being over, a large steamer called the Zebra had just arrived and in moving about to take her station or position managed to place herself on the piles forming the coffer dam and consequently was very much damaged. The owners brought an action against the Metropolitan Board of Works and the firm jointly for £20,000. As a rule and if at all possible, I knew the firm always avoided law, but this case was much too serious to accept defeat and Hargrove, Fowler and Blunt, Solicitors had the management of my father's defence. The Counsel were Mr Hawkins QC with Horace Lloyd as his junior. Baron Martin was the Judge and the leading Counsel of our opponents was Mr Bovill QC, afterwards Chief Justice Bovill; the name of his junior I can't remember.

To me it seemed perfectly clear, there could be no claim for damages against the contractors as they were simply the agents and acted entirely on the instructions of the Board's Clerk of Works and this was admitted by the Clerk in cross examination. But it was a curious and unfortunate matter altogether so far as this action went as Bovill was far too clever in cross examination and he turned our witnesses inside out in a very short time. Poor George Hall, a man as a rule always knowing what he was talking about and well able to take care of himself - in a very few minutes he evidently did not know whether he was standing on his head or his heels. In Bovill's cross examination he gave such extraordinary replies that we had hard work to restrain my father from interfering, he got so excited. In the meantime, Hargrove was rushing about trying to find our leading Counsel, Hawkins, to come in but he was unable to do so at the moment, I believe, as he was engaged on a consultation in another case. Horace Lloyd the junior was in the Court, but asleep; someone said he had been playing cards all night. However, the verdict went against us and in favour of the Zebra.

The firm could not of course accept this and were, I believe, advised by Counsel to bring an action against the Metropolitan Board and notice was sent to that effect. But after some little time, I suppose seeing the error of their ways they gave the firm notice through their solicitors that, after considering the matter, they would take the responsibility on their own shoulders and release the firm from the case altogether. This was a very righteous judgement and I need not say a great relief in my father's mind.

Continuing my efforts to find another home for us I would mention that towards Twickenham there was another meadow and then a somewhat distinguished looking house with a large lawn facing the river called Radnor Lodge, tenanted by a Mr Chillingworth, a well-known wine merchant in the City. Then there was an ordinary looking house built with white brick and immediately opposite the Grotto Tavern. Opposite the entrance to Radnor Lodge another road started and went by the entrance to Strawberry Hill and continued along to the back part of Teddington into Bushey Park. On one side of the gates there was the gatekeeper's lodge, and on the other a pretty cottage; very small but so prettily situated. There was a piece of garden stolen from the Park with only an iron open railing to separate it from the park itself. Going over it, I found there really was not sufficient accommodation for my family, but very nice for a bachelor or student. Going back again on the road to Twickenham, I found a board had been put up, that it was to be let or sold and apply to Mr Maw, Aldersgate Street in the City.

At this time young Mr Quick had become quite a feature in our lives; when boys we were like brothers. He was now at work, assisting his father in the engineering department of the Southwark and Vauxhall Water Company and my dear old friend Mr Fraser had become the engineer of the Grand Junction. Young Joe, however, found time beyond this to exercise his talents in various speculations, particularly in house property and building societies. Mr Quick himself, I remember, was speculating in this way in addition to his position with the water companies, for many years, while Joe was still too young to do the same. But as Joe became older, he seemed to turn into it quite naturally and it was a very pleasant and practical way for any man to increase his income.

They were small properties, improved and put in better order and sometimes sold again at a very handsome profit. I can very well recollect one of these properties where my wife and the children spent many happy days as each summer came round. This was at Bognor, then the most primitive seaside place, I should think, in the Kingdom. Mr Quick, in that early period had bought a row of small buildings called Rock's Buildings which were all stuck together forming a crescent. Numbers 1 to 9 or 10 were painted a bright green and I think faced with flints, looking immediately on to the sea. There was no sea wall and only a rough beach and in fact the houses were almost in the sea – what I call real seaside. If you walked about half naked, no one would notice.

There was only a little high street with a few shops, the butcher and fishmonger being difficult to find. There were also a few interesting and very old-fashioned villages and houses round about outside – Aldwich, South Bersted and Feltham. At the first named place, just beyond Rock Buildings, was a large house and grounds facing the sea belonging to Baron Grant, a well-known financier. Neither Mr Quick nor his son Joseph (when he was old enough) appeared in these matters; they were arranged and let by a man named Restler, the Foreman of the turncocks in the Southwark and Vauxhall Water Company's offices, who arranged everything about Rock Buildings. I would have to see him and take one of them for two guineas a week. During Goodwood Races, the whole row, and nearly all the private houses, were taken by the officers of several of the crack cavalry regiments at an immense rent for that week. I need not say that my family were not there for that noisy time as I believe these gentlemen, after dining, played old harry with the furniture and everything round about for which, of course, they had to pay. They would also have their own four in hands and drive to the course each morning. Indeed, I heard of some extraordinary tricks and amusements they were always up to during Goodwood week.

When my family were staying at Bognor, I could only join them from the Saturday to the Monday, but even that little bit of rest was very enjoyable in that quiet place with what I always so loved - real sea air.

Writing now of the house at Twickenham that I thought might suit my family, I one day asked young Joe, as he was then, to go with me to Aldersgate Street and interview Mr Maw, whose board I had seen. We did so, but a "maw" disagreeable man I never met! He had evidently not made up his mind whether to sell or let and as it was only a waste of time, we very soon retired. Some months after this the house was occupied by a Mr Child, who had some business in the City, but whether he had purchased or whether he was renting it, I never asked or cared to know.

These repeated failures to find some other business or occupation to work up by the time the lease of the Kingston gas works expired made me feel very anxious, although there were still some three or four more years of the lease to run. Time now seemed to race along too quickly, the end would soon come round and in all probability I would have to provide for a larger family.

Although I had done so well and would probably do even better, as it was extraordinary how every year the rental increased, I could not shake off a nervous feeling that I might tumble into some business that might prove unfortunate and lose all the savings for which I had worked so hard. Thinking it over quietly I came to the decision that it would be better not to be in too great a hurry, but by keeping my eyes and ears open, perhaps something really substantial might turn up. Considering I was only about 35 years old, I certainly had nothing to complain of so far.

The three companies of which I was a director gave about £400 a year – the Crystal Palace District Gas, the Gas Product Company and the Angelsea Slate Company. It was on this Board that I first met Makins, then a young man and nephew of Mr Lightly Simpson, Chairman of the Great Eastern Railway and well known in the railway world. This nephew I knew better late on as Colonel Makins MP and Chairman of the old Chartered Gas Company.

My few investments were, I think, safe and now and then, by watching some movements on the market, I sometimes added a little more by what I suppose must be termed a bit of a gamble; though with my shrewd friend H. P. Stephenson by my side, it was almost sure to turn up trumps. One of these little specs was for some shares in the Great Southern Buenos Aires Railway Company and another in the Anglo Austrian Bank; I made a profit on both of these.

The Engineer of the first named was a Mr Thomas Rumbold and I made his acquaintance in Hargrove's office one day. The firm John Aird and Son had just completed a large contract for a gas works and its main laying in the city of Bahia in the Brazils. James Eagle had charge of the work and I remember on his return he brought back with him two large cases with a glass front in each, containing a most magnificent collection of the largest and most beautiful moths and butterflies from that part of the world. Each specimen had been carefully pinned, forming a most interesting and beautiful group of colour. He very kindly presented these cases to my mother and she afterwards had them placed in a large gilt frame and standard forming a splendid pair of fire screens and then gave them to me, much to my surprise and delight.

Mr Rumbold, on his return to England brought with him a concession for a gas works in Para, the next largest city in Brazil. Knowing Hargrove, he induced him to go into the matter and met Mr Quick, my father and brother John. After several meetings a company was formed, the Directors being Sir W Clay MP, Chairman, Mr Henry Whiting, Mr Frederic Wigan, Mr Quick, another, whose name I cannot now remember, but an influential man and with much knowledge of that country. There was also Mr Alfred Rumbold (Mr Thomas Rumbold's brother) and myself. When the subject of my becoming a director of this company was mentioned, I could not help thinking it might perhaps cause some unpleasantness to the Chairman, Sir W. Clay, when finding that one of this Board (myself) was some seven or eight years ago one of his subjects in the Grand Junction Water Company office. It was, however, arranged that Mr Quick should prepare him for the shock.

I had heard that it was the practice of Sir William, on the Board day of the water companies of which he was Chairman, to go first into the Secretary's room and receive from Mr Quick some idea or knowledge of the several matters to be transacted by the Court, and this same practice was always followed in this case. I can so well remember that day all the other directors were present waiting, the door opened and Sir William and Mr Quick entered. If he had been the king, he could not have assumed a more dignified and solemn manner, then he bowed very gravely to one and all, sat down and said a few introductory words about it being our first meeting and so on. I certainly need not have troubled about being present, for before the Board broke up I received a most gracious smile of recognition.

As time went on and the Brazilian gas works was completed, it became rather a responsibility on my shoulders, exporting the proper coals and attending to the reports from the manager who had been sent out to that country, for I was the only gas man on the Board.

I have just noticed that having named the several members of the board of this company I have forgotten to mention the Secretary, Mr Arthur Blunt, a younger brother of William Blunt of Hargrove's firm and a very clever and pleasant young fellow. I remember several years before this when I was with the Grand Junction Water Company, the Secretary there, Mr E. O. Coe, who resided above the offices at the corner of South Molton Street and Brook Street, was very fond of amateur theatricals. The premises, having been recently considerably enlarged, then gave him some very nice rooms for such a purpose. I can also recollect that Mr Coe had these new rooms furnished by a Mr Maple, who then had a small, insignificant shop in the Tottenham Court Road. What a difference in that shop and business now!! As the Aird family were very friendly with the Coe family, I was often invited to be present at some of the entertainments he got up and very successfully, sometimes even writing the play. This young fellow Arthur Blunt was always a success at acting, mostly in a comic part, and I shall never forget him in the farce *Betsy Baker.*

Things were now going on much the same; the fee from this additional directorship increased my income a little more, apart from the pleasure of meeting such colleagues every fortnight.

At Kingston, by way of a little change for some time past, I had been very much troubled by some members of the Council headed by that wretched little baker Wenman from the London Road, the problem being the price charged for the public lamps. The new railway, now being completed, had caused many new roads to be opened up and mains and services laid in order to light up the additional lamps on them. Houses were being built, but of course it would be some little time yet before they were occupied. The lighting of public lamps is never very profitable, there being so much waste of gas with them. However, after a lot of pro and con letter writing in the local newspaper, the Surrey Comet, and a lot of tall talk on the Council about another company, the subject as usual seemed to end in nothing and the contract was renewed on the same terms for another three years which would last my time. In answer to their request for a reduction in the charge for the public lamps I remember remarking how could it be possible when a re-assessment of the Borough has just been made and the valuation of the gas works doubled. But in all matters of business, particularly when dealing with a corporation, there must now and then be some worry and trouble.

I am writing these words 40 or 50 years later, but the Kingston on Thames Gas Company is still there; still the only gas works in Kingston on Thames and still paying their ten per cent. Their Engineer and Manager is still a Packham, the son of the boy Henry Packham whom I took on to help me in my little office when he was only 12 or 13 years old. I think it would be difficult to find another instance such as this of successful management.

In the Times newspaper of this day, 23rd July 1910, I notice an account of the Annual Meeting of the Gas Engineers and Managers Association. How different in every respect to old days and times. In the first place, where are all the gas engineers and managers I knew when I was working out the lease of the Kingston Gas Works? All gone, and younger men come to the front whose names I do not know. The two or three left I can remember are: Corbett Woodall, Henry Hack, C. Hunt and C. Carpenter. [Woodall eventually came on the Glover and Main Board on which Charles Aird's son, Kenneth, was Chairman; the latter had served his own indentures under Hack, then Chairman of Birmingham Gas].

There were many interesting papers read on the subject of gas, which seems to be holding its own in spite of the competition from electric lights; but what great and wonderful changes in the manufacture. In my days, if we got 9000 cubic feet from a ton of good gas coal, we did well and with an illuminating power of 16 candles (the government standard). Now I see they can get 13,000 cubic feet per ton and by the use of these new incandescent burners and mantels can defy the electric light. Also, now there is very little manual labour involved, nearly all done by machinery and to this old man it is wonderful how primitive the manufacture of gas must have been in my time.

Having been so very busy lately with one thing and another at the works at Kingston, as it was becoming difficult to keep up with the demand for gas since the new railways were completed, the new roads formed and new buildings erected, I seemed to have forgotten all my kind friends. I could only rush up to London to attend the Board meetings of the companies I was connected with, but fortunately, except for one, the Anglesea Slate Company, the meetings were in the afternoons at 3 o'clock. This gave the whole morning to do a lot of preparation and see that things were all in order.

I had heard that my kind friend Mr Innes and his niece Miss Mynott had left their pretty house and garden at West Wickham and had taken a lease of a large place near Horsham in Sussex, plus the farm adjoining – some three or four hundred acres I believe. This residence was formerly the home of the Shelley family where the well known poet had resided. It seemed difficult for me to understand this, as the position he occupied with the old Phoenix Gas Company was quite as much as any living man could manage properly. But Mr Innes was a wonderful man and I made up my mind to call upon him the very next time I went up to London and explain that I had been too pressed with work.

The offices of his company were at Bankside, Southwark facing the river, with wharves of one sort or another (but mostly iron and coal), between Blackfriars and Southwark Bridges. I always knew where to find him. Going a little way up the yard through the large entrance gates by the side of the office buildings, there was a little window looking onto the great yard and retort houses. If Mr Innes was in his room, by looking up to that window I would see him sitting at his desk. If alone, he would smile immediately, jump up and open the door, opening into the yard. Ascending two or three steps, I would find myself in his office, about the size of a lavatory, say 10 or 12 feet square, smaller than my own office at Kingston. If he should happen to be engaged with anyone, I would stroll about the yard until he would open the door and call me; we quite understood one another. I have often wondered how many times I had occasion to go into that little office. There was, of course, another door from it leading into the general offices. The Secretary of the company then was Mr Isaac Crookendon (a very nice fellow) who lived in the rooms above.

After this perhaps unnecessary description, we had a little friendly chat, more particularly about his recent move to Horsham and as he really seemed to wish that my wife and I should spend a day or two there with them and have a look round, it was so arranged. We would accompany him home the following Saturday and we were able to carry out this arrangement. Minnie was delighted to see us for I am sure the poor girl must have felt it awfully dull and depressing having to live alone the best part of each week in such a large place. Field Place was the name of the house and a most interesting place it was; very quiet with a lot of large rambling rooms, but very comfortable. The gardens and grounds were very singular in their arrangements and I had never seen anything like them before.

The farm seemed to be of large extent and quite close to the house. I could not help being amused to see Mr Innes, the moment we arrived, rush off to his bedroom and in a few moments return in what he called his farming suit and then rush off again to assist or join in any farming operations that were going on. But as it was very nearly 9 o'clock before he came in again for dinner, which had been arranged as usual for 7 o'clock, this was no joke for his hungry guests. I remember Minnie telling us afterwards that this was quite a common occurrence.

How well I remembered all this some 12 years later on. However, in spite of the overcooked dinner, we had a very good evening; chatting in the most cosy and comfortable drawing room I ever sat in.

The following morning (Sunday) we walked to the pretty little church and after luncheon Mr Innes and I had a good long walk over the farm; the weather, I remember, being very fine and it was very enjoyable. In the evening after a very pleasant dinner (at 7o'clock this time) we had another friendly chat in the cosy old-fashioned drawing room.

In the course of conversation I can recollect Mr Innes asking whether we would like to drive home instead of returning by rail the next morning; he had a pony in the stable, and as he expressed it, eating its head off. I first had been obliged to decline his very kind invitation to stay a few days with Minnie, being so very busy, and it would rather trespass on my own time. However, I thought the drive might please my wife and as ill luck had it, assented.

In the morning Mr Innes left early for London and at about 10 o'clock I think the pony carriage was brought round, but the pony certainly did not impress me; I should fancy it must have been born in Shelley's time. My wife also did not seem quite happy in her mind about something and poor Minnie seemed distressed at our leaving, but we tried to cheer her up by promising to stay a little longer the next time.

The weather being very fine, it should and would have been a very pleasant and beautiful drive had not the wretched animal commenced to play the fool. It had evidently not been doing any work for some time past and was therefore very fresh and consequently required all my attention to keep the creature going. To make matters worse, my wife became nervous and imagined we were on our way to kingdom come, now and then making futile attempts to clutch the reins: this did not improve matters. I have never professed to have much knowledge or experience with horses, but I can often remember the great delight and pleasure it was to me when in Canada, during the winter with the thermometer several degrees below zero and nothing but snow everywhere, driving on a Sunday (the only day we had to ourselves in Toronto). We would hire a sleigh and span of horses with the slight kind of harness they used to wear then, covered with little bells and, with one of my colleagues, generally Quill or Kent or with one of the stable men, we would drive many miles in the back part of the country.

This unfortunate little trip makes one smile to think of the past. However, after lunch and a drink, we started again. Perhaps seeing the error of his ways, the pony behaved better and we finally arrived at Surbiton, where I was only too glad to leave the animal at the stables of the Southampton Arms hotel with instructions to have it well looked after and sent back in a day or two to Field Place. We then jumped into one of the station cabs and drove home, stopping a little while on our way at the gas works to see if all was going on well.

Before leaving Field Place on this first visit, I remember noticing, when walking over the grounds, two large expanses of water, really large enough to be termed lakes and in some places very deep. On my return, the idea occurred to me of having a little boat made and sent as a present to Minnie as there was nothing of the kind in their possession.

Just by our home at Teddington Lock, there was a well known boat builder, James Messenger, who had been the champion oarsman of the Thames for several years and now the Chief of the Queen's Bargemen with their gorgeous scarlet and gold uniform. Early one morning, I therefore walked round to his yard and saw him about making a boat. After some two or three weeks, it was finished. A pretty little thing, properly equipped and with her name in fanciful gilt letters on each side of the outer bow. I had also taken the precaution to have it made with a good, broad beam to make it as safe as possible. It was then sent down by road and on arrival, I heard it caused no end of surprise and pleasure.

Although we visited Field Place many times later on, there were no more returns home by road. Sometimes on these little visits Miss Betsy Goring would be staying there, as she and Minnie were always great friends.

At this time there were not many holidays, occasionally a visit here when taking an extra day off now and then and strolling about the place with a gun or little rook rifle to pot a few rabbits. I have often thought what a silly fool a hare is, to stand up, stare at you and get shot!

During the summer, my wife and children would go down to

Bognor and stay at Rock Buildings for a few weeks and on the Saturday I would join them, what a primitive little place it then was.

15

The purchase of Pope's Villa, and Charles, aged 35, joins John Aird and Son as a Partner.

I am now approaching two events so astounding, so extraordinary and undreamt of, that the whole future of my career was at once changed. Up to that time my success had been wonderful, but I had worked hard for it. I had been content with a simple, quiet life and the few dear, real friends I had made and valued. I never cared for society or gaiety and was always uncomfortable when meeting strangers.

This strange and miserable failing has never left me. Throughout my life I have never been able to seem or feel natural as other men do. How often have I compared myself with my brother John, who in such cases, at once and without any effort, quite naturally assumed the position of senior and mastermind. How often have I wondered at his ease and manner, when I became more closely associated with him later on. I have sometimes thought, looking back on some of these pages that there seemed a desire to bring myself forward too much and, as they say, blow my own trumpet. If such an idea was really conceived, it would grieve me very much for there is really no man living who has less conceit or opinion of himself than this old fellow.

Some few weeks after our return from this first visit to Field Place, I received a letter from Mr Chancellor, the house agent at Richmond, who had been trying to find another home for us. He said that a place had been put in his hands that day and he would like to see me about it. I replied, any evening after 7 o'clock, so he came the very next evening. To my great surprise and astonishment, he stated that the owners and occupiers (Mr and Mrs Balfour) of the well known Pope's villa had placed it in his hands to dispose of, with the seven acres of garden grounds attached to it. They also very much wished it should be disposed of privately if possible, and not by advertisement or public auction, and they further stated their price should be £10,000.

They had been obliged to take this step with much sorrow, as they loved the place, having spent so many happy years in it, but Mr Balfour's failing health made this necessary. He was a silk merchant in the City and now found the journey to and fro' each day, although short, too much for him.

Well, when Mr Chancellor had said all he had to say on the matter, I really could not help laughing; it did seem so absurd that such a proposal should be made to a young fellow like myself, situated as I then was. I no more dreamed of such a thing than of flying up to Heaven. However, I thanked him for his kindness (for he really was a very kind and obliging man) in giving me the first opportunity, but for many reasons I could not accept it. Whatever my means might be, it would be bad form to put myself in such a prominent position and might do me much mischief in my future progress.

It crossed my mind afterwards that Mr Chancellor knew the man he was hoping to do business with better than the man himself. Before leaving, he suggested that before finally making up my mind to decline the offer altogether, he thought it would certainly be worth my while to take the opportunity of going over the place to see what it really was like. Instead of persisting in my refusal, I weakly assented to do so, thereby showing that I must be one of those unfortunate men who have no self-denial or strength of mind when tempted. He then left, promising to let me know what day and time it would be convenient for them to let my wife and I go over the property. The arrangement he made was, I think, for the following morning or the one after. I think the latter after 11 o'clock; it is becoming rather difficult for me to remember, precisely, days and dates.

My wife and I therefore went round at about half past ten and were received most kindly by Mr and Mrs Balfour who, in a few introductory remarks, repeated the reasons for their leaving their pretty home; so much so that we felt quite at our ease at once. Mrs Balfour, a very pleasant and buxom lady, took the lead and started with us on our little survey of the villa and the grounds. Mr Balfour, not feeling well enough, begged to be excused.

Before starting on this inspection, my memory went back to days long past and gone, the days when my brother John and I were little lads, living at the gas works at Greenwich where my father was then the Manager and Engineer. I have talked already of the day when, wishing to give us a day's pleasure, he engaged an open cab for the day from the public stand to drive him and his two boys to the then well known races on Moulsey Hurst, better known to all cockneys as the 'appy 'ampton races, and almost as crowded as Derby day at Epsom. My father, knowing the coachman to be a decent, steady fellow thought he might trust him. It was rather a long drive for one poor horse, but I believe he assured my father that if we started early, so that the horse have a frequent "bait" (bite), he would do it beautifully. That was certainly one of our red-letter days and how pleased my father always was now and then to refer back to it. Why I refer again to that incident was that on the drive to those races we passed Pope's villa.

I could remember quite well the almost frantic excitement of my brother John and I when passing it, never having seen such a curious house before and asking my father whether Chinese lived there, as it looked so much like a Chinese Pagoda. I can remember his stopping the vehicle that we might look at it while he told us the particulars.

Again, I have talked of another little incident many years later on when I was engaged with the Grand Junction Water Company, and going one Sunday to Twickenham with a chum in the same office (Frank Carter) for a day's fishing. Having told the fisherman to fix his punt in a good place, he did so, almost immediately opposite Pope's villa saying it was the best pitch on the river. It may have been so, but I know I must have lost many good fish by admiring the beautiful place and velvet lawn instead of watching my float. How little I then knew what was to come.

I will now endeavour to give some idea of the place as my wife and I viewed it under the kind guidance of Mrs Balfour. The houses right and left on the river side of the road opened direct from the pavement and on the same level, but with Pope's villa the entrance door was set back some 12 feet, still level with the road and with no steps. There was a wide, gravel path curving to the entrance door from two double gates, one each side of the villa's grounds. This door opened into a large hall, about 20 to 25 feet square, with a quaint, handsome fireplace on the left and a door on the same side, just beyond, opening into the drawing room. This was a very pretty, comfortable room, not grand in any way and with no gilt about, but all in good taste.

This room extended the whole length of the house from the road front to the river front and was 18-20 feet wide. At the end, facing the river there was a very handsome stained glass window with a very clever medallion portrait of Pope (head and shoulders, life size) painted in the glass above; its two doors were of plate glass and opened onto a balcony of carved wood. Looking from this balcony to the lawn below, the difference between the road level and the river side level was quite eight feet.

There was no window in this room at the road side end, but in front of the door from the hall there was a glass door opening to a portion of the garden. It was on the same level and the same length, exactly as the drawing room, perhaps a little wider and about 28 feet. On the left of this piece of ground the wall separating it from the road was covered with ivy and other creepers, with a narrow little border at the bottom with flowers. On the wall opposite (the division wall between Pope's villa and the grounds of the Misses Young) there was a small conservatory, well stocked and in good condition and with a little sunken pit and boiler. Fixed to the wall above there was an old leaden cistern with the date 1646 on it, which at once reminded me of days gone by of the old dirty cistern in the drawing office of No. 10 Walbrook where the best years of my life were wasted! There was also a little door in the road wall for the gardener's use.

On the right of this piece of garden ground there was a very ornamental balustrade from the corner of the house and drawing room to some stone steps, with the same balustrade going down with them and covered with pillar roses. These steps led down to the river side lawn and a wide gravel path like a terrace running along the whole river side front of the basement with its thick vines and many creepers again. On the left of this pathway, or terrace as it might be called, there were a number of glorious standard roses; what a soil that was for roses, I can remember. Along the two division walls separating Pope's villa from its two neighbours there were several very beautiful trees or flowering shrubs such as rose acacia, Thorn's white or red wisteria, lilac and laburnum. The lawn was like a piece of green velvet sloping down to the river side where there was a strong timber wooden shed to protect it. I can also remember a very beautiful wisteria climbing and spreading over the front, as high as the bedroom windows on the first floor.

Continuing along this terrace or wide pathway, there was an open space or passage, about five to six feet wide with a tile floor which ran from the path the whole length of the house to meet some stone steps coming down from a back door entrance from the road for the tradespeople. About midway along this open passage there was a door leading into a passage or corridor in the basement. On the left it went to the kitchen and all the other usual offices. On the right there was a door opening into a small cosy little sitting room with windows looking out on to the beautiful riverside lawn and the standard roses.

Two or three yards further along the terrace we came to the grottoes. Here there was a sort of recess about 18 inches, or perhaps a couple of feet, deep and in this niche or corner there was a little marble figure of a child playing by the side of a pool of water, roughly worked out with pieces of stone or rock about 18 inches in diameter. We were told this little basin was always kept supplied by a tiny spring of water which had been found there. Now we were at the entrance to the grottoes and became much interested. There was a tunnel that ran right through under the main road about 6 feet high, glittering with all sorts of spar, quartz and specimens of different ores, fossils and no end of interesting objects, all moulded together with some judgement.

At the entrance right and left of this tunnel there were two chambers or cells, on each side, about the same height or perhaps a little higher, say seven or eight feet with a curved roof and faced all over in the same manner. They were circular, about ten feet across and in one of these chambers there was a large limb of a willow tree, said to be a portion of Pope's tree. In the other there was a pretty little marble figure of a boy and we were told that strangers, generally Americans, frequently sent in their cards when passing by and asked permission to view these things.

Having now come to the end of this wide pathway or terrace and to the division wall of this side of the property, there were some rustic steps on the right leading to another pretty little bit of garden above the grottoes, level with the road as on the drawing room side. There was no gravel path leading to the river on this side, but the division wall was quite hidden by shrubs and trees all about.

When reading these pages that attempt to give some idea of the place, one would almost imagine we had been wandering about nearly the whole day, whereas it really had not occupied so far, much more than half an hour. After seeing the grottoes, it was suggested we should walk down the side gravel path from the steps from the drawing room to the boat house where the gardener, David Trussler, would take us in hand and show us round the garden and grounds on the other side of the road. Mrs Balfour, who was a largely made lady and not accustomed to much walking about, would return to the house and, before we left, show us over the bedroom floors.

Up to this point I had been much pleased and interested, the place being altogether so different to the several ordinary houses I had been wasting so much time in looking over lately. But when we came to the boat house I became a different being altogether and this feature of the property seemed to take absolute possession of me. Curiously, my first thought was a glance on the river. Yes, there was a punt I had noticed on the very same pitch that Frank Carter and I used to visit. When I was a happy young bachelor living in my little cottage at Surbiton during the summer months, after a good day's work I would look forward to a pull up the river in my boat as far as Hampton Court. After my dinner (generally cold) and having changed into my flannels, I would stroll quietly along to Parker's boathouse only a few yards off. Phil Parker would (weather permitting) always be on the look out for me and my boat, of which I was very proud, would be out of the shed and floating ready in the water. Tossing my jacket aside, I soon put my back into it and arriving at the bridge, if feeling thirsty, would haul the boat up and run across the road to the Mitre and get a bottle of seltzer water. Having the stream with me on returning, I would quietly paddle back and think; always thinking, the poor brain never at rest. There were several boating clubs, but I never cared or wished to join them. When married, of course all this sort of happiness vanished, as I then believed, never to return so I gave my pretty boat to my young friend Arthur, Mr Simpson's youngest son. How all this comes back to me once more with a possibility, yes a possibility, that we might live here. I cannot put this aside, even trying my best to do so, and find myself asking where is the wonderful self denial that I think I possess?

The little boat house, as it then appeared to me was so ideal, so perfect, so far beyond anything of the kind I could conceive that I must try and give a little sketch of it. The Balfours, having no family made no use of it themselves, but sometimes I remember Mrs Balfour remarking that some young friends came and stayed a little while and a boat was hired for their use and amusement. Anyhow, everything now seemed to be in perfect order and condition.

In the first place, there was always a sufficient depth of water, even at low tide, for boats to float, both in the little dock and alongside. Two boats in the dock would have lots of room and the roof was sufficiently high for a man to stand up in one. It was made safe inside by a pair of easy shutting lattice wooden gates or doors. There were also three little wooden steps by the side of the river for a boat to lie alongside or to embark. There was a wooden lattice door at the side of the gravel path opening into the dock and an iron ladder inside fastened to the wall to enable anyone to descend into the boats or come up from them.

In fact, all the arrangements were simply perfect. At the garden end of the boat house outside there were a few wooden steps up to a pretty room or summer house, above the dock and in perfect condition. This had a little table and some easy chairs and even some coloured prints on the walls and a double window at the end looking on to the river.

Having taken in all this, we then went through the grotto or tunnel under the road, then up two or three stone steps to a gravel path which went straight on some distance until it came to the wall of the kitchen garden. The gardener's pretty cottage was like the villa itself, set back about 12 feet from the road. There was a path of coloured tiles leading to the door behind the cottage where there was a little yard, enclosed with a coach house and a couple of stalls, but evidently these had not been used as stables for many years. The yard was now being made use of as a chicken run. I quite approved of this for throughout my life, I have always thought and said that as I had a good pair of legs, I would never use a carriage even if well able to do so.

Behind this yard there was a very large and long lawn, somewhat narrow in proportion to its length and on it near the cottage were two immense, handsome cedar trees, their flat branches spreading out and giving most grateful shade on a hot day. They must have been of a great age. A little further along the same lawn there were two very large quince trees, such fruit not being very common, but a slice of quince in an apple tart, what flavour it gives!

All this portion of the grounds were enclosed by two roads; on the left by one named Pope's Grove and on the right by a narrow lane that no one seemed ever to use and which joined another lane running along the end of the property. All along this path on the right were a number of splendid standard roses again. There was no path on the other side, only a pretty shrubbery hiding the wall.

About half way along on this side there was a door in the wall which opened into a meadow just about the same length and width as the lawn we were on. This also had a gravel path leading from this door to the end of the meadow and another door into the kitchen garden. On the left of the path was an iron fence or railing. On the right was a border along the wall separating this meadow from the garden grounds and on which strawberries were growing in abundance. On that side with the wall facing south there were a lot of fruit trees planted and on the other side of the meadow were some gates opening into Pope's Grove.

Returning again into the garden, the lawn continued a little distance further on and then came to a beautiful hedge four feet high; entirely of sweet briar, carefully clipped and running from one side of the lawn to the other. There was a little border at the base full of dwarf roses and odds and ends of other flowers, but there were no formal beds about the gardens anywhere. One more step or two to the other side of this hedge and we came to the most perfect bowling green I had ever seen, as level as a billiard table and not a weed on it. Here there was a very old wall from one side to the other which, with the briar hedge quite enclosing it and with some old fashioned garden seats about to make it perfect, it reminded me of some of the pictures of Marcus Stone, that well known artist.

In the centre of this old wall there was a door, opening into a large kitchen garden or more properly speaking, fruit garden. The four walls were covered with every fruit tree that could be named, many of great age. There were espaliers round all the paths, also standards everywhere. There was also a great growth of vegetables but fruit predominated. I could not help thinking that a little money ought to be made out of all this abundance and I can well remember it reminding me of the wild fruit growing in the backwoods of Canada. Also, when staying with my brother William at the Cage Farm in Tunbridge, how it used to grieve and worry me to see in the orchards so many beautiful apples lying on the ground, rotting.

The whole of this part of the country for miles round on the Middlesex side of the river was very flat and given up entirely to market gardens. Isleworth, Hounslow, Hanworth, round Kneller Hall and as far as Brentford. It was a sight in the early morning to see the stream of wagons packed high up with all kinds of vegetables and fruit (no finer strawberries could be grown anywhere) travelling along through Brentford and Hammersmith to Covent Garden.

One thing I was glad to notice in this garden was that there was no glass, no expensive hot houses to keep up; the desire evidently being throughout to let nature have its run as much as possible, but always kept in order. There was another door here on the left, leading into the other end of the meadow.

Having now seen all the grounds, we hurried back to the villa again, crossing the road this time and entering by the front door into the hall. I do not know why, but it then seemed to me that I was more impressed with it than I was an hour ago. The cornice, with its clever carving or moulding and the quaint, old fashioned marble chimney-piece. On the right there was a window high up to give proper and sufficient light. This had a heavy frame of carved wood and curved at the top with stained glass round the margin and embossed glass in the centre. This wall on the right ended just half way and a recess formed about 10 or 12 feet square for the stairs, one set going up, the other set going down. Those going up were the most important looking and very nice, being wide with easy treads and a somewhat handsome carved balustrade.

The smaller set of stairs went only to the basement. The stairs were not seen at all until one advanced a step or two beyond where the wall seemed to end. It therefore kept the hall perfectly square, making it quite a comfortable room to sit in now and then. I should mention that this recess in which the stairs were built went right up and formed the tower.

Passing by the stairs, we came to a door facing us which opened into the dining room, a well formed comfortable room (all the rooms on the ground floor were of good height). The cornice of the hall and dining room was cleverly done, each different and perhaps a copy of some of Grinling Gibbons' work. I noticed also that the walls of both the hall and the dining room were painted and panelled in some dark colour with some sort of varnish to protect it; the effect was very good. I remember Mrs Balfour remarking at the time that it must have been in that state for many years and it really looked as if it would remain so for many years more.

There was a large marble mantlepiece in the room and I think a window at the other end, but the window at the side facing the front garden and the river was striking. It was very large and divided into three panels with heavy carved frames; no stained glass in it, but heavy plate glass, prettily engraved round the margin of each panel. There was no balcony outside as in the drawing room and it seemed to be unusually set back, outwards, giving a very wide sill or slab, quite 12 inches wide, almost wide enough to sit upon, but it was too high and therefore not possible. The great feature and charm of this window was that, although the glass in each of these three panels was so heavy and massive, by putting one's little finger to the bronze handle at the bottom, each window could be lifted wide open with the greatest ease, they were so well balanced. It seemed to me that such work was much better then than it is now. Whoever built this house, there could be no question about the standard of work or materials.

Leaving the dining room, we went up stairs to the first floor. Here there were, I think, three or four bedrooms – not very large, but very comfortable. The principal one facing the river had a dressing room and there were the same number of rooms on the next floor. The stairs continued up to the tower where a beautiful view was obtained of the country on the opposite side of the river - Ham Common and the Petersham meadows, with Richmond Park at the back and not a house to be seen to spoil it. The same pretty view could be seen from all the windows below on the riverside, but not of course to the same extent as from the tower.

I think we had now seen everything and having thanked Mr and Mrs Balfour for their kindness and said how pleased and interested we had been, we then left promising to write.

The proposition Mr Chancellor had made that I should invest in the property was a very formidable one, and much too soon to think of such a thing. However, I had been much impressed by the simplicity and quiet homeliness of the place and putting aside for the moment the beautiful grounds and the situation, it seemed to me it would not be more expensive to keep up than any other suitable home for us of moderate character. I then returned to my work at Kingston.

In the evening after dinner I had a further talk with my wife about the villa and the future. She naturally seemed a little nervous about it all, and I was anxious about it all myself, and wished Chancellor had never thought of bringing it to my notice. I now only sincerely wished to do that which was best and try and not make a fool of myself. However, after a little quiet thought and having very carefully looked into my position as regards finance and having also conceived an idea that might be useful, I decided to write a line to young Joseph Quick asking if he could conveniently run down the next day and stay the night, as I very much wanted to speak with him on a rather important matter. Always a good-natured fellow, he came as wished.

When Mr Chancellor had mentioned the sum of £10,000 at our recent interview, I could not really help laughing and Joe, when I mentioned that amount, positively roared with laughter. Having calmed down a little, however, I proceeded with the main point I had in view and told Joe what had passed during the last few days. In answer to Mr Chancellor I had told him distinctly and plainly that such a proposition to a man like myself for many reasons was absolutely and entirely out of the question and I no more dreamt of such a position than of flying to Heaven. But, as I had further to explain, "When Mr Chancellor was leaving he remarked that it would do no one any harm and might perhaps very much interest me to go over the place first. To this view I had weakly assented; a day with the owners was arranged by him and yesterday my wife and I went all over the property and I need not say we were much interested and pleased; indeed so much so that I began to look at things a little differently." I then gave Joe a description of the house and grounds, as well as I could, particularly the boat house. Finally, I came to the idea I had conceived, upon which I had ventured to trespass upon his kindness.

The idea I had in my mind and the question I wished to ask him was this. Would it be possible, entirely in the way of business, and supposing after all that I could see my way to make them an alternate offer? Of course £10,000 was absolutely out of the question, but having looked carefully into money matters the last day or two, I could make an offer of £8,000 without breaking my back. Could the Building Society, with which he was connected and in which I had a few shares, see their way to assist by advancing any portion of the amount or the whole sum, if they preferred. Also if this could be spread over the next three or four years with their usual interest on such dealings, it would of course entirely relieve me of any pressure during that time and make matters quite easy for me. Having been so successful in my profession up to the present time, I could see no reason why, if I kept my health, I should not do even better than I had yet done. From the rapid extension of the town and increased demand for gas, there was every prospect of it being so. Of course, it meant plenty more hard work but that did not disturb me, the more the better. I had mentioned to Joe that there yet remained three or four years before the lease of the gas works expired.

After some little pause, Joe was evidently surprised to find or rather to hear that I had been so successful during the last few years and very kindly said a few words expressing how pleased and glad he was to find it was so. He also thought that the Building Society would be very glad to take up the matter as it should be to be advantage of both parties.

The first thing they would require to do would be to send the Society's surveyor to the villa and make a general report of the building and the grounds and if the report was satisfactory, the matter would proceed at once and be concluded in a few days. Directly he returned to London he would see his colleagues and put the matter in hand. In the meantime it would be well, as I no doubt would wish the matter kept private, to let him have a few lines from me telling the Society he was acting for me and that the owners could give him certain further information on my behalf. I at once wrote such a letter and in the morning, on leaving, Joe promised to let me know the result as soon as possible. I also wrote a line to Chancellor saying I hoped to write and make him an offer in a few days.

I need not say that during all this time I felt very worried and anxious, but my work always kept me steady. However, after about a week, so far as I can recollect, I received a letter from the Secretary of the Society informing me that their surveyor, having viewed the property, had reported most favourably upon it. Therefore, the Society was now quite willing and prepared to advance the whole amount or any lesser proportion I might require, to be spread, capital and interest, over the next four years.

I also had at the same time a few lines from young Joseph Quick saying there had been no difficulty about the matter and that the Society would be very glad to have a few more clients of similar character. I can very well remember that about this time he was mixed up in some way with the well-known Henry Labouchere and some other men in the same field whose name I cannot now remember, but it began with the letter C. I do know, 12 years later it was still more curious when I sold the house to Henry Labouchere himself.

Having simply acknowledged these letters, I also wrote to Mr Chancellor to the effect that I was now prepared to give £8,000 for the property as it then stood: the sum of £10,000 as I sure he would know with his experience was quite out of the question. Also, that I very much regretted that my offer was so different from the amount named by his clients, but I could not, whatever my own desires might be and in justice to my own family, offer a larger sum. It was a sincere offer and I honestly thought a fair one. I would therefore beg that a reply, simply yes or no might be sent as soon as convenient as there could not possibly be any increase and the worry and anxiety of the matter was very much interfering with my work.

My letter and offer to Mr Chancellor was something of this kind, but I hope very much better. With all my failings, faults and sins, at that time I was not a bad man of business and could, I think, draft and write a letter as well as any man (now I am old and sadly rusty).

I think it was one day within the week that I received Mr Chancellor's reply stating that my offer of £8,000 had, after some discussion with his clients, been accepted. However, plus another £500 for a slip of land alongside which had curiously been forgotten and omitted in the particulars of the property given to him in the first place. For the life of me (putting aside for the moment this additional £500) I could not make out whether I was pleased or sorry at my offer having been accepted. However, pulling myself together a bit, I tried to make out what this additional £500 meant.

It appeared that some few years after the Balfours had purchased the villa, the owners of the land opposite to Mr Maw's house (a little to the west of the Pope's villa grounds) decided to make a new road alongside to be called Pope's Grove. At the corner of this road on the main road they, whoever 'they' were, intended also building a tavern to be called the Grotto tavern. Hearing of this the Balfours and other residents nearby, the Countess of Waldegrave, Mr Freake, the great London builder, residing then at Cross Deep House, Mr Chillingworth of Radnor Lodge, Mr Laming and Mr Robert Morley and I suppose also Mr Maw, made such a bother about it that they were obliged to build it some distance back from the main road. At the same time, there happened to be alongside Mr Maw's bit of garden between the road and the river, a narrow strip of ground, but not large enough to build a house on. The Balfours, knowing this and fearing the tavern people might get hold of it to make a communication between the river and the tavern to allow boating parties to land and visit the tavern for refreshment, set their solicitor to work and bought it. Although I should have preferred that this £500 had not to be added, I thought under the circumstances it would be better to accept and the matter therefore was concluded.

I wrote a line then to young Joseph Quick telling him my offer of £8,000 had been accepted and that the Society might proceed at once to prepare the necessary papers for the sum agreed upon. As he was aware I very much wished the matter be kept as quiet as possible and as private as possible, it might perhaps simplify matters if Mr Rutter, the Society's solicitor would also act as my solicitor in both cases. This being arranged, when next in London I called upon that gentleman at his offices in the

Temple and was much pleased and impressed with him; a shrewd, clever old gentleman. I cannot quite remember what interval occurred for the completion of all the necessary papers or deeds and the final handing over of the money, but probably some three weeks. Twelve years later on I was glad to avail myself of his services once more.

During my career up to that time I had never required the services of any solicitor. The only bit of law I had ever indulged in was to ask my brother-in-law, Mr Hargrove, to send a letter to my then manager, Strachan, at Kingston threatening that if he did not refund certain monies he had collected by a certain time, proceedings would be taken at once and I was glad to find it had the desired effect.

Under ordinary circumstances I should have asked Hargrove's firm to act for me in this matter, but feeling so anxious and so nervous that remarks might be made if it were generally known that a man like myself should put himself in such a prominent position, I thought the arrangement I had made would be best and as it turned out, I think it was.

At last, the first of the two great events had come and gone and Charles Aird, then 35 years old stood there, the owner of Pope's villa; how wonderful and how bewildering it did seem to me then. But the other event, how can I possibly describe it?

It was not an event, it was really a miracle coming at such a moment. Since the final completion of the matter I had been very busy for some two or three weeks arranging for the removal of all our goods and chattels from one house to the other, when a letter came from the firm of John Aird & Son. Little dreaming what its contents might mean to me I opened it in my usual way, when I suddenly felt all my blood rushing to my head with the immense surprise and astonishment at the contents. These were to this effect: that the business now had grown so large and was increasing still, it had occurred to them that with my knowledge of construction and gas work I might be of much assistance and they therefore would be glad if I could see my way clear to become a partner. Now, was this not a miracle, coming at such a moment? Neither my father, brother or any close friends then knew anything of my purchase of Pope's villa, excepting young Joseph Quick. He had faithfully promised me not to mention it to anyone for the reasons I have already given; and yet, all in a moment as it were, without lifting a little finger, all my fears of reproach and trouble at what I had so recently accomplished would at once disappear. No one would or could express any surprise at a partner of the well-known firm of John Aird and Sons taking such a step.

Should this wonderful, this undreamed of proposal be arranged, what happiness, what a future – so different to the uncertain and precarious prospect I had before me only an hour or two ago. This would neutralise at once any question as to my imprudence in buying the villa, indeed it seemed to me that it would rather act the other way. The more I tried to realise it all, the more astounding it seemed. There would be no more wandering about to find further occupation for the future, when the lease of the Kingston Gas Works expired. If the whole world could have been offered to me to select an occupation, nothing could have given me greater happiness than the interesting work of the firm and to think of the still further great happiness of meeting and sitting with my dear father and brother each day.

Well, the day I received that letter was indeed an eventful one. So far as my work was concerned, it was really a *Dies Non* for I could not apply my mind to anything; but I managed to write a few lines in answer, saying I could not express in writing how deeply I felt about their great kindness, but if all was well, I would be at Belvedere Road the following morning between 11 and 12 o'clock.

The following morning must have found me a different sort of being, not quite knowing whether I was standing on my head or my heels, but after an early breakfast and a couple of hours work at the gas works, I went up to London to Belvedere Road and met my dear father and brother. For a few minutes I really could not say a word, my feelings were so intense, but presently I controlled myself sufficiently to do so. They no doubt could see how deeply I felt their great kindness and after a few general remarks, explained a little more fully their reasons for suggesting the idea of partnership. I said a few words to try and express my gratitude for such undreamed of happiness. It is always so difficult with me to find suitable and proper words to convey my feelings sufficiently. When I had been fully acquainted with the new position I was to hold, it would possibly make me a stronger character in the future and of greater help to the firm.

I think I then endeavoured to give them an idea of how I then stood financially. As I proceeded I could see they were greatly surprised and when I came to the recent purchase of Pope's villa and describing how I had in some sort of accidental way tumbled into it, I thought they both seemed a little alarmed; but when I had concluded, there came a distinct look of pleasure and perhaps a little bit of pride. But there were serious matters yet to be discussed and arranged.

First of all as regards the Kingston Gas Works, did I see my way to continue to manage them as I had done, at the same time acting as a partner in the firm? Yes, without the least hesitation as I anyway should have had to do this had I succeeded in finding another business which I should have worked up, or tried to work up and continued after the Kingston Gas lease had expired. With a little feeling of pride, I had no hesitation in saying that the Kingston works had been so worked up it was now like a clock, only requiring to be wound up occasionally. Half a day each week, morning or afternoon as most convenient, would be all that would be required. The half yearly accounts I should of course prepare myself and there would be no difficulty about that as I should keep the books at home and work at them now and then after dinner. The Directors' meetings were always held in the evening at 8 o'clock. Really, now the new railways had been completed, it would not make much more than half an hour's difference whether I went to Belvedere Road by way of Kingston or straight from Twickenham.

My two assistants, Watson and Packham, I would trust with everything I possessed and knew well how they worked their best to keep things going satisfactorily. This knowledge and opinion of them relieved me altogether of any anxiety on that score and from that time to the very day the lease of the gas works terminated, it was always the same with them. The result of our working up to this time had indeed been a success and from the way new roads were being formed in all directions (thanks to the new railways) and houses springing up like mushrooms, the results of the remaining three or four years should even be greater.

In endeavouring to give some idea of this important meeting

(certainly as regards myself) I have done my best. After so many years it of course has been impossible to describe our discussions exactly and I trust that some idea can be formed from the words my memory has helped me to write.

Having thus explained all matters so far regarding myself, I ventured to speak of a subject which I considered to be of great moment and for the future happiness of us all. Would it not be possible to arrange that some contracts or works now in hand might be selected by the two senior partners and placed under my entire control and responsibility? I of course assumed that having been invited to join, lend a hand and even to enlarge the operations, they believed in me and that I was capable of doing so. I had a very strong feeling that no work could prosper with two or three masters giving different orders; it must eventually upset the men and create trouble. I did not for one moment suggest any hard and fast line that neither of the other partners should come on the work – that of course would be absurd; speaking for myself, nothing would give me greater pleasure than tc go round with one or both and hear their comments. After a little further consideration it was consented to and a short interval was to occur so that the necessary partnership deeds should be prepared and also that I should explain to the Directors of the gas works and to my two assistants Watson and Packham, the change in my position. The conversation I had subsequently with them was very gratifying and I assured them how much I appreciated and valued their services during the past.

I cannot remember who prepared the partnership deeds, but it must have been either Mr Sturmy of Southwark, my father's first solicitor who lived at Blackheath, or the firm of H F Blunt.

Charles Aird with his family. From Left to right: *standing* Maggie (m: Dent), Charlie, Nellie, Lily; *sitting* Edith, Charles Aird, Maye (m: Jörgensen) Kenneth, Ellen Aird; *kneeling* Mabel (m: Duval) Winnie, Sybil.

Pope's Villa, Twickenham. Bought and
enlarged by Alexander Pope, in 1717.

Epilogue

So far my younger daughter and I have transcribed six volumes of my grandfather Charles's diaries; the first five are beautifully handwritten, but in the last the writing is obviously declining. His final volume, No. 7 is a complete jumble of words and sentences, some being repeats of what he had already written about the family firm, Pope's villa, of which he was greatly proud, and interspersed with domestic notes. "Are cherries eaten? Bread and cutlets; my boat, why small fish? Maple catalogue; bed rest; change hall picture again" and the like. He was obviously failing. The following extract which we did manage to dig out of volume 7 and was written in 1910 probably explains it all.

"It is not more, I think, than two years ago that I conceived the idea of writing my career or life; not that there was anything to be proud of, but perhaps it was a singular and curious one. I started with the idea of a sort of sketch, but it went on and I found it interested and amused me and took me away now and then from my great suffering. In spite of all this I have jogged along with it, never being able to write more than one hour a day, but frequently unable to continue for a week. My memory of the past is wonderful, but the difficulty of expressing myself clearly and distinctly, both in grammar and spelling is at times very distracting. I have become rusty in that respect; when in harness there was no man in the busy world who could put me aside".

Charles died in 1910 aged 79. His father John had already died in 1876 and his brother John died in 1911. These three produced the real power and drive in the firm, although Charles had to retire in 1876 having suffered a bad stroke, but not before he had fathered 11 children.

It is sad that nothing has come down about his time as a partner at John Aird's. He was in that position with them for twelve years before his stroke and what he might have written about the people he met and the work he undertook would have made fascinating reading. The firm's greatest achievement, the Egyptian Aswan Dam and its barrages down the Nile was not started until 1898 and was carried out under the direct control of his brother John, who for that task received a baronetcy. This huge work to hold back the waters for 140 miles would provide more than 800,000 acres of fertile land and was completed in 1902, a year before the stipulated time. Aird's received no payment until the dam was finished and the resultant crops started to produce a return.

Charles sold Pope's villa to Henry Labouchere in 1876 after his stroke. The family then went to The Hollands, a regency house at Langton Green, Tunbridge Wells, for about a year before finally moving to The Grove at Eastbourne. After Charles died, his widow Ellen and their unmarried daughters moved to Engalee at East Grinstead. She died in 1945 aged 102.

Peter Stirling-Aird
April 2004

FAMILY TREE OF JOHN AIRD, 1801-1876

David Aird (b. about 1740) of Teanauld, near Evanton
Married (mid 1760's) Roberta Munro of Foulis

Katherine	Elizabeth	Margaret	Robert	Jane	Janet	Georgina
(b. 1767)	(b. 1768)	(b. 1771)	(1774-1817)	(b. 1776)	(b. 1778)	(b. 1780)

(All christened at Rossleen)

Robert:
Married Elspet MacRitchie (1773-1857)
dau: of John MacRitchie (1709-1787) and Jane Junor (1747-1845)
ftom the Black Isle and formerly Templand

Janet	Jean	Elspeth	George	Alexander	John*
(b.1791)	(b. 1793)	(b. 1794)	(b. 1796)	(b. 1799)	(1801-1876)
Avoch	Avoch	Rosemarkie	Rosemarkie	Rosemarkie	Rosemarkie
				dau: Ann	Married 1831 Agnes Bennett (d. 1869) Dau: of Charles Bennett, Lambeth

(Christening village)

Charles	John	Alexander	Agnes	William	Jessie	Joseph	Henry	Walter	George
(1831-1910)	(1833-1911)	(b. 1836)	(1837-1939)	(1840-1917)	(1842-1918)	(1845-1906)	(1849-1914)	(1852-1905 or before)	(1855-1856)
						Wellington Tube Works	Emigrated to Australia		
See separate Family Trees		married Janet Keith	m. 1864 Henry Gill	m. 1867 Blanche Stock	married James Hargrove		m. 1873 1. Margaret Rutherford 2. 1879 Margaret McCrae	m. 1876 Leah Bailey	
				For 3 generations were the family lawyers					
		7 or more children. Lived in Germany	4 children. Lived in Potsdam	5 children. Farmed in Kent Then emigrated to Australia	8 children		3 children	Emigrated to USA 1882. 3 children	

Index of Surnames

C

F

G

I

J

Q

R

S

T

V

W

YZ

Index of Place Names

D

E

F

G

H

IJK

L

OPQ

R

UV

W

Printed in the United Kingdom
by Lightning Source UK Ltd.
107552UKS00001B/85